REEDS MARINE ENGINEERING AND TECHNOLOGY

MOTOR ENGINEERING KNOWLEDGE
FOR MARINE ENGINEERS

REEDS MARINE ENGINEERING AND TECHNOLOGY SERIES

12

REEDS MARINE ENGINEERING AND TECHNOLOGY

MOTOR ENGINEERING KNOWLEDGE
FOR MARINE ENGINEERS

Revised by Paul A Russell

Thomas D Morton

Leslie Jackson

Anthony S Prince

REEDS

LONDON • OXFORD • NEW YORK • NEW DELHI • SYDNEY

REEDS
Bloomsbury Publishing Plc
50 Bedford Square, London, WC1B 3DP, UK
Bloomsbury Publishing Ireland Limited,
29 Earlsfort Terrace, Dublin 2, D02 AY28, Ireland

BLOOMSBURY, REEDS, and the Reeds logo are trademarks of Bloomsbury Publishing Plc

First published in Great Britain 1975
Second edition 1978
Third edition 1994
Fourth edition 2013
Fifth edition 2018
This edition published 2025

A catalogue record for this book is available from the British Library

Library of Congress Cataloguing-in-Publication data has been applied for.

ISBN: PB: 978-1-3994-1446-3; ePub: 978-1-3994-1447-0; ePDF: 978-1-3994-1445-6

2 4 6 8 10 9 7 5 3 1

Typeset in Myriad Pro 10/14 by Newgen Knowledge Works Pvt Ltd, Chennai, India
Printed and bound in Great Britain by CPI Group (UK) Ltd, Croydon CR0 4YY

MIX
Paper from
responsible sources
FSC® C013604

To find out more about our authors and books visit www.bloomsbury.com
and sign up for our newsletters
For product safety related questions contact productsafety@bloomsbury.com

CONTENTS

PREFACE

The objective of this book is to prepare students for the Motor Engineering Knowledge part of the Certificates of Competency for marine engineering officers. This text is based on UK practice and has a central focus on the Maritime Coastguard Agency (MCA) in the United Kingdom. However, the information holds good for other Flag State administrations, and the book also covers the International Maritime Organization's (IMO) requirements for engineers, which are detailed in Chapter III of the Standards of Training, Certification and Watchkeeping for Seafarers (STCW). This edition of *Motor Engineering Knowledge* includes the most up-to-date information relating to the requirements of IMO's STCW and MARine POLution regulation MARPOL Annex VI, which sets out the agenda for reducing the emissions from ships' engine exhaust gases.

The text is intended to cover the groundwork required for the examinations at the different levels of Engineering Officer of the Watch, Second Engineering Officer and Chief Engineering Officer. The syllabuses and engineering principles involved can be similar for all examinations but questions set for the Chief Engineering Officer examination require a more detailed answer than those set at Second Engineering Officer level. It is extremely important for the student preparing for the Officer of the Watch examination to concentrate on the safety procedures and practices of marine engineering. While it is not acceptable for the Engineering Officer of the Watch (EOOW) to keep answering a question with 'I will ask the Second or Chief', it should be remembered that responsibility does lie with the Chief and she/he is available to consult if all other options fail. The Chief, on the other hand, has no one to fall back upon although she/he can consult technical manuals.

The book can now also be considered as more than a specific examination guide and will be useful to superintendent engineers wishing to have a general guide to the latest trends, from which they can seek more detail. Engineering knowledge is delivered via several different academic pathways from the Scottish Qualifications Authority's (SQA) Maritime Studies Qualification (MSQ) through Higher National Diplomas (HND) to foundation degrees and full honours degrees. The drawings are still intended to have direct relevance to the examination requirements but it is left to the student to practise his/her own versions.

The best method of study is to read carefully through each chapter, practising the drawings, and, when the principles have been mastered, attempting the few examples at the end of the chapter.

It is also important that the information contained within this text is related back and linked with the student's own practical experience. Flag State examiners will be looking for detailed answers to the written questions set or the oral questions asked. Answers such as 'Complete the pre-start checks given in the safety management system' without giving information about what those checks are will result in failure.

Don't forget that the examination is also a chance for the candidate to show the examiner the extent of his/her knowledge and understanding of the machinery concerned.

Finally, the miscellaneous questions at the end of the book should be worked through methodically. The best preparation for any examination is to work on the examples; however, this is difficult in the subject of engineering knowledge as no model answer is available, nor indeed any one textbook to cover all the possible questions. As a guide, it is suggested that the student finds his/her information first and then attempts each question in the book in turn, basing their answer on either a good descriptive sketch or writing a description covering about a page and a half of A4 paper. Try to complete this exercise in half an hour. I have found it particularly useful to use an artist's sketch pad, fill it with relevant drawings and practise them so that they can be reproduced as required in the examination.

Use of a small ruler that can be quickly moved around to give your sketches straight line is a particularly useful technique to master. This technique gives a much more 'professional' look to the final presentation of your sketches.

ACKNOWLEDGEMENTS TO SIXTH EDITION

I wish to acknowledge the invaluable assistance given by the following bodies in the revision of this book:

ABB Turbo Systems Ltd

Rolls-Royce

MAN Diesel & Turbo

Krupp-MaK Maschinenbau GmbH

Dr. -Ing Geislinger & Co

Wärtsilä Corporation

The Institute of Marine Engineering, Science and Technology (IMarEST)

Scottish Qualifications Authority (SQA)

Merchant Navy Training Board

I also wish to extend my thanks to colleagues in Maritime Education in the United Kingdom.

Paul A Russell

BASIC PRINCIPLES

Definitions and Formulae

Boyle's law

Boyles law states that the volume of an ideal gas multiplied by its pressure equals the same value over a wide range of pressures and volumes and all at a constant temperature. Please note the word 'ideal' as this law does not apply to all gases under all values of pressure and volume. The reason is that some gases experience internal chemical reactions during the pressure change. (For more details about the Ideal Gas Laws please see chapter 5 of *Reeds vol 3: Applied Thermodynamics for Marine Engineers*.) Students will also be able to note that to have the temperature remain constant heat will need to be added or removed as the compression changes.

Charles's Law

Charles's law explains the relationship between the volume of a given mass of gas and its temperature. Here the pressure is assumed to remain constant and therefore the volume of a gas varies directly with its temperature measured in degrees Kelvin.

Combined Gas Laws

Both Boyle's law and Charles's law assume that one of the three properties remains constant. However, it is more likely that in the real world all three of the properties will change. Therefore, combining the two laws we find that for a perfect gas under ideal conditions the Pressure multiplied by the Volume and divided by the Absolute Temperature is constant. The equation becomes.

$$P_1 \times V_1 / T_1 = P_2 \times V_2 / T_2$$

More details about the gas laws and how they are used in calculating the working properties of gases in Internal Combustion engines can be found in Chapters 5 and 6 of volume 3 (*Applied Thermodynamics for Marine Engineers*) of the Reeds Marine Engineering Series.

Working from these theoretical gas laws came the working engine thermodynamic cycles of the Internal Combustion engine.

Isothermal operation (*PV* = constant)

This is an ideal, reversible process carried out at constant temperature. It follows Boyle's law, requiring heat addition during expansion and heat extraction during compression. It is, however, impractical due to the requirement of very slow piston speeds.

Adiabatic operation (*PV^y* = constant) (where *y* = gamma = the adiabatic index *Cp/Cv*)

This is also an ideal and reversible process but with no heat addition or extraction and therefore the work done is equivalent to the change of internal energy. It is again impracticable due to the requirement of very high piston speeds.

Polytropic operation (*PV^n* = constant)

This is close to a practical process where the value of the index *n* usually lies between unity and gamma.

Volumetric efficiency

This is a comparison between the mass of air induced per cycle and the mass of air contained in the stroke volume at standard conditions. This term is usually used to describe four-stroke engine and air compressor operation. Due to the restrictions of practical engine design, typical values range between 86% and 92%.

Scavenge efficiency ('scavenging' is the term used to describe the air exchange process)

This is similar to volumetric efficiency but is used to describe two-stroke engines where some exhaust gas from the previous stroke may be included with the induction air at the start of compression. Both efficiency values are reduced by high revolutions (less time for the exchange process) and high ambient air temperature (less weight of incoming air). The introduction of exhaust gas recycling will change the values recorded on modern engines.

Mechanical efficiency

Mechanical efficiency is a measure of the mechanical perfection of an engine. It is numerically expressed as the ratio between the indicated power (power available from burning the fuel) and the brake power (power measured at the flywheel).

Uniflow scavenge

With uniflow scavenging, the two-stroke engine is designed to have the exhaust at one end of the cylinder (top) and scavenge air entry at the other end of the cylinder (bottom) so that there is a clear flow traversing the full length of the cylinder (see figure 1.1) This design means that the scavenge air does not have to travel up the cylinder and down again, as with the other designs, to purge the exhaust gas from the previous cycle, hence the name UNIflow. Due to the increased efficiency, all modern engine designs are now based on this arrangement.

Loop scavenge and cross scavenge

This is the traditional two-stroke design where the exhaust gas exit and scavenge air entry are at one end of the cylinder (bottom); examples are the older Sulzer (now Wärtsilä) RD RND and RL engines. This general classification simplifies and embraces variations of the sketch (figure 1.1) in cases where air and exhaust are at different

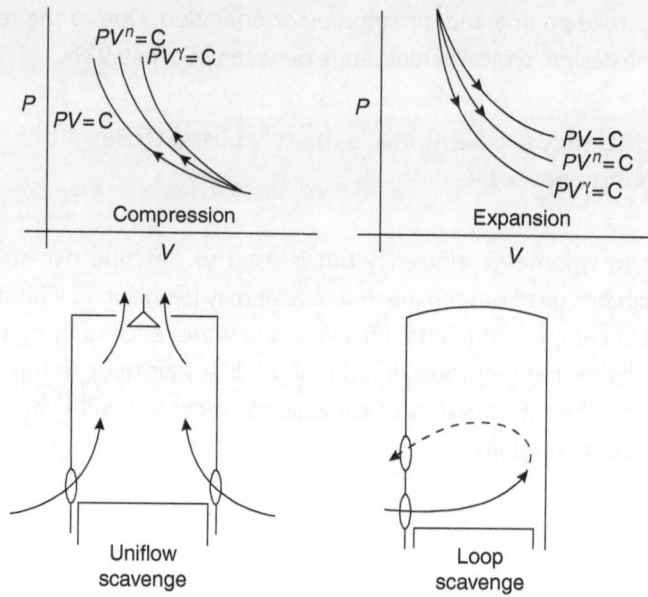

▲ Figure 1.1 *Compression, expansion*

sides of the cylinder with and without crossed flow loop (cross and transverse scavenge).

Brake thermal efficiency

This is the ratio between the energy developed at the flywheel, or the output shaft of the engine, and the energy supplied from burning the fuel. Traditionally this was measured by placing a 'load' or 'brake' on the output shaft, hence the term brake thermal efficiency.

Specific fuel oil consumption (SFOC)

SFOC is the fuel consumption per unit of energy at the cylinder or output shaft, kg/kWh (or kg/kWs); 0.38 kg/kWh would be normal for measurement at the shaft for a modern engine. However, the current general practice is for the manufacturers to quote a consumption figure measured at the cylinder and expressed in g/kWh and not kg/kWh. Therefore, a typical fuel consumption figure for a modern two-stroke diesel main engine would be quoted as being between 160 and 185 g/kWh.

Compression ratio (CR)

CR is a measurement of the ratio of the volume of air at the start of the compression stroke to the volume of air at the end of this stroke (measured between top dead centre (TDC) and bottom dead centre (BDC)). Usual value for a compression ignition (CI) oil engine is about 14:5 to 20:1, that is, clearance volume is 7.5% to 5% of stroke volume.

Fuel–air ratio

Depending upon the type and quality of fuel, the amount of air required to give enough oxygen to completely burn all the fuel is about 14.5 kg for each kg of fuel. However, engines supply excess air to the combustion process and therefore the actual air supplied varies from about 29 to 44 kg/kg fuel. The percentage of excess air is about 150 (36.5 kg for each kg of fuel).

Performance curves for fuel consumption and efficiency

The initial design considerations for main engines powering merchant ships will be for optimised thermal efficiency (and minimum specific fuel consumption) to occur at the power conditions required to maintain the chosen service speed of the vessel. Marine practice is to quote the minimum specific fuel consumption at a given percentage of engine service load but maximum speeds are occasionally required when the specific fuel consumption will be much higher. Modern tonnage is often required to operate at speeds other than the design service speed. The practice of 'slow steaming' isn't uncommon and this means that the main engine will be required to operate at loads well below its service maximum continuous rating (SMCR). Engine manufacturers have responded and produced modern engines that have a much improved efficiency when running continuously at part load.

The design of internal combustion (IC) engines driving electrical generators is arranged so that the peak thermal efficiency is at approximately 70% maximum load because this is the engine unit's average load during normal ship operation.

Manufacturers published performance curves that are useful in establishing principles, such as:

1. The fuel consumption (kg/s) increases steadily with load. However, the fuel consumption is not reduced by 50% if the load is reduced by 50% as certain essentials consume fuel at no load (eg heat for cooling water warming through, etc).

2. Mechanical efficiency steadily increases with load as friction losses are almost constant and therefore become a smaller percentage of the total losses.

3. Thermal efficiency (brake, for example) is designed to be at maximum at full load.

4. Specific fuel consumption is therefore a minimum at 100% power. Fuel consumption on a brake basis increases more rapidly than indicated specific fuel consumption as load decreases due to the friction losses being almost constant.

Heat balance

A simple heat balance is shown in figure 1.2. There are some factors not considered in drawing up this balance but as a first analysis this serves to give a useful indication of the heat distribution for the IC engine. The high thermal efficiency and low fuel consumption obtained by modern diesel engines is superior to any other form of engine in use at present.

1. The development of waste heat recovery systems gives the marine plant an efficiency gain as this is heat that would otherwise be lost to the environment.

2. The recent efficiency increases of exhaust gas-driven turbochargers not only contribute to high mechanical efficiency, by taking no mechanical power from the engine, but they also take a smaller percentage of the exhaust gas output to drive the charge air compressor. This means that more gas is left over to drive turbogenerators, exhaust gas boilers and other waste heat recovery systems.

3. Cooling loss includes an element of heat energy due to generated friction.

4. Propellers do not usually have propulsive efficiencies exceeding 70%, which reduces brake power according to the output power.

5. In the previous remarks, no account has been taken of the increasingly common practice of utilising a recovery system for heat normally lost in coolant systems.

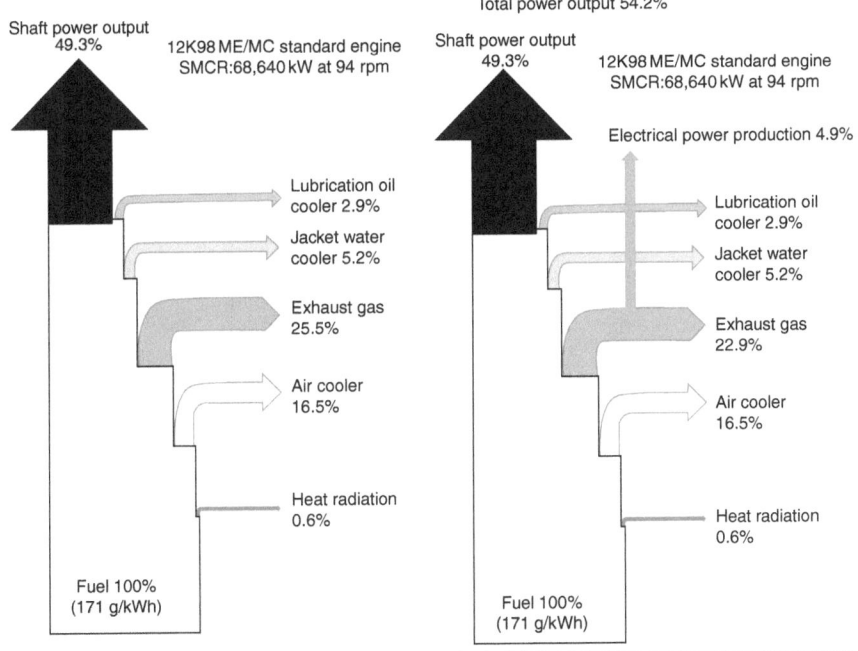

▲ Figure 1.2 *Heat balance including waste heat recovery (courtesy of MAN Diesel & Turbo)*

Analysis of the simplified heat balance shown in figure 1.2 reveals two important observations.

1. The difference between the indicated power and brake power is not only the power absorbed by the friction losses as some power is required to drive engine components such as camshafts, pumps, etc, which means a reduced potential for brake power.

2. Friction also results in heat generation, which is dissipated by the various fluid cooling media, that is, oil and water, and hence the cooling analysis in a heat balance equation will include the frictional heat effect as an estimation.

Engine load diagram showing different combinations

Figure 1.3 shows a typical load diagram for a slow-speed two-stroke engine. It is a graph of brake power and shaft speed. Line 1 represents the power developed by the engine on the test bed and runs through the maximum continuous rating (MCR) point. The lines running parallel to line 2 represent constant values of mean effective pressure (P_{MEP}). Line 3 shows the maximum shaft speed, which should not be exceeded. Line 4 is important since it represents the maximum continuous power and MEP; at a given

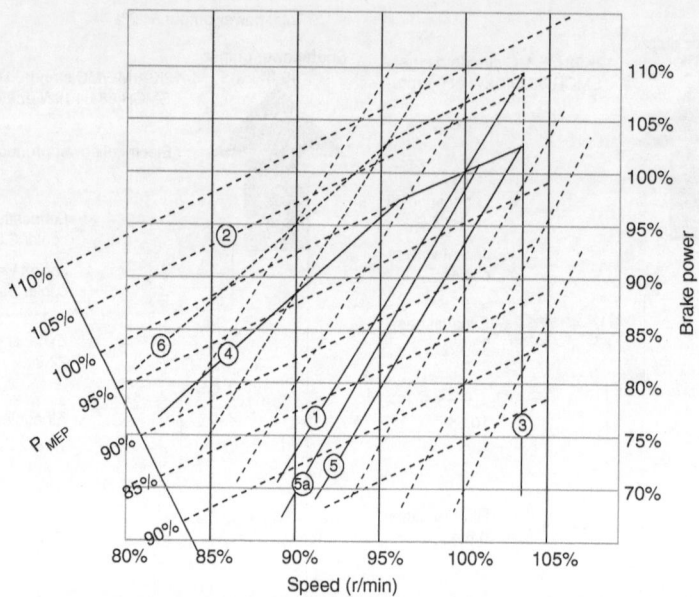

▲ **Figure 1.3** *Standard engine load diagram*

speed, this will depend upon an adequate supply of charge air for combustion. Line 5 represents the power absorbed by the propeller when the ship is fully loaded and has a clean hull. The effect of a fouled hull is to move this line to the left as indicated by line 5a. In general a loaded vessel will operate between lines 4 and 5, while a vessel in ballast will operate in the region to the right of line 5. The area to the left of line 4 represents overload operation.

It can be seen that the fouling of the hull, by moving line 5 to the left, decreases the margin of operation and the combination of hull fouling and heavy weather can cause the engine to become overloaded, even though engine revolutions are reduced. Following on from this diagram, the engine manufacturer will calculate the most efficient operating point for the engine. The operational requirements of the owner will determine the design speed and power for the normal running point of the engine (see layout information below).

Engine layout points

In designing engines for different types of duty, the specific consumption minima may be at a different load point. As quoted earlier, this could be about 70% for engines driving electrical generators. The effective output power of a diesel engine is proportional to the MEP and engine speed measured in revolutions per minute. When a vessel is running with a fixed pitch propeller, the relationship between the power required to operate the ship at a given speed and the efficiency of the propeller can also be plotted on a graph. If the two functions are combined in the layout and load diagrams for diesel engines, then when logarithmic scales are used, the result is a simple diagram with straight lines (see figure 1.4).

Engine layout diagram

An engine's layout diagram is limited by two constant MEP lines L1 to L3 and L2 to L4, and by two constant engine speed lines L1 to L2 and L3 to L4 (see figure 1.4). The L1 point refers to the engine's nominal MCR. However, within the layout area the vessel designer has the freedom to select the engine's actual specified MCR point, which would be designated as point M, and relevant optimising point designated as point O, which is the optimum combination for the ship and the operating profile. However, the lowest SFOC for a given optimising point O will be obtained at 70% and 80% of the power at point O for electronically and mechanically controlled engines, respectively.

Based on the best propulsion and engine running points, drawn up by the designer, the layout diagram of a relevant main engine may be drawn up. The specified MCR point M must be inside the limitation lines of the layout diagram. The optimised layout point of the engine is the rating at which the engine, timing and CR are adjusted to work most efficiently with the scavenge air pressure of the turbocharger.

However, engines without variable injection timing (VIT) fuel pumps cannot be optimised at part load. Therefore, these engines are always optimised at point L1.

Other information might also be included in these graphs by the engine manufacturers, such as:

- Propeller curve through an optimised point
- Layout curve for engine line
- Heavy propeller curve due to fouled hull and/or heavy seas
- Speed limit line
- Torque/speed limit
- MEP limit
- Light propeller curve clean hull and calm weather layout curve for propeller
- Limit for continuous running
- Overload limit
- Sea trial speed limit
- Constant MEP.

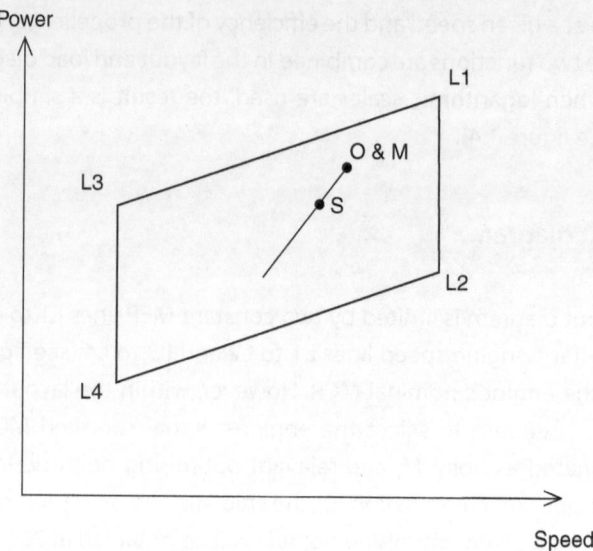

▲ **Figure 1.4** *Engine layout points L1–L4*

Ideal Thermodynamic Cycles

Thermodynamic cycles are a series of operations carried out by a machine manipulating a substance. During the process, heat and work are transferred by varying temperatures and pressures and eventually returning the system to its original state. The ideal thermodynamic cycles form the benchmark for reference against the actual performance of IC engines. In the cycles considered in detail, all curves are regarded as frictionless adiabatic, that is, isentropic. The usual assumptions that are made are that constant specific heats and mass of charge are unaffected by any injected fuel, etc and hence the expression *air standard cycle* may be used. There are two main classifications for reciprocating IC engines: (a) spark ignition (SI) such as petrol engines and (b) CI such as diesel and oil engines. Liquid natural gas (LNG) is beginning to find favour with the main engine manufacturers due to its potential for producing less harmful emissions from the exhaust. Early engines designed to run on gas operated using the Otto (SI) process; however, MAN Diesel & Turbo have recently conducted trials of a two-stroke engine operating on the Diesel process. Engines designed to run on 'alternative fuels' are being developed and refined in the light of manufacturers' research. Methanol and ammonia, together with LNG, are spearheading the change.

Older forms of reference used terms such as light and heavy oil engines but this is not very explicit or satisfactory. Four main air standard cycles are first considered, followed by a brief consideration of other such cycles less often considered. The cycles have been drawn using the usual method of P–V diagrams. Research into reducing the exhaust emissions from marine diesel engines has led the manufacturers to develop engines operating using the 'Miller' cycle. The primary reason for this is that the highest temperatures of combustion are avoided and therefore the harmful nitrous oxides (NOx) are not produced to then be released into the atmosphere through the exhaust.

Otto (constant volume) cycle

The Otto cycle was named after Nicolaus Otto, the inventor of the first efficient working IC engine working on the four-stroke cycle. The Otto cycle now forms the basis of all SI and high-speed CI engines. The four non-flow operations combined into a cycle are shown in figure 1.5.

Air standard efficiency = work done/heat supplied

$$= \frac{(\text{heat supplied} - \text{heat rejected})}{\text{heat supplied}},$$

referring to figure 1.5.

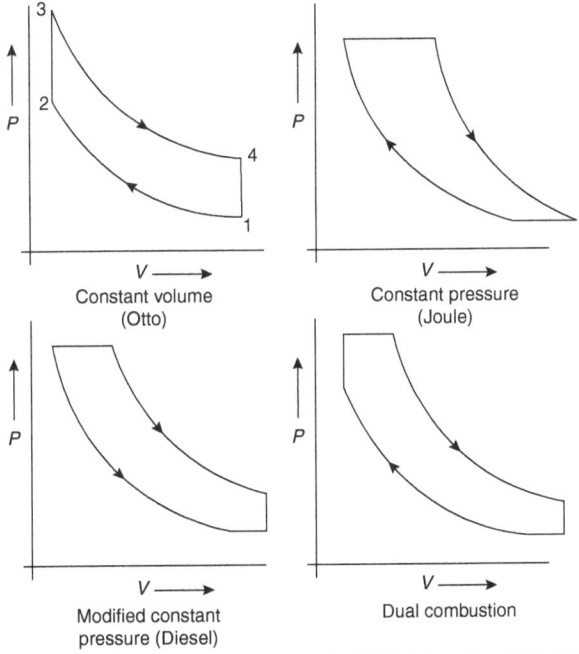

Constant volume
(Otto)

Constant pressure
(Joule)

Modified constant
pressure (Diesel)

Dual combustion

▲ **Figure 1.5** *Theoretical (ideal) cycles*

Air standard efficiency $= 1 -$ heat rejected/heat supplied

$= 1 - MC(T_4 - T_1)/MC(T_3 - T_2)$, where M is the mass and C is the specific heat capacity of the substance

$= 1 - 1/(r\gamma^{-1})$

[using $T_2/T_1 = T_3/T_4 = r\gamma^{-1}$, where r is the CR].

Note: Efficiency of the cycle increases with an increase in the CR. This is also true for the other four cycles.

Diesel (modified constant pressure) cycle

This cycle is more applicable to older CI engines utilising long periods of constant pressure fuel injection in conjunction with blast injection. Modern engines do not in fact aim to follow this cycle, which in its pure form requires very high CRs. The term semi-diesel was used for hot bulb engines using a CR between that of the Otto and the Diesel ideal cycles. Some very early Doxford engines utilised a form of this principle with low compression pressures and 'hot spot' pistons. The Diesel cycle is also sketched in figure 1.5 and it should be noted that heat is received at constant pressure and rejected at constant volume.

Dual (mixed) cycle

This cycle is applicable to most modern CI reciprocating IC engines. Such engines employ solid injection with short fuel injection periods fairly symmetrical about the firing dead centre. The term semi-diesel was often used to describe engines working close to this cycle. In modern turbocharged marine engines the approach is from this cycle almost to the point of the Otto cycle, that is, the constant pressure period is very short. This produces very heavy firing loads but gives the necessary good combustion.

Joule (constant pressure) cycle

This is the simple gas turbine flow cycle. Designs at present are mainly of the open cycle type, although nuclear systems may well utilise closed cycles. The ideal cycle *P–V* diagram is shown in figure 1.5 and again as a circuit cycle diagram in figure 1.6, in which intercoolers, heat exchangers and reheaters have been omitted for simplicity.

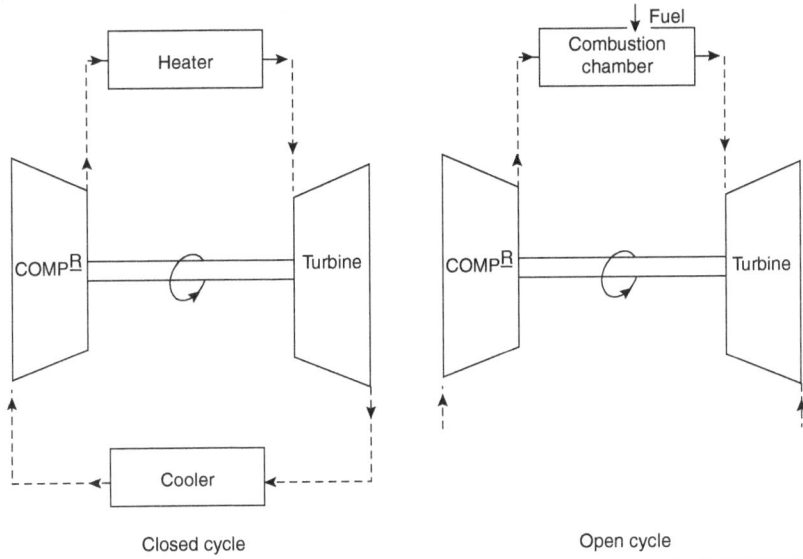

▲ **Figure 1.6** *Gas turbine circuit cycles*

Other cycles

The efficiency of a thermodynamic cycle is a maximum when the cycle is made up of reversible operations. The Carnot cycle of isothermals and adiabatics satisfies this condition and this maximum efficiency is, referring to figure 1.7, given by $(T_3 - T_1)/T_3$ where the Kelvin temperatures are maximum and minimum for the cycle. The cycle is practically not approachable as the MEP is so small and CR would be excessive. All the four ideal cycles have efficiencies less than the Carnot. The Stirling cycle and the Ericsson cycle have equal efficiency to the Carnot. Further research work is being carried out on Stirling cycle engines in an effort to utilise the high thermal efficiency potential. The Carnot cycle is sketched on both P–Y and T–S axes (figure 1.7). However, as with all these theoretical cycles, the reality of producing a practical working engine running on one is very difficult. Therefore, actual engines are always a compromise.

Miller cycle

Modern engines are expected to become much more fuel efficient in the near future. Explanations surrounding the circumstances about why and when this will happen appear elsewhere in this volume. However, it is generally accepted that diesel engines will not reach the reduced values for NOx emissions unless the Miller cycle is used.

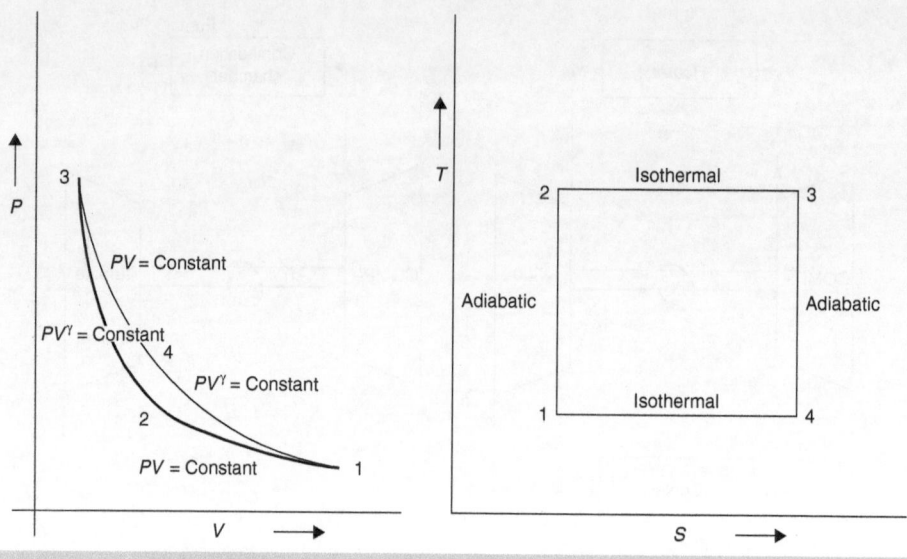

▲ **Figure 1.7** *Theoretical (ideal) cycles*

This engine cycle cannot be used unless an engine has full electronic control of both the fuel injection process and the ability to vary the valve operating timing. The temperature peaks during combustion are responsible for over 90% of NOx formation. Therefore, manufacturers use 'primary' combustion measures to eliminate the peak temperatures in the combustion chamber without incurring fuel consumption penalties or, if possible, at improved fuel efficiency. To achieve this, a range of engine modifications have been used, including:

- further cooling of the charge air
- improved re-entrant piston bowls
- low swirl inlet ports
- higher CRs
- higher fuel injection pressures and improved injector nozzle spray patterns
- revised fuel injection timing
- a combination of revised 'Miller cycle' valve timing and high-efficiency, high-pressure turbocharging.

The Miller cycle involves the early closure of the inlet valve, causing the air entering the cylinder to expand and cool. This cooling action reduces the temperature peaks during combustion, which is the major cause of NOx production during combustion. However, the shorter inlet valve opening period would mean a lower mass of combustion air entering the cylinder and hence reduced engine power and torque. To counter this effect, higher-pressure turbocharging ensures that an equal – or in the case of MAN Diesel's new technology package – or greater amount of air can enter the cylinder in the shorter time available. During trials using an intensive Miller cycle under full load conditions and turbocharger pressure ratios of 6.5–7, MAN Diesel has recorded reductions in NOx of over 30%, reductions in fuel consumption as great as 8% and a 15% increase in specific power output.

Combustion of Alternative Marine Fuels

Designers of Internal Combustion engines that are to be run on fuels other than traditional fossil fuels are currently developing engines to explore the best operating format. This can be either the Otto (Constant Volume) cycle, where the fuel is pre-mixed with the air before compression, or alternatively the Diesel (Constant Pressure) cycle where the fuel is introduced under higher pressure, which is after the air has been compressed. Currently the ignition source is a pilot injection using a conventional fuel, such as Diesel Oil. The challenge with these alternative fuels is achieving the correct timing and pressure of fuel injection. Modern injectors and fuel delivery systems have enabled manufacturers to achieve what would not have been possible just a few years ago.

Engineering Officers should pay close attention to the specification of the fuel being delivered to the ship. The significant reason is that the calorific value of the 'alternative' fuels may not be the same or as stable across deliveries as it is with traditional fuels.

Reviewing the latest Marine Oil Specification (ISO 8217), students will be able to see that marine diesel oil can be delivered with up to 7% (volume) content of Fatty Acid Methyl Ester (FAME). These are under the column labeled as (DF) in the ISO specification of marine distillate fuels. However, it will be of note that the category DMB/DFB is a marine diesel oil with a higher density and lower Cetane number than regular marine diesel oil.

The use of alternative fuels on existing ships remains quite a challenge for the older technology. For example, an engine with fully variable injection and valve timing will be able to compensate for any slight differences in fuel quality. This will be especially true if the engine is fitted with electronic load/speed management systems. However, the older engines with only basic fuel injection and timing control will not be able to adjust their settings, making them more prone to malfunction.

Marine fuels termed 'biofuel' and containing a percentage of FAME might have a lower calorific value than the equivalent fossil fuel. The chief marine engineering officer should be able to ensure that the fuel has been tested in the correct way using the 'bomb calorimeter' method and not just calculated from theoretical equations. See the testing of liquid fuel and oils in *Reeds vol 8: General Engineering Knowledge*.

A further complication from burning LNG via the 'otto' cycle means that if the LNG is premixed with air before the combustion, then concentrations of methane can be found to accumulate in the exhaust (so called methane slip). The inclusion of Exhaust Gas Recirculation can reduce the effect of methane slip. In addition, IMO Tier III compliance can be achieved with the inclusion of the Selective Catalytic Reduction (SCR).

Engineering officers onboard ship should be aware that a symptom of poor quality fuel would be an extended ignition delay and combustion period, poor heat release resulting in poor or incomplete combustion. This will all lead to reduced performance and a possible increase in fuel consumption.

Actual Cycles and Indicator Diagrams

There is a correlation between the real IC engine cycle and the equivalent air standard cycle, as is shown by the similarities in the $P–V$ diagrams, but how does this help the marine engineer working in remote places away from any substantial support?

For many years, engineers recorded the pressure in each cylinder by the use of a mechanical cylinder pressure indicator apparatus (figure 1.14). These were small hand-held units made up of a piston that was able to move up and down in a cylinder. The machine was connected to the engine via a cock on the engine called the indicator cock. Opening the cock allowed the full force of combustion through to the indicator equipment's cylinder. The pressure from the combustion acts upon the piston operating within the cylinder. Movement of the piston is restricted by a spring that can be changed to match the different combustion pressures of different engines. The vertical movement of the piston drives an arm, at the end of which is a pointer that is used to draw the vertical line on a 'card' corresponding to the pressure in the cylinder. The horizontal movement of the card is achieved by rotating the drum that the card is attached to, in time with the movement of the piston. The movement was achieved by using a cord attached to a roller on the camshaft of the engine. Using the equipment shown in figure 1.14, an actual diagram was produced as shown in the lower drawing of figure 1.8. This allowed the engineer to assess the efficiency of the combustion process. Only one cylinder could be reviewed at a time but the different cards could then be compared alongside each other.

The differences between the cycles can now be considered and for illustration purposes the drawings given are of the Otto cycle. The principles are generally the same for most IC engine cycles.

1. The actual compression curve (shown as full line in figure 1.8) shows a lower final pressure and temperature than the ideal adiabatic compression curve (shown as

dotted). This is caused by the time lag in the heat transfer taking place, variable specific heats, a reduction in γ due to gas-air mixing, etc. The resulting compression is not adiabatic and the difference in vertical height is shown as x.

2. The actual combustion gives a lower temperature and pressure than the ideal due to dissociation of molecules caused by high temperatures. These twofold effects

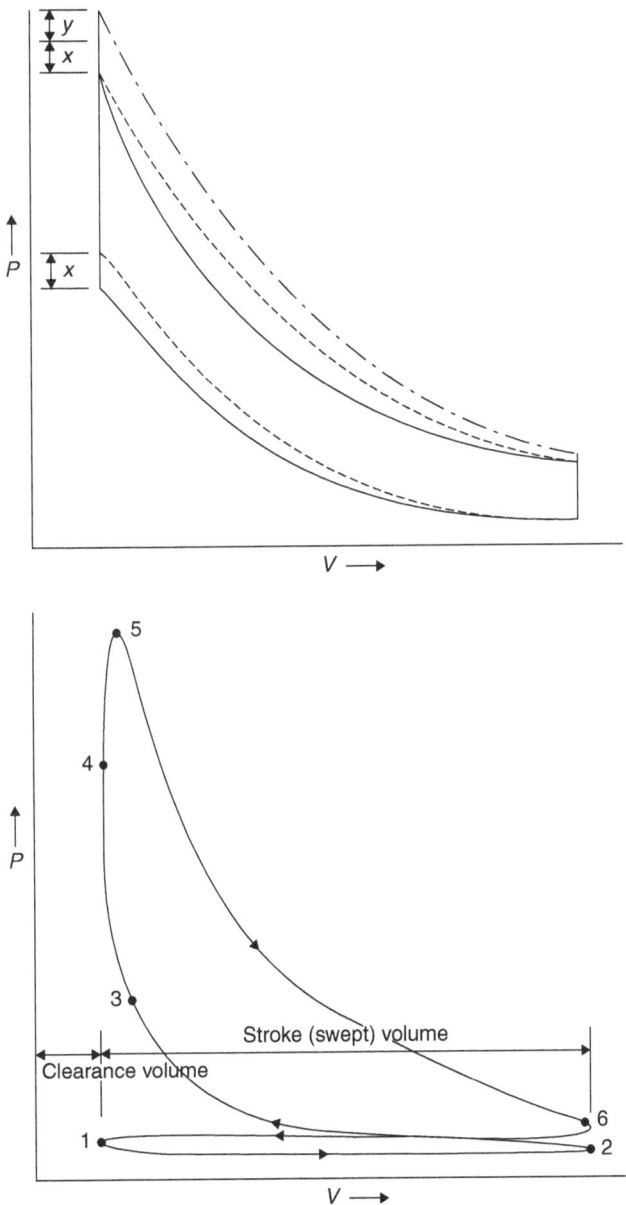

▲ **Figure 1.8** *Actual cycles (Otto basis)*

can be regarded as a loss of peak height of $x + y$ and a lowered expansion line below the ideal adiabatic expansion line. The loss can be regarded as clearly shown between the ideal adiabatic curve from maximum height (shown as chain dotted) and the curve with initial point $x + y$ lower (shown as dotted).

3. In fact, the expansion is also not adiabatic. There is some heat recovery as molecule recombination occurs but this is much less than the dissociation combustion heat loss in practical effect. The expansion is also much removed from adiabatic because of heat transfer taking place and variation of specific heats for the hot gas products of combustion. The actual expansion line is shown as a full line on figure 1.8.

In general, the assumptions made at the beginning of the section on ideal cycles are worth repeating, that is, isentropic, negligible fuel charge mass, constant specific heats, etc plus the comments above such as, for example, on dissociation. Consideration of these factors plus practical details such as rounding of corners due to non-instantaneous valve operation, etc mean that the actual diagram appears as shown in the lower sketch of figure 1.8.

Typical indicator diagrams

The power and draw cards are shown in figure 1.9 and should be studied closely. These examples are for two-stroke and four-stroke CI engines and the typical temperatures and pressures are shown on the drawings where appropriate.

The draw card is an extended scale picture of the combustion process. They have been given the name 'draw cards' because in early marine practice the indicator card was drawn by hand. The later practice was for an 'out-of-phase' (90°) cam to be provided adjacent to the general indicator cam. Incorrect combustion details are highlighted by taking the draw card. There is no real marked difference between the diagrams for two-stroke or four-stroke. In general, the compression point on the draw card is more difficult to detect on the two-stroke as the line is fairly continuous. There is no induction – exhaust loop for the four-stroke as the spring used in the indicator is too strong to discriminate on a pressure difference of, say, 1/3 bar only.

Compression diagrams are also given in figure 1.10, with the fuel shut off expansion and compression appearing as one line. Errors would be due to a time lag in the drive or a faulty indicator cam setting or relative phase difference between camshaft and crankshaft. Normally such diagrams would only be necessary on initial engine trials unless loss of compression or cam shift on the engine was suspected.

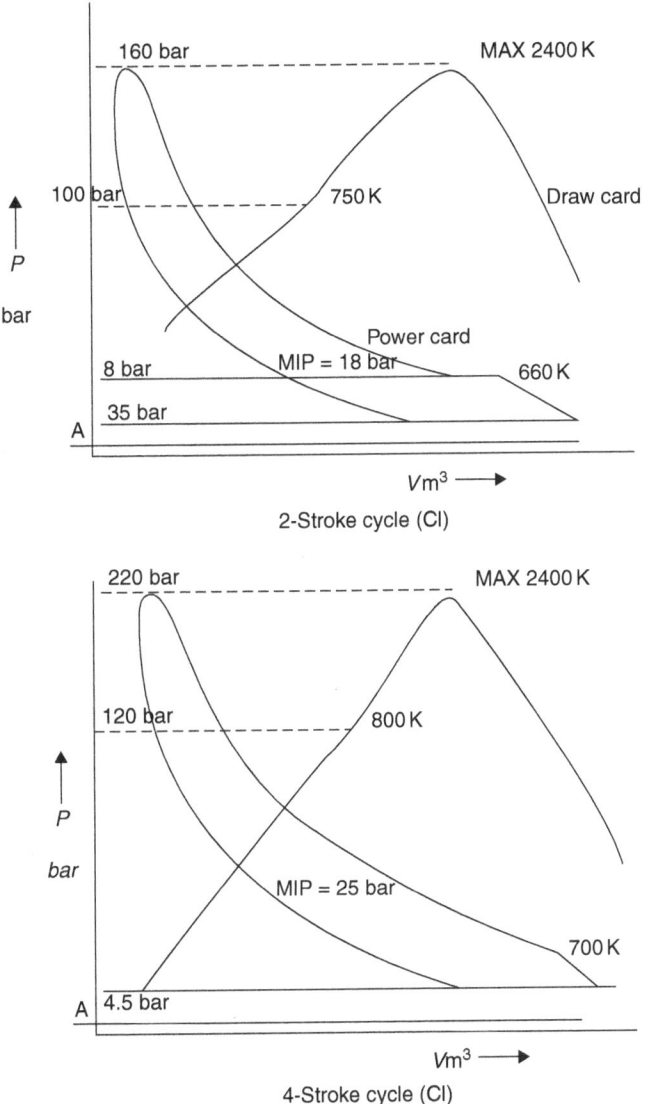

▲ **Figure 1.9** *Typical indicator (power and draw) diagrams*

Figure 1.11 shows the light spring diagrams for CI engines of the two-stroke and four-stroke types. These diagrams are particularly useful in modern practice to give information about the exhaust – scavenge (induction) processes as all main engines are now turbocharged. The turbocharge effect is shown in each case and it will be observed that there is a general lifting up of the diagram due to the higher pressures.

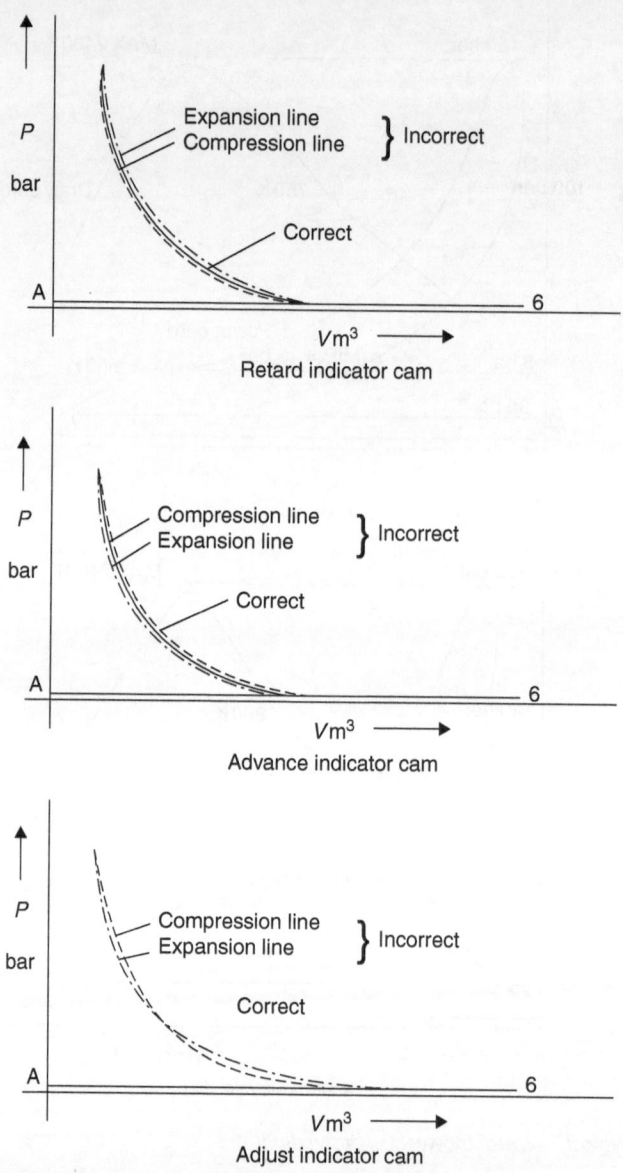

▲ **Figure 1.10** *Compression diagrams*

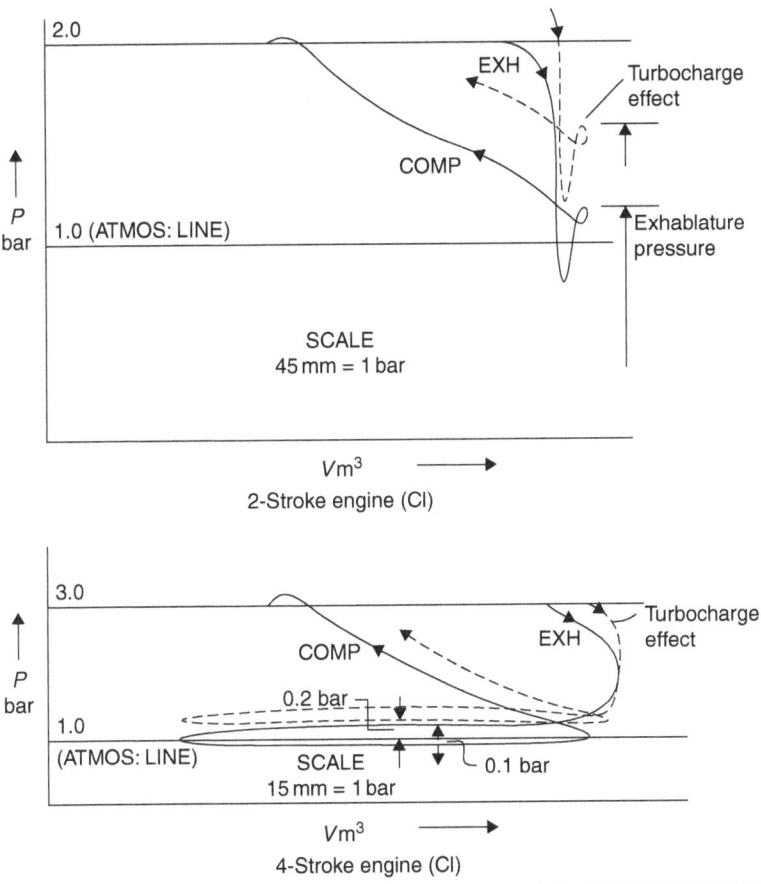

▲ **Figure 1.11** *Typical indicator (light spring) diagrams*

Other Related Details

Fuel valve lift cards are useful to obtain characteristics of injectors when the engine is running. A diagram given in figure 1.12 shows the outcome of such a card. Electronic system can easily be set up to obtain such a result even on high-speed engines. This is an example of the advantage of the electronic system over the older mechanical ones and is a common question asked by the Flag State examiner.

Typical diagram faults are normally best considered in the particular area of study where they are likely to occur. However, as an introduction, two typical combustion faults are illustrated on the draw card of figure 1.12. Turbocharge effects are also shown in figure 1.11 and compression card defects are shown in figure 1.10. It should perhaps

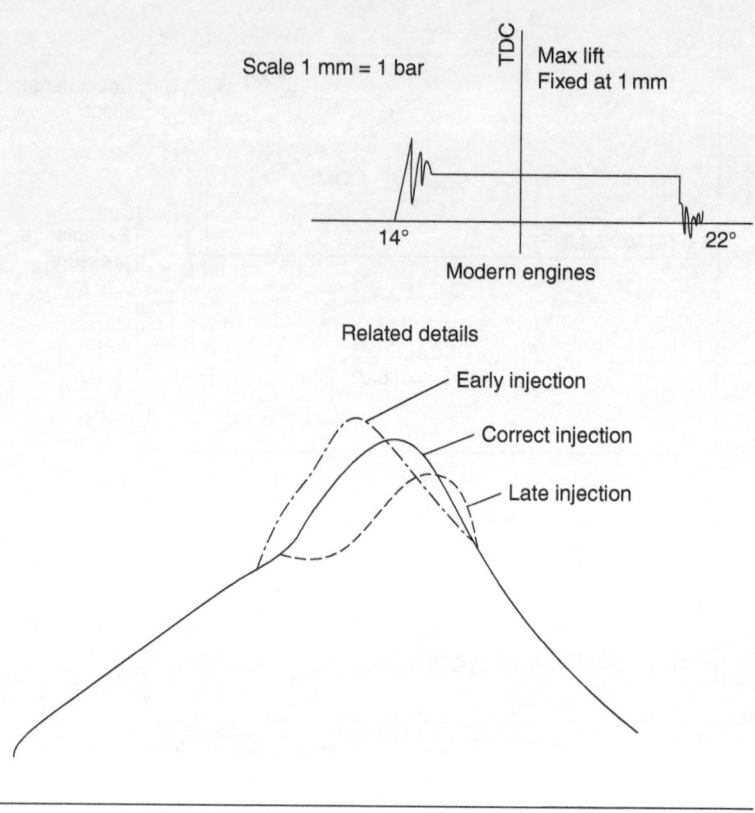

Scale 1 mm = 1 bar

Max lift
Fixed at 1 mm

TDC

14°

22°

Modern engines

Related details

Early injection

Correct injection

Late injection

Typical faults shown on draw card

▲ **Figure 1.12** *Fuel valve lift diagrams*

be stated that before attempting to analyse possible engine faults it is essential to ensure that the indicator itself and the drive are free from any defect.

CR has been discussed previously and with SI engines the limits are pre-ignition and detonation. Pinking (knocking) and its relation to octane number are important factors as are anti-knock additives such as lead tetra-ethyl, $Pb(C_2H_5)_4$. Factors more specific to CI engines are ignition quality, diesel knock and cetane number, etc. In general, these factors, plus the important related topics of combustion and the testing and use of lubricants and fuels, should be particularly well understood and reference should be made to the appropriate chapter in Volume 8 of the Reeds series. This is especially true for the modern practice of using low sulphur fuels (LSFs) or biofuels. Modern electronic systems rely on being able to measure the knocking condition and alter the injection timing to remove this harmful effect.

Accuracy of indicator diagram calculations is perhaps worthy of specific comment. The area of the power card is quite small and therefore the errors introduced by the measuring device used to determine the area, planimeter, are therefore significant. Multiplication by high spring factors makes errors in evaluation of mean indicated pressure (MIP) also significant and certainly of the order of at least ±4%.

Further application of engine constants gives indicated power calculations having similar errors. Provided the inaccuracies of the final results are appreciated then the real value of the diagrams can be established. Power card comparison is probably the most vital information to be gained from indicator diagrams. However, modern practice using mechanical devices would perhaps favour maximum pressure readings, equal fuel quantities, uniform exhaust temperature, etc for cylinder power balance and torsion meter for engine power. The draw card is particularly useful for compression and/or combustion fault diagnosis and the light spring diagram for the analysis of scavenge – exhaust considerations.

Turbocharging

This subject is considered in detail in Chapter 4; however, one or two specific comments relating to timing diagrams need to be made at this point. The start of the exhaust process is required much earlier in turbocharged engines to drop exhaust pressure quickly before the induction of the next air charge. The time allowed for the discharge of the greater gas mass needs to be longer than for naturally aspirated engines. As the air induction phase is slightly longer, the two-stroke cycle exhaust is open from 76° before BDC to 56° after (unsymmetrical by 20°) and scavenge 40° before and after. For the four-stroke cycle, air is open as much as 75° before top centre for 290° and exhaust is open 45° before bottom centre for 280°, that is, considerable overlap.

Actual timing diagrams

Figure 1.13a–d shows examples of actual timing diagrams for four types of engine. It will be seen that in the case of the poppet valve type of engine, the exhaust opens at a point significantly earlier than on the loop scavenged design. This is because the exhaust valve can be controlled, independently of the piston, to open and close at the optimum position. This means that opening can be carried out earlier to effectively utilise the pulse energy of the exhaust gas in the turbocharger. The closing position can also be chosen to minimise the loss of charge air to the exhaust. With the loop scavenged engine, however, the piston controls the flow of gas into the exhaust with the result that the opening and closing of these ports are symmetrical about the bottom centre. To minimise the losses of charge air to exhaust, the choice of exhaust opening position is dictated by the most effective point of exhaust port closure.

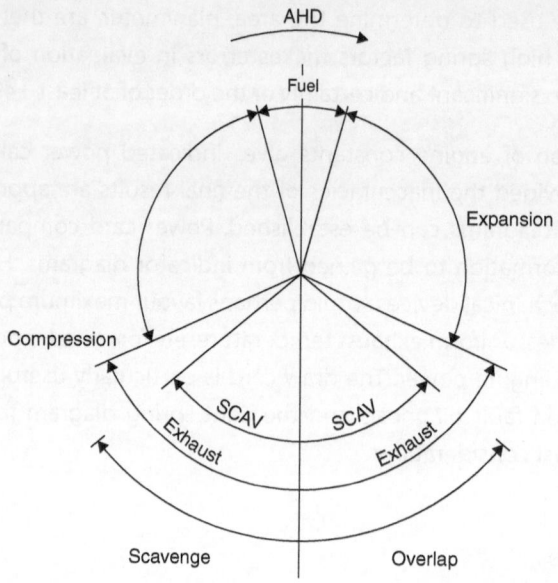

▲ **Figure 1.13a** *Crank timing diagram for two-stroke loop scavenged turbocharged engine. Exhaust and scavenge symmetrical about bottom dead centre.*

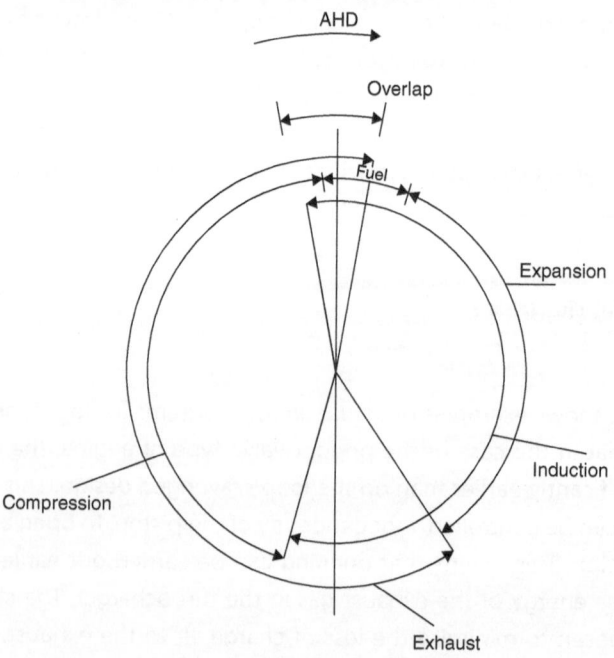

▲ **Figure 1.13b** *Four-stroke naturally aspirated engine. Note the difference of overlap between turbocharged and naturally aspirated four-stroke engine*

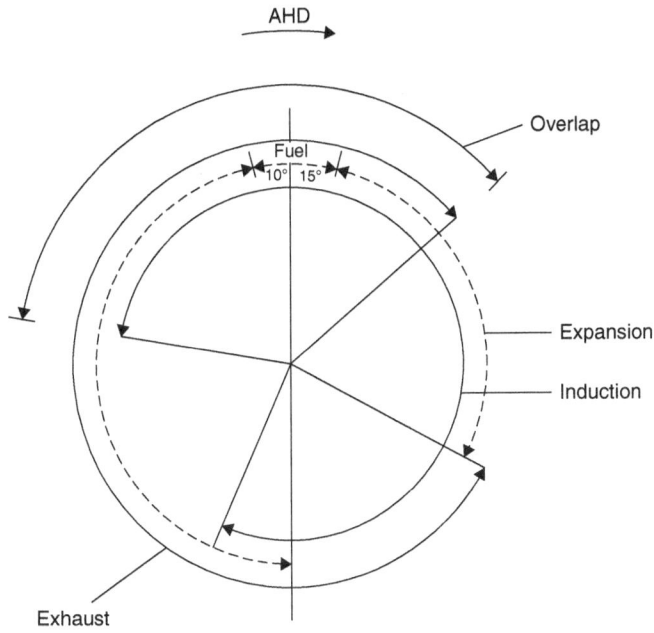

AHD

Overlap

Fuel
10° 15°

Expansion

Induction

Exhaust

▲ **Figure 1.13c** *Four-stroke turbocharged engine*

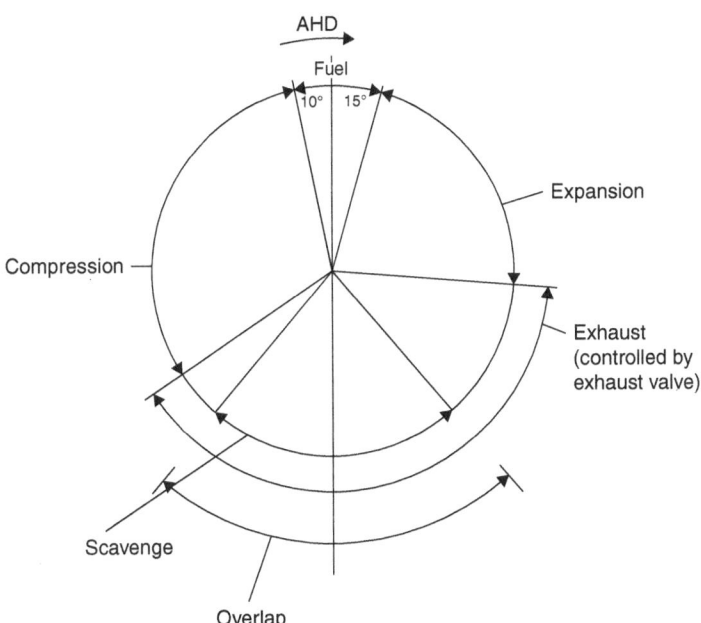

AHD

Fuel
10° 15°

Expansion

Compression

Exhaust
(controlled by
exhaust valve)

Scavenge

Overlap

▲ **Figure 1.13d** *Crank timing diagram for two-stroke turbocharged engine (uniflow scavenge; exhaust controlled by exhaust v/v in cylinder cover)*

Comparison of the crank timing diagrams of the naturally aspirated and turbocharged four-stroke diesel engine shows the large degree of valve overlap on the latter. This overlap, together with turbocharging, allows more efficient scavenging of the combustion gases from the cylinder. The greater flow of air through the turbocharged engine also cools the internal components and supplies a larger mass of charge air into the cylinder prior to commencement of compression.

Types of indicating equipment

Conventional indicator gear is shown in figure 1.14; its precise details and manufacturer's descriptive literature can be found on the LEHMANN & MICHELS GmbH website. For high-speed engines an indicator of the 'Farnboro' type is often used. Maximum combustion pressure and compression pressure can be taken using a peak pressure indicator. The fuel is shut off to the cylinder being measured to obtain the compression pressure.

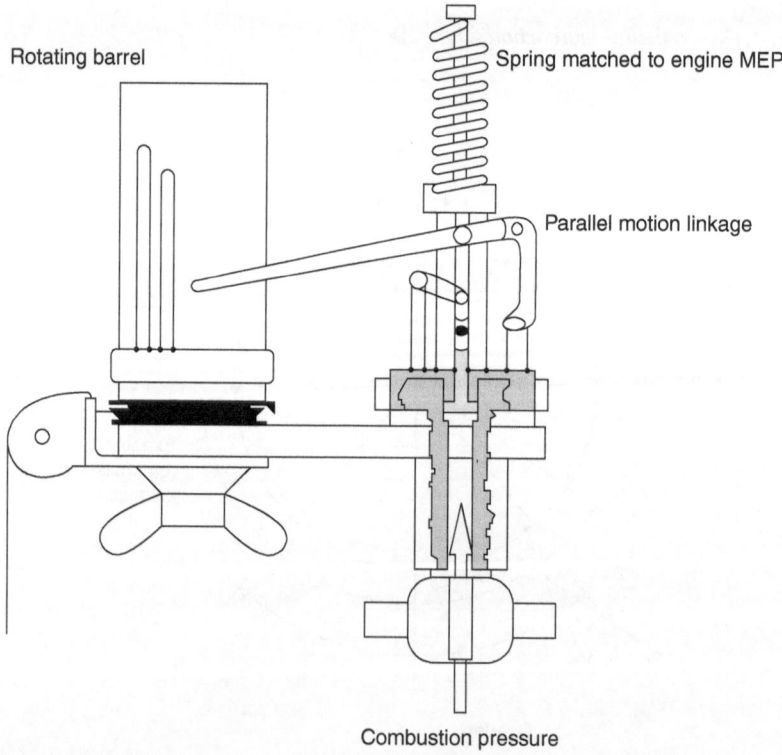

▲ **Figure 1.14** *Engine indicator (mechanical)*

Electronic indicators

The limitations of mechanical indicating equipment have been overtaken by the need for greater accuracy and a higher level of detail as engine powers have risen and especially now as electronic combustion control has become normal practice. With outputs reaching 5500 hp per cylinder, inaccuracies of ±4.0% will lead to large variations in indicated power and therefore attempts to balance the engine power by using mechanical indicating equipment will only have limited success. The inaccuracies stem from the friction and inertia of mechanical equipment transmitting errors in recording as well as the inaccuracy of measuring the height of the power card.

Modern practice utilises electronic equipment to monitor and analyse the cylinder peak pressures and piston position and presents the information on a display, which could be part of the analyser or could be the display on a computer. The cylinder pressure is measured by a pressure transducer, which can be attached to the indicator cock. Engine position is detected by a magnetic pick-up in close proximity to a toothed flywheel. The information is fed to a microprocessor, where it is averaged over a number of engine cycles, before calculations are made as to indicated power and MEP (figure 1.15). The advantages of this type of equipment are as follows:

1. It supplies real-time dynamic operational information. Therefore, the injection timing can be measured while the engine is running, which is a more accurate method of measuring the timing, and if sensors are placed on the crankshaft at each unit, as with Wärtsilä's 'intelligent combustion' system, it allows for crankshaft twist while the engine is under load, unlike other methods, which do not.

2. It can compare operating conditions with optimum performance, which leads to improvements in fuel economy and thermal efficiency.

3. It produces a load diagram for the engine, clearly defining the safe operating zone and optimum performance zones for the engine.

4. It can produce a trace of the fuel pressure both rise and fall in the high-pressure lines, which is invaluable information when diagnosing fuel injection faults.

Early operational experience with this type of equipment pointed to problems of unreliability with the pressure transducers when connected continuously to the engine. To overcome this problem, manufacturers briefly experimented with alternative methods of measuring cylinder pressure, such as attaching a strain gauge to the cylinder head stud of each cylinder. Since the strain measured is a function of cylinder pressure, this information can be fed to the microprocessor. However, modern production techniques have improved the situation and there are now a number of different sensor technologies used for detecting and transmitting a high pressure that is also fluctuating at a high frequency, but the most common technique is the resistive measurement method because it has the advantage of referencing the measured pressure to the ambient conditions.

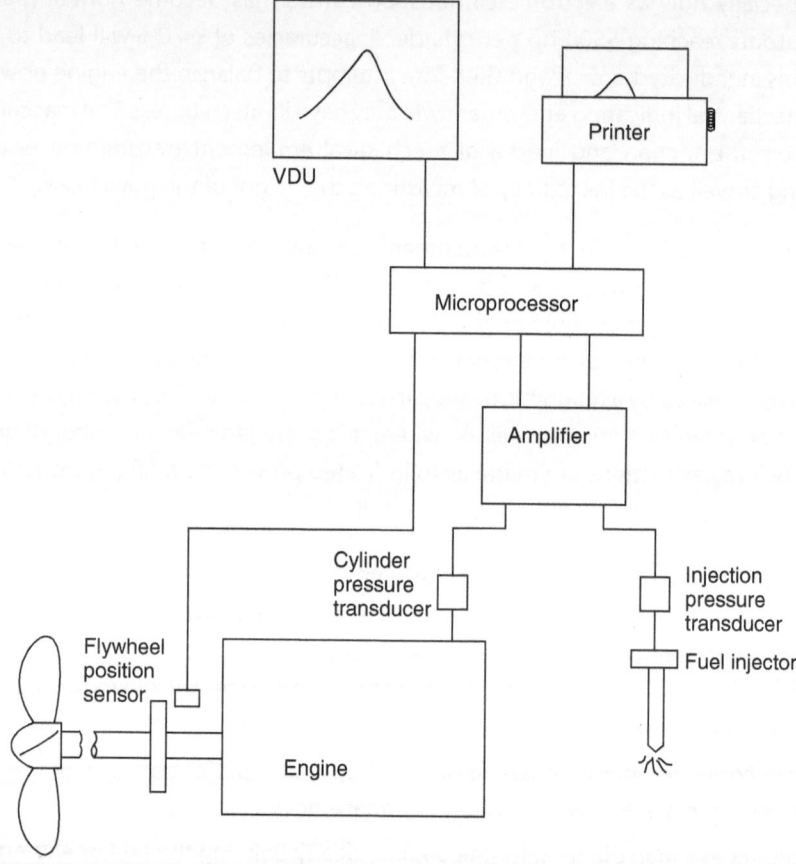

▲ **Figure 1.15** *Basic electronic combustion indicator equipment*

One construction method for this type of pressure sensor consists of using two coating techniques. The first step is to electrically insulate the active sensor components from the body of the device by using a thin silicon dioxide (SiO_2) layer, which is placed on the sensor using a chemical vapour deposition (CVD) process. The insulating layer is capable of withstanding in excess of 500 V a.c. Resistors made of nickel and nickel-chrome alloy are sputtered directly onto the SiO_2 surface to form a Wheatstone bridge circuit. In subsequent process steps, conventional lithographic techniques are used to create the conductor paths typically used in strain gauges.

Instrumentation and control

The accuracy and usefulness of any control system depends upon the measured data that is collected. Automation, instrumentation and therefore control of machinery and dynamic 'systems' has moved at a rapid pace since the development of electronic, solid state, sensors and transducers.

Engineers and operators of ships have been assisted for many years by the measurement of parameters such as temperature, pressure, movement, level and flow. However, the instruments in the past were usually mechanical, electrical, pneumatic or a combination of these.

The instruments were also expensive as they had to be manufactured to fine tolerances to be accurate and reliable. They often needed adjustment and they were also subject to wear, especially in the harsh marine environment.

With modern materials and manufacturing techniques coupled to electronic components, the latest measuring devices, or sensors, are very small, robust and reliable. This has helped fuel the explosion of information available for control systems, diagnostic systems and safety systems.

The design of modern sensors is helped by the use of finite element modelling (FEM). FEM analysis can be used to investigate sensors under different load conditions and their performance is evaluated so that designs can then be modified accordingly.

For example, electrical connection from the strain gauge on the cantilever beam to a printed circuit board (PCB) is made with the help of wire bonds in order to mechanically decouple the sensor element from the electronics and the output signal from the Wheatstone bridge, which is generated by the very slight stretching of the diaphragm under the action of the pressure that is being measured.

The accuracy of modern sensors is now well below ±1% and in some cases the error in a pressure sensor amounts to a few bar over the operating range of 2500 bar and they also have to conform to the marine quality standards for the equipment.

The quality process could include highly accelerated lifetime testing such as subjecting the sensors to a series of tests designed to exceed the sensor specifications. The vibration test, for example, involves sensors with a specified vibration tolerance being tested outside their design conditions and at a higher temperature range than expected in service, over a long time period.

The low cost of sensors means that tracking can be achieved that was not possible even a few years ago. For example, by using the latest, economically priced sensors to monitor fuel injector needle lift and cylinder pressure, the combustion process in each cylinder can be individually optimised, thereby allowing operators to reduce emissions, increase fuel efficiency and monitor engine health more closely.

Combustion control

A purely mechanical system of introducing and shutting off fuel during the control of the combustion process within the cylinders of a diesel engine is very crude when compared with computer controlled systems.

For example, the highly reliable 'scroll-type' fuel pump (see page 134) is manufactured to high precision but will only normally alter the end of injection depending upon the load on the engine. Therefore, optimising the timing for fuel injection will be compromised with this system. The fast-acting computer controlled 'common rail' systems (see page 144) will accomplish optimisation over a much broader range of conditions.

The increased reliability of these systems allows combustion and fuel injection monitoring equipment to be permanently installed, giving real-time data to inform the engineering staff about engine performance and maintenance.

It has always been the case that 'trend analysis' of performance data has been used to good effect to inform the engineering officers about the internal condition of their equipment.

However, with the increase in the data available and the development of the subject of 'mechatronics' (see page 234 for more details), trend analysis is now much more sophisticated and therefore any faults can be detected early.

Variable valve timing (VVT)

With the modern electronic control and hydraulic/electrical actuation of components affecting engine combustion, manufacturers can now accomplish so much more than in the past. VVT is one of the areas of combustion control that has had a big influence on the operation of engines.

The problem with mechanical operation using camshaft control of the inlet and exhaust valve is that it is difficult to change the cam profile and influence the time of opening and closing for different engine speeds.

VVT enables the use of the Miller cycle, as described on page 13, without the disadvantage of the very poor part-load operation. If the Miller cycle is used at part load then the combustion is starved of air and a large amount of smoke is produced by the engine. The use of VVT technology overcomes this and allows the inlet and exhaust valve timing to be optimised for all engine loads. It also allows the Miller cycle to be switched on and off to match the engine operating conditions.

Fatigue

Fatigue is a phenomenon that affects materials that are subjected to cyclic or alternating stresses. Designers will ensure that the stress induced in a component is below the yield point of the construction material as measured on the familiar stress/strain graph (see Volume 2 of the Reeds series). However, if that component is subjected to constant cyclic stresses, it may fail at a lower value due to fatigue. The most common method of displaying information on fatigue is the S–N curve (figure 1.16). This information is obtained from fatigue tests usually carried out on a Wohler machine in which a standard specimen is subjected to an alternating stress due to rotation. The specimen is tested at a particular stress level until failure occurs. The number of cycles to failure is plotted against stress amplitude on the S–N curve. Other specimens are tested at different levels of stress. When sufficient data have been gathered, a complete curve for a particular material may be presented.

It can be seen from figure 1.16 that, in the case of ferrous materials, there is a point known as the 'fatigue limit'. Components stressed below this level can withstand an infinite number of stress reversals without failure. Stress is equal to the load divided by the cross-sectional area (CSA) of a given material; therefore, we can deduce that reducing the stress level on a component involves increasing the CSA, resulting in a weight penalty or reducing the load. In marine practice the weight implications are generally regarded as secondary to reliability and long life and so components are usually stressed below the fatigue limit. This is not the case in, for example, aeronautical practice where weight is a major consideration. In this situation the component designer would compromise between weight and stress levels and from the S–N curve would calculate, with the addition of a safety margin, the number of cycles the component could withstand before failure occurs. A marine example is that the working life of four-stroke medium-speed diesel bottom end bolts are calculated in this way. Indeed, recent investigations by Lloyds Register into engine failures concluded that cyclic stresses induced in bottom end bearings can cause ovality of the shell bearings that, if left unchecked, would lead to big end bolts failure. This aspect of the bearing structure should be checked at the appropriate service interval.

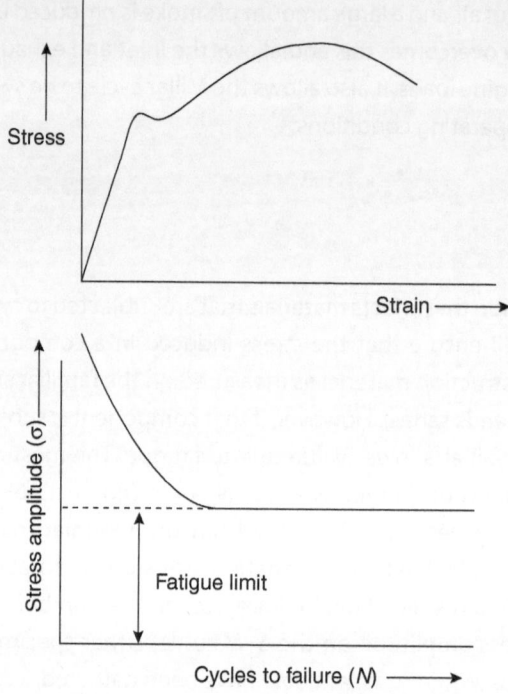

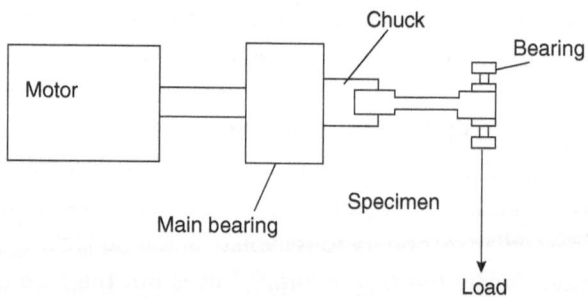

▲ **Figure 1.16** *Wohler machine for zero mean stress fatigue testing*

Engine combustion control systems

The advanced abilities of electronic control systems have been the key component in the improvements in diesel engine efficiency and reduced emissions of harmful chemicals. Engine design and developments in material science are the other factors that contribute to improved engine efficiency.

Fig 1.17 shows a basic combustion monitoring and control system where the traditional mechanical camshaft and associated equipment is retained. The effective control of engine combustion will depend upon the collection of data in real time. Fig 1.17 shows how the combustion control module, local to the engine, receives information from the pressure sensor on each cylinder. This will enable a comparison of the load in each unit to take place and if the system has the link to control the fuel to each cylinder then, under steady state conditions, the engine can be balanced in real time by the system. If the data collection system is only used to monitor the combustion in each cylinder, then the engineers will need to review the data and make manual adjustments to the fuel supply and injection system.

Data relating to the position of the crankshaft is essential for the correct timing of the fuel injection. Input from the scavenging process will then help to optimise the engine operation, especially at part or transient loads, during manoeuvring for example.

The electronic control of the equipment that affects the combustion of the diesel engine means that fine adjustments can be made in a very short time scale. However, this does rely upon a fast and efficient communication system – see page 236 for more information about these systems.

More advanced control systems can be designed to have a combustion control module dedicated to each unit of the engine and on engines without a mechanical camshaft will also take on the timing of the exhaust valve and the air start valve.

It can be appreciated by marine engineering students that those responsible for the safe operation of these engines will need an in-depth knowledge of the capabilities of the control system. The correct operation of mechanical machinery is only one part of the functioning system. Engineering officers should also be on the lookout for incorrect functioning of faulty sensors. For example, an incorrect voltage from the crankshaft positioning sensor could trigger an engine shutdown. The cause of the failure could be straightforward, such as vibration causing sensors to fail or come loose.

The success of the system relies upon each controller or device being able to interpret the information moving around the system. Therefore, standards must be used and students should be aware that the National Marine Electronics Association (NMEA) has developed the standard, MNEA 2000, for the data communication for devices used in ship's control systems. Based on the original SEA J1939 standard, the MNEA has also included some marine specific instruction sets.

Other standards that are used in industrial control engineering are Fieldbus, Profibus and HART. Please note that security from outside interference is important and manufacturers are working all the time to make systems more secure. Therefore, regular updates of software are important.

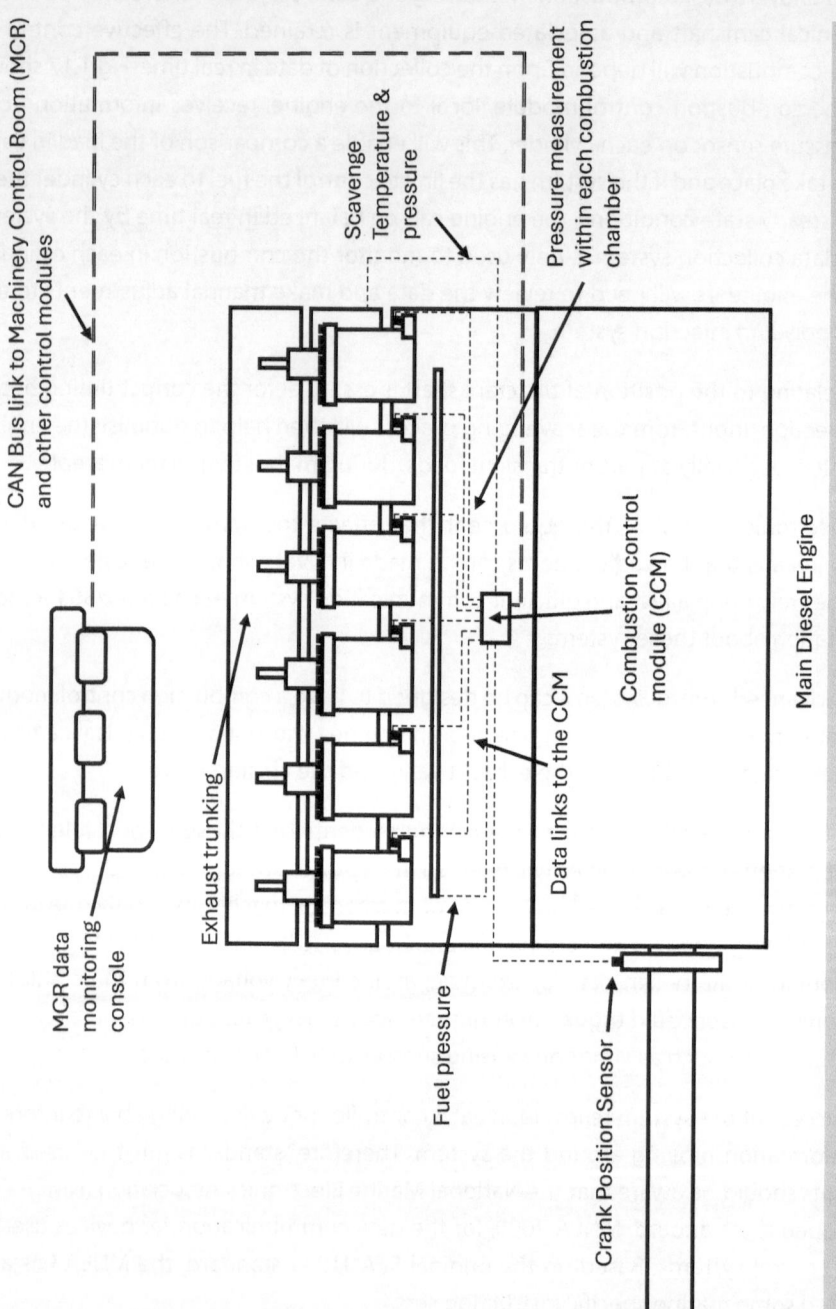

CAN Bus link to Machinery Control Room (MCR) and other control modules

MCR data monitoring console

Scavenge Temperature & pressure

Pressure measurement within each combustion chamber

Exhaust trunking

Combustion control module (CCM)

Data links to the CCM

Fuel pressure

Main Diesel Engine

Crank Position Sensor

▲ Figure 1.17

Engine Emissions

Where countries have signed up to environmental agreements, they have made a commitment to reduce the harmful effects that humans were having on the world's climate, which includes the pollution that ships create.

The International Maritime Organization (IMO) is continually updating its marine pollution convention (MARPOL) Annex VI, which is designed to limit the air pollution from ships.

The main aim is to reduce the oxides of nitrogen and sulphur and particulate matter that are contained in the exhaust gases of marine diesel engines. Also, by improving the efficiency of the power plant, better fuel consumption is achieved and therefore less carbon dioxide (CO_2) is produced to complete the same task of running and driving a ship of a given size and duty.

The term NOx and SOx has come about; these are not direct chemical terms but are a general 'collective' term used to describe the different compounds of oxides that are under consideration.

Page 328 onwards gives more details about the strategies being used in engine and propulsion systems design and management, to reduce the air pollution from ships. It is important to note here that the development of electronic sensors and digital control systems have enabled designers to build and control engines much better than in the past.

More detail about electronic sensors is contained in this volume starting on page 230 and more information about 'control systems' will be found in Volume 10 of the Reeds series.

2

STRUCTURE AND TRANSMISSION

Introduction

Designers are continually refining the transmission of power to drive the ship while making the newer versions more efficient than previous ones.

To enable marine engineering officers to completely understand the transmission of power it is necessary for them to know the strengths of the different systems and where power is 'lost' on its way from the prime mover to the propeller. This will help with monitoring the efficiency of the plant and determining where any faults may be developing.

Diesel engines are not very efficient when transitioning from one load to another. In addition, two stroke direct drive engines need to operate over a range of speeds as well as a range of loads. This means that changing speed and load when manoeuvring is a particularly inefficient time for this type of engine.

Thinking about this characteristic, students can then see that where a vessel is expected to be used on long ocean voyages and where the percentage of manoeuvring the vessel is a small part of its duty, then the large two stroke, direct drive engine will have the advantage due to its efficient 'steady' state operation.

The smaller four stroke indirect drive engine is better at handling transient loads, as well as not needing to stop and reverse during the manoeuvring period. Therefore, these will be the preferred power source for vessels such as ferries where a much larger proportion of the vessel's duty is devoted to manoeuvring the vessel in and out of port.

This is especially so with Diesel-electric plants, where the power from the diesels is also used to supply electricity to the rest of the vessel as well as electric propulsion motors. With this system, it means that the engines are not running on very light loads for long periods.

Manufacturers have now started integrating batteries or super capacitors into the mix and therefore these can be used to support the engines at the time when they are most inefficient. Tug boats are particularly suited to this system as they are required to increase and decrease the load on the engines drastically and at short notice.

Some systems have a battery bank which is so powerful that the vessel can be run in 'battery only' mode when entering and leaving port. The batteries can be charged by the diesels when at sea or by plugging into the mains while in port.

The structure of the large two-stroke, low-speed marine engine is quite different from the structure of the smaller higher-speed four-stroke medium-speed engine. Both types of engine must sit on a solid foundation called the bedplate but the two-stroke engine has a triangular 'framebox' or 'A' frame that sits between the bedplate and the structure that carries the cylinders, which MAN Diesel & Turbo call the cylinder frame but is sometimes termed the entablature. In modern designs the frame section must have the following fundamental requirements and properties.

Strength is necessary since considerable forces are set up within an engine as it is operating. These may be due to out-of-balance effects (which might differ depending upon how the ship is loaded), vibrations and variations in the gas load forces transmitted and gravitational forces (due to the motion of the vessel).

Rigidity is required to maintain the correct alignment of the engine running gear. However, a certain degree of flexibility will prevent high stresses that could be caused by any slight misalignment, although with modern manufacturing techniques and quality control, misalignments should be few and far between.

Lightness is important as it may enable an increase in the power to weight ratio; also, less material would be used, bringing about a saving in cost. Both are important selling points as they would give increased cargo capacity or reduced fuel consumption.

Toughness in a material is a measure of its resilience and strength; this property is required to enable the material to withstand the fatigue conditions that prevail.

Simple design – if manufacture and installation are simplified then a saving in cost will be realised with reduced maintenance time and less downtime.

Access – ease of access to the engine transmission system for inspection, maintenance and installation is a fundamental requirement.

Dimensions – ideally, these should be as small as possible to keep engine containment to a minimum, which will give either more engine room space to work in or more cargo carrying ability of the vessel. One of the drawbacks of the two-stroke engine is the empty space required above the cylinder head. This is due to the requirement for the piston rod to be withdrawn vertically at the same time as the piston when the piston needs maintenance, such as the replacement of the piston rings. In response to this problem, MAN Diesel & Turbo have designed the 'double jib crane', which allows the two-stroke piston to be extracted from the engine when there is a low height above the engine tops.

Seal – the engine and transmission system container must effectively seal in the oil and any vapours generated from the rest of the engine room.

Manufacture – increasingly, modern engines are manufactured in prefabricated sections with the sections brought together for the final assembly. This is a much easier method than trying to build the whole engine in one place section-by-section. The general arrangement of the engine is shown in figure 2.1.

The designers use computer programs to plan the dimensions of the components and overall structure of a new engine. Computer-aided drawing (CAD) software will enable the designer to ensure that the engine fits into the vessel and other software called finite element analysis (FEA), will run simulation models to work out if the engine can withstand the operational conditions for which it was designed.

FEA is a computerised method of calculating all the loads, mechanical stresses, thermal stresses and cyclical fluctuating stresses in complex structures. The algorithms that make up the computer models can be in two dimension (2D) or three dimension (3D). The 2D models are less complex and can be run on powerful personal computers (PCs). The 3D models, on the other hand, are very complex and would have to be run on mini or mainframe computers. These techniques have enabled a substantial increase in the designers' ability to design accurately stressed engines and indeed whole vessels.

The structure of the modern two-stroke engine must be rigid and able to withstand the tremendous forces imposed by the combustion pressures set up by these high-performance engines. Full 3D stresses and bending moments are calculated on each engine design before it is put into production or a prototype is built.

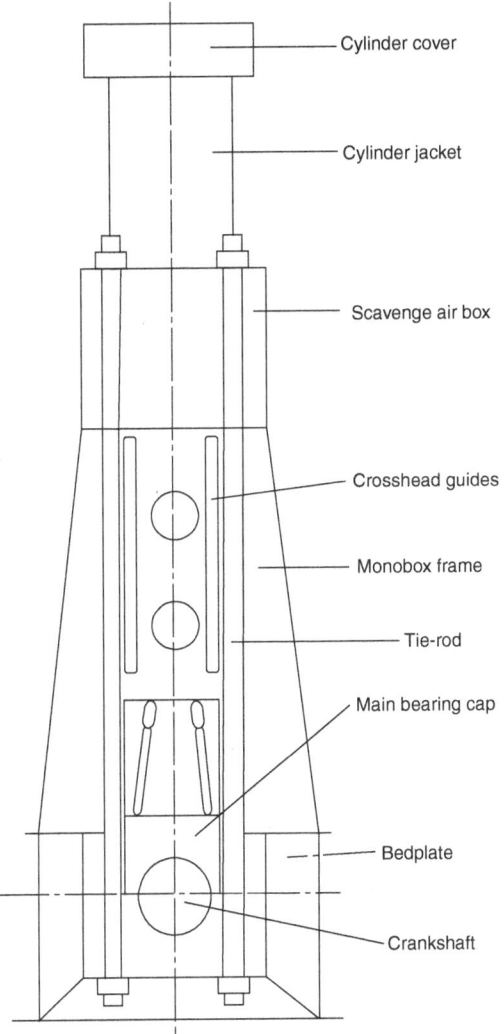

▲ **Figure 2.1** *General arrangement of engine structure*

Bedplate

The engine bedplate is a structure that was traditionally made of cast iron or cast steel because of the materials' excellent resistance to compressive forces. However, cast metals can add a lot of weight to the structure and, especially on the large two-stroke engine, manufacturers are using prefabricated steel, or a hybrid arrangement of cast steel and prefabricated steel, as alternative methods of construction.

Cast iron, one-piece structures are generally confined to the smaller high- or medium-speed engines rather than the larger slow-speed crosshead type engine. This is due to the problems that are encountered with the quality control of large castings. These problems include poor flow of material to the extremities of the mould, poor control of the grain size, which leads to a lack of homogeneity of strength and soundness, and poor impurity segregation. In addition to these problems, cast iron has poor performance in tension and its modulus of elasticity is only half that of steel, hence for the same strength and stiffness a cast iron bedplate will require to be manufactured from more material. This results in a weight penalty for larger cast iron bedplates when compared with a fabricated bedplate of similar dimensions. Cast iron does, however, enjoy certain advantages for the construction of smaller medium- and high-speed engines, which are as follows:

- Castings do not require heat treatment.
- Cast iron is easily machined and is good in compression.
- The master mould can be reused many times, which results in reduced manufacturing costs for a series of engines.
- The noise and vibration damping qualities of cast iron are superior to that of fabricated steel.
- As outputs increase, nodular cast iron, due to its higher strength, it is becoming more common for the manufacture of medium-speed diesel engine bedplates.
- Modern cast iron bedplates for medium-speed engines are generally, but not exclusively, a deep inverted 'U' shape, which affords maximum rigidity for accurate crankshaft alignment. The crankcase doors and relief valves are incorporated within this structure. In this design the crankshaft is 'underslung' and the crankcase is closed with a light unstressed oil tray (figure 2.2).
- As outputs of medium-speed engines increase, some manufacturers choose the alternative design in which the crankcase and bedplate are separate components, the crankshaft being 'embedded' in the bedplate (figure 2.3).

Now that welding techniques and methods of inspection have improved and larger furnaces are available for annealing, the switch to prefabricated steel structures with their saving in weight and cost has been made where the advantages are realised. It must be remembered that the modulus of elasticity for steel is nearly twice that of cast iron, hence for similar stiffness of structure roughly half the amount of material would be required when using steel (Young's modulus of elasticity is given the symbol 'E' and is a measure of the stress to strain ratio of a material) – cold draw steel = 200 GPa and grey cast iron = 124 GPa.

Early designs were entirely fabricated from mild steel but radial cracking due to cyclic bending stress imposed by the firing loads was experienced on the transverse members, especially in way of the main bearings. The adoption of cast steel, with its

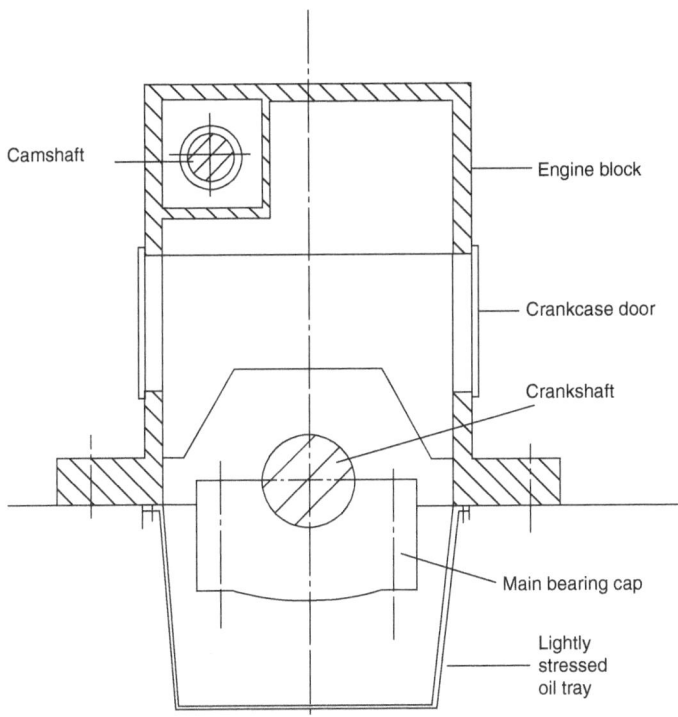

Camshaft

Engine block

Crankcase door

Crankshaft

Main bearing cap

Lightly
stressed
oil tray

▲ **Figure 2.2** *Section through engine block of medium-speed engine with underslung crankshaft*

greater fatigue strength, for transverse members has eliminated this cracking. Modern large-engine bedplates are constructed from a combination of fabricated steel and cast steel. Modern designs consist of a single-walled structure fabricated from steel plate with transverse sections incorporating the cast steel bearing saddles attached by welding (figure 2.4). To increase the torsional, longitudinal and lateral rigidity of the structure, suitable webbing is incorporated into the fabrication.

It is modern practice to cut the steel plate using automatic contour flame cutting equipment. Careful preparation is essential prior to the welding operation because:

- It is necessary to prepare the edges of the cut plate and therefore it is also necessary to make an allowance for this during the initial cutting phase.
- The equipment needs to be set correctly to ensure the smallest heat-affected zone (HAZ).
- Welding consumables are stored and used correctly to prevent hydrogen contamination of the HAZ, which could lead to post-annealing hydrogen cracking.

Following the welding operation the welds are inspected for surface cracking and subsurface flaws. The surface inspection is carried out by the dye-penetrant method or

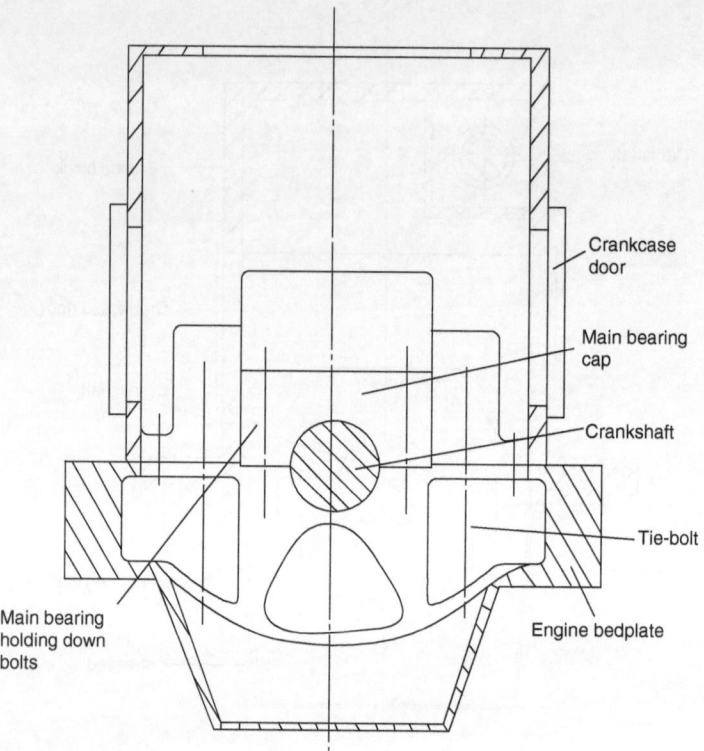

Crankcase door

Main bearing cap

Crankshaft

Tie-bolt

Engine bedplate

Main bearing holding down bolts

▲ **Figure 2.3** *Medium-speed engine bedplate with embedded crankshaft*

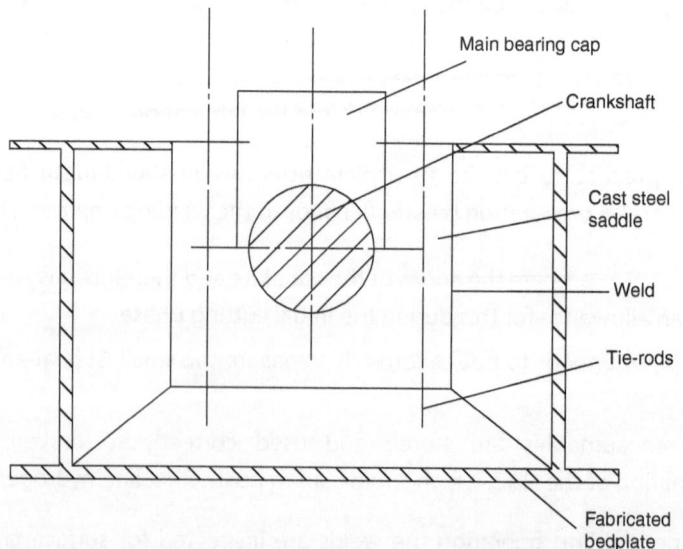

Main bearing cap

Crankshaft

Cast steel saddle

Weld

Tie-rods

Fabricated bedplate

▲ **Figure 2.4** *Modern fabricated single-walled bedplate with cast steel bearing saddle*

the magnetic particle method, while the subsurface flaws are inspected by ultrasonic testing. Any flaws found in the welds would be cut out, re-welded and tested.

The bedplate is then stress relieved by heating the whole structure to below the lower critical temperature of the material in a furnace and allowing it to cool slowly over a period of days. When cooled, the structure is shot-blasted and the weld is again tested before the bedplate is machined.

In order to minimise stresses due to bending in the bedplate, without a commensurate increase in material, tie-rods are used to absorb the combustion forces. Two tie-rods are fitted to each transverse member and pass, in tubes, through the entire structure of the engine from bedplate to cylinder cooling jacket (entablature). They are pre-stressed at assembly so that the engine structure is under compression at all times (including when the engine is in operation). Initially, engines that utilised the opposed piston principle had their combustion loads absorbed by the running gear and therefore did not require tie-rods. Modern engines employ tie-rods or stay bolts, as MAN prefer to call them, and to minimise bending moment forces across the main bearing, these are placed as close as possible to the crankshaft centreline. In some cases this has led to the use of 'jack bolts' as shown in figure 2.5. The RT84 flex, for example, uses this arrangement.

Figure 2.5 shows diagrammatically the arrangement used in the larger-bore Wärtsilä Sulzer engines. The idea was first used on engines before the merger with Wärtsilä. Employing jack bolts, under compression, to retain the main bearing caps in position allows the distance x to be kept to a minimum. Hence the bending moment Wx, where W is the load in the bolt, is also kept to a minimum.

Owing to their great length, tie-rods in large slow-speed diesel engines may be in two parts to facilitate removal. They are also liable to vibrate laterally unless they are restrained. This usually takes the form of pinch bolts that prevent any lateral movement. Although tie-rods are pre-tensioned to their correct value during assembly, they should be checked at intervals. This is accomplished by:

- Connecting both pre-tensioning jacks to two tie-rods lying opposite each other (figure 2.6).
- Operating the pump until the correct hydraulic pressure is reached. This pressure is maintained.
- Checking the clearance between the nut and intermediate ring with a feeler gauge. If any clearance exists then the nut is tightened onto the intermediate ring and the pressure is released. If no clearance is found then the pressure can be released and the hydraulic jacks can be removed.
- Figure 2.7 shows the two different arrangements on the MAN 'MC' series of engines.

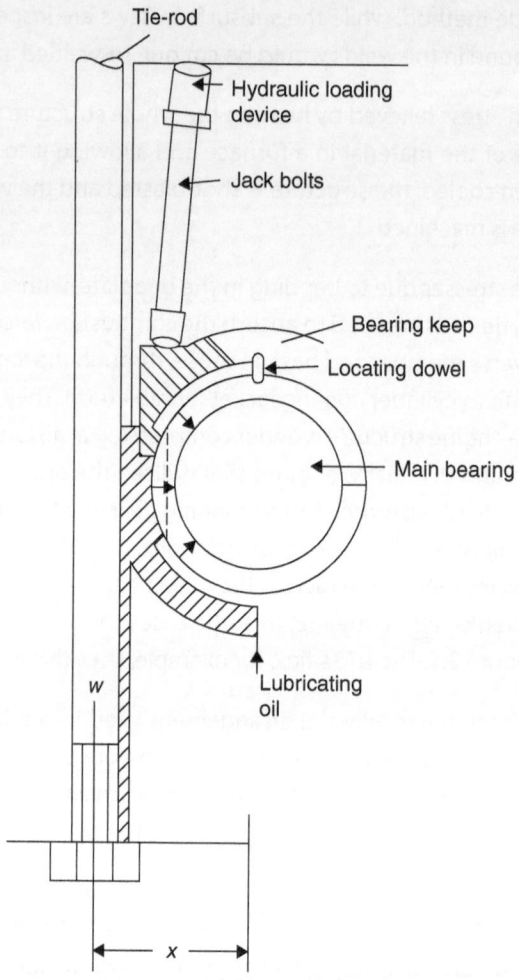

▲ **Figure 2.5** *Use of main bearing 'jack' bolts*

When hydraulic tensioning equipment is specified for use on the main engine bolts, it is essential that all the equipment is maintained in good order and the accuracy of the pressure gauges is checked regularly. If the hydraulic equipment will not exert the necessary pressure on the tie-rod then there will be great difficulty in releasing the nuts from the rod as, invariably, the nuts are round, to fit within the design of the hydraulic jack (see Fig 2.6 – upper tie-rod nut and tommy bar). With this arrangement, conventional spanners will not fit; this is also true for the cylinder cover bolts.

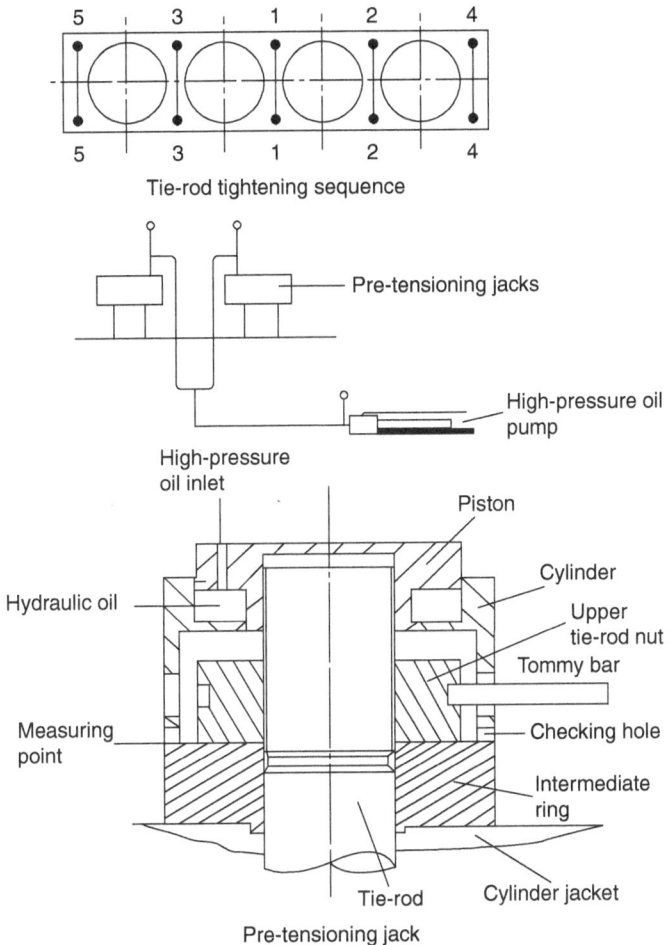

▲ **Figure 2.6** *Tie-rod tightening sequence and pre-tensioning jack*

If when inspecting the engine it is found that a tie-rod has broken, it must be replaced immediately. If the breakage that occurs is such that the lower portion is short and can be removed through the crankcase, the upper part can be withdrawn with relative ease from the top. If, however, the breakage leaves a long lower portion, it may be necessary to cut the rod and remove it in sections through the crankcase.

The latest Wärtsilä Sulzer RTA small-bore engine has moved away from this arrangement and uses bolts on the main bearings. The tie-rods are slightly further apart but this is not a problem for the modern materials used in the construction of the latest engines. The design of the latest MAN engines is to incorporate a shorter 'twin' tie or stay bolt, as MAN prefer to call them. This arrangement sees the tie-bolt stop short of penetrating through the bedplate (figure 2.7). Students undertaking

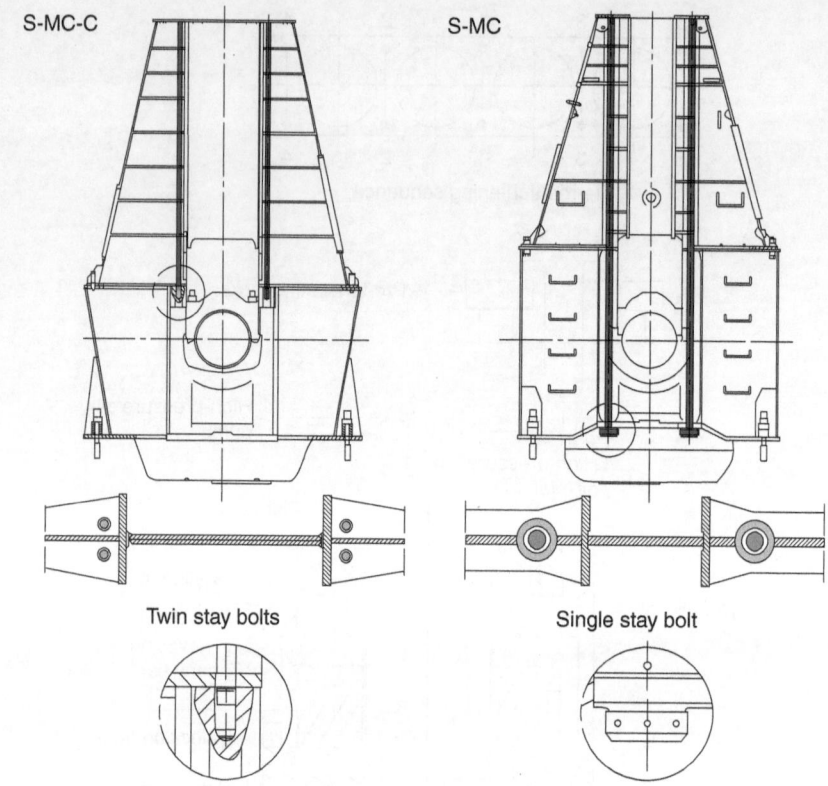

S-MC-C

S-MC

Twin stay bolts

Single stay bolt

▲ **Figure 2.7** *Arrangement of the tie-rods on the MAN S-MC series of engines*

the Flag State exams must have a good knowledge of both these systems and the reasons for their construction, as the student may be required to serve on a vessel with these engines fitted following qualification.

'A' Frames or Columns

The advent of the super-long-stroke and now the ultra-long-stroke – MAN Diesel & Turbo call their ultra-long-stroke engine the G-type, where G denotes Green – slow-speed diesel engine has resulted in an increase in lateral forces on the crosshead guides. This is due to the use of a relatively short connecting rod, which is used to reduce the overall height of this type of engine, and that results in an increased angle between the connecting rod and the guides and consequently the higher lateral force component generated. Figures 2.8(a,b) and 2.9a and 2.9b show an example of the actual arrangement of the two main engines that are currently in production.

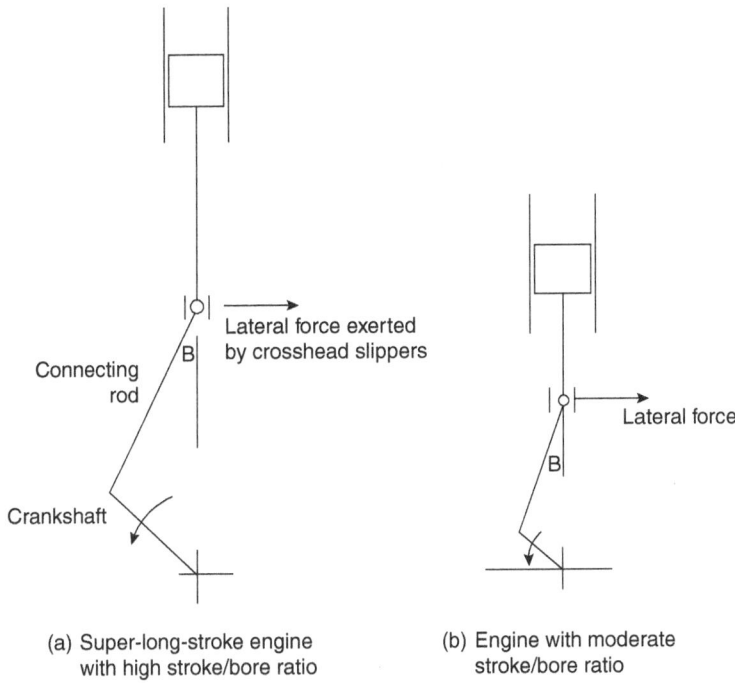

(a) Super-long-stroke engine with high stroke/bore ratio

(b) Engine with moderate stroke/bore ratio

▲ **Figure 2.8** *Increased connecting rod angle giving higher lateral forces*

In order to maintain structural rigidity under these conditions, designers tend not to utilise the traditional 'A' frame arrangement, preferring instead the 'monoblock' or 'cylinder frame' structure (figure 2.10), which consists of a continuous longitudinal beam incorporating the crosshead guides. The advantages of the monoblock design are:

- greater structural rigidity
- more accurate alignment of the crosshead
- forces are distributed throughout the structure, resulting in a lighter construction
- improved oil tightness.

In the G-type engine, the standard cylinder frame is predominately a cast construction; however, a welded frame is available upon request. The welded construction offers increased rigidity and when the scavenge air receiver is integrated into the frame a weight saving of up to 30% can be achieved. The options are made available by MAN because a large number of the engines are made under licence and the different licensees have different manufacturing capacities.

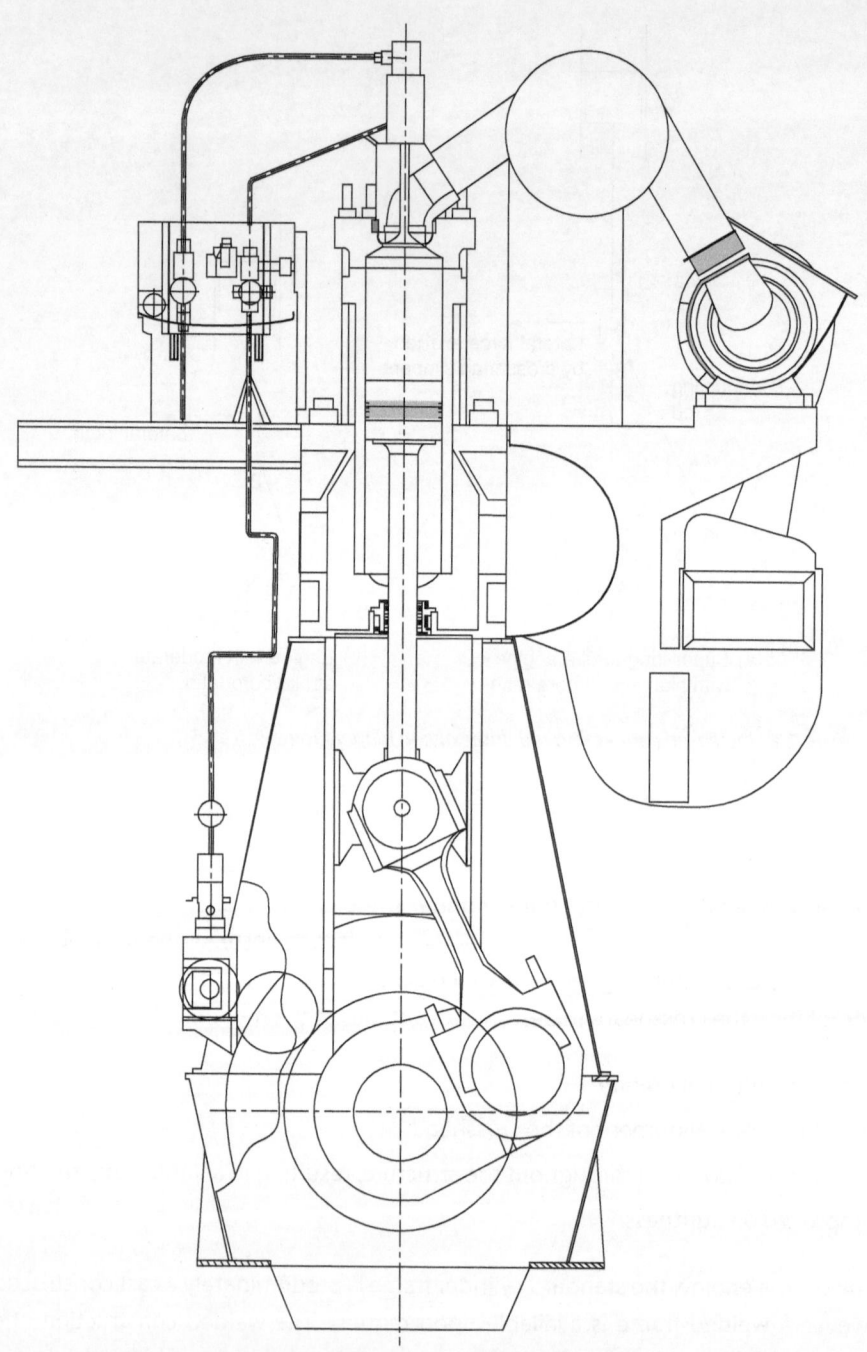

▲ **Figure 2.9a** *Wärtsilä RTA engine cross section*

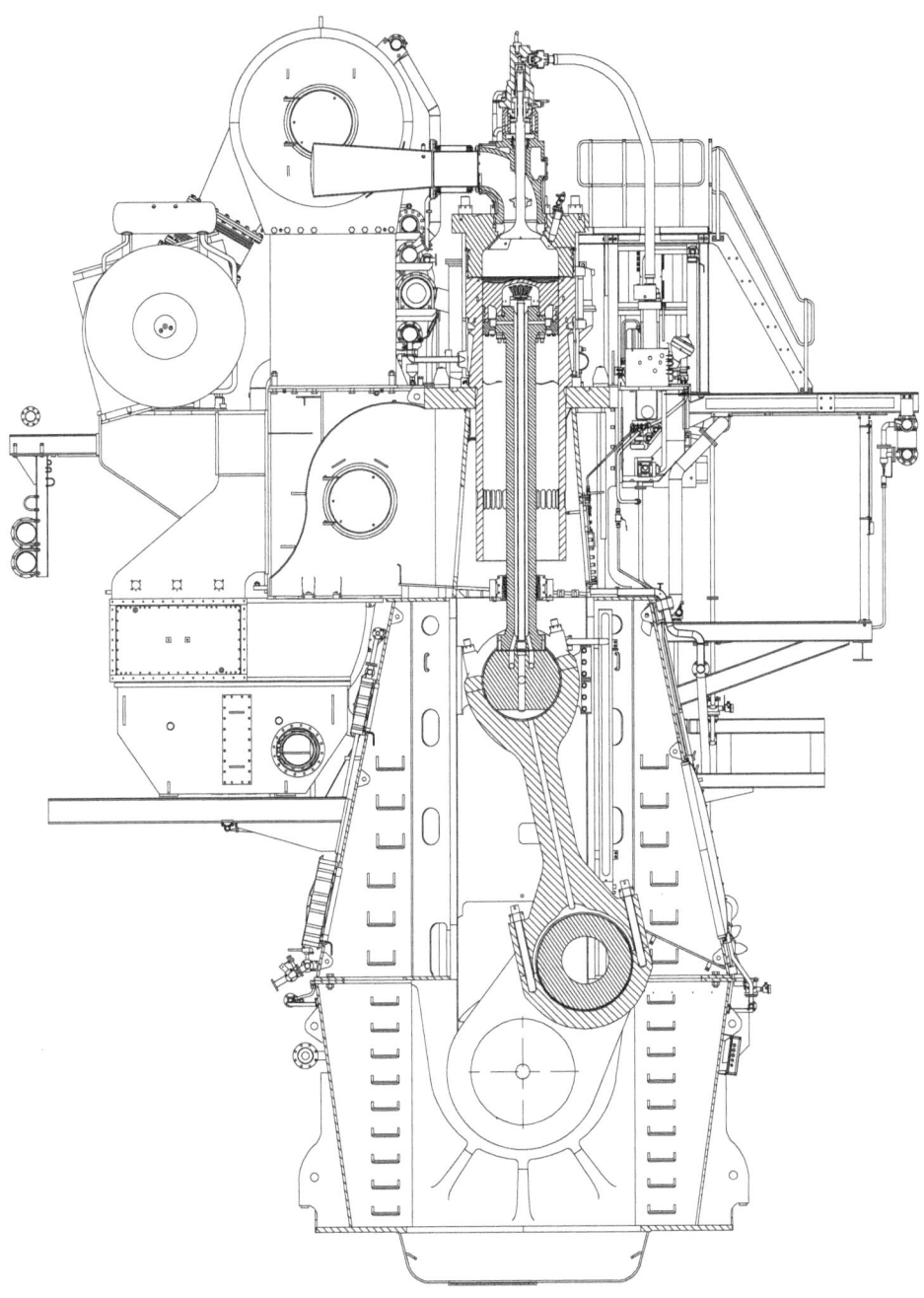

▲ **Figure 2.9b** *Cross section of MAN two-stroke engine*

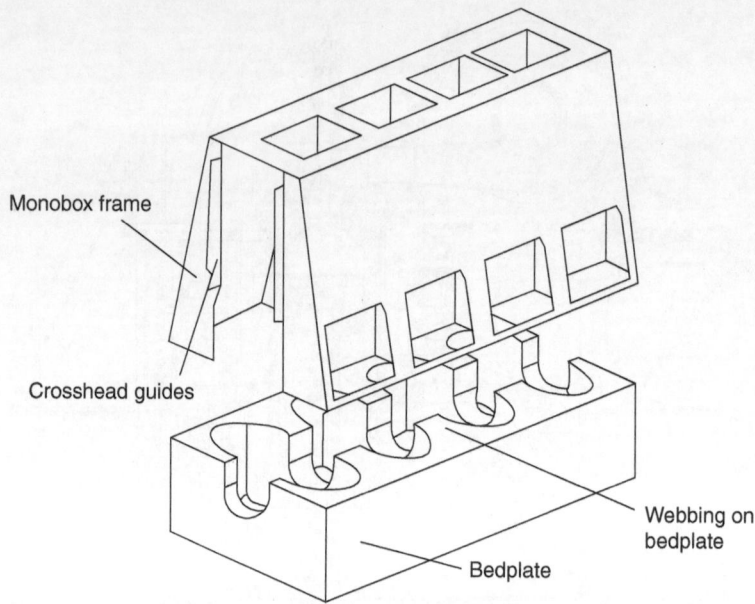

Monobox frame

Crosshead guides

Webbing on bedplate

Bedplate

▲ **Figure 2.10** *Modern monobox construction*

Holding Down Arrangements

The engine must be securely attached to the ship's structure in such a way as to maintain the alignment of the crankshaft within the engine structure. There are two main methods of holding the engine to the ship's structure.

1. By rigid foundations fixed onto the ship's structure.
2. Mounting the engine onto the ship's structure via resilient mountings.

Rigid foundations

In this more traditional method, fitted chocks are installed between the engine bedplate and the engine seating on the tank top. The holding down bolts pass through the chocks. During installation of the engine, great care must be taken to ensure that there is no distortion of the bedplate, which would lead to crankshaft misalignment. In addition, great care must be taken to correctly align the crankshaft to the propeller shaft. The engine is initially installed on jacking bolts, which are adjusted to establish its correct location in relation to the propeller shaft. When the engine is correctly positioned, and crankshaft deflections indicate no misalignment, the space between the bedplate and seating is measured and chocks are manufactured. To facilitate the fitting of chocks, the top plate of the engine seating is machined with a slight outboard-facing taper. The chocks, usually made from cast iron, are individually fitted and must bear the load over at least 85% of their area. The surface of the bedplate and the underside of the top plate that will make contact with the holding down bolt and the nut faces are machined parallel to ensure that no bending stresses are transferred to the bolt. As the holding down bolts and chocks are installed, the jacking bolts are removed.

Holding down bolts for modern slow-speed installations tend to be the long-sleeved type and are hydraulically tensioned (figure 2.11). This type of bolt, because of its greater length, has greater elasticity and is therefore less prone to fatigue cracking than the superseded short un-sleeved bolt. The bolts are installed through the top plate and a waterproof seal is usually used with 'O' rings providing the seal. Fitted bolts are installed adjacent to the engine thrust block.

The holding down bolts should only withstand tensile stresses and should not be subjected to shear stresses. The lateral and transverse location is maintained by side and end chocking. The number of side chocks will depend upon the length of the engine (figure 2.12a,b).

It is extremely important that the engine is properly installed during building. The consequences of poor initial installation are extremely serious since it may lead to the fretting of chocks, the foundation and/or the bedplate, resulting in a slackening and breakage of holding down bolts and ultimately in misalignment of the engine. To maintain engine alignment, it is important to inspect the bolts for correct tension and the chocks for evidence of fretting and looseness. There are a number of factors that can cause a holding down bolt to look as if it is tight when in fact it has become stretched. Corrosion or the nut binding on the threads may mean that the nut is held fast, appearing to be firmly in place.

One method of testing the holding down bolts is to place your thumb or finger on one side of the nut, so that you can feel any movement between the nut and the surface of the bedplate that it is supposed to be firmly against. If you then carefully tap the opposite side of the nut with a hammer, any movement felt at your thumb will indicate that the bolt has become stretched and is not contributing to holding down the engine. In this case it is very important that the bolt is replaced because it will have become elongated beyond its elastic limit and is therefore outside its design specification. If you can reach, the use of feeler gauges would also be a good way of checking for any clearance between the nut and the bedplate.

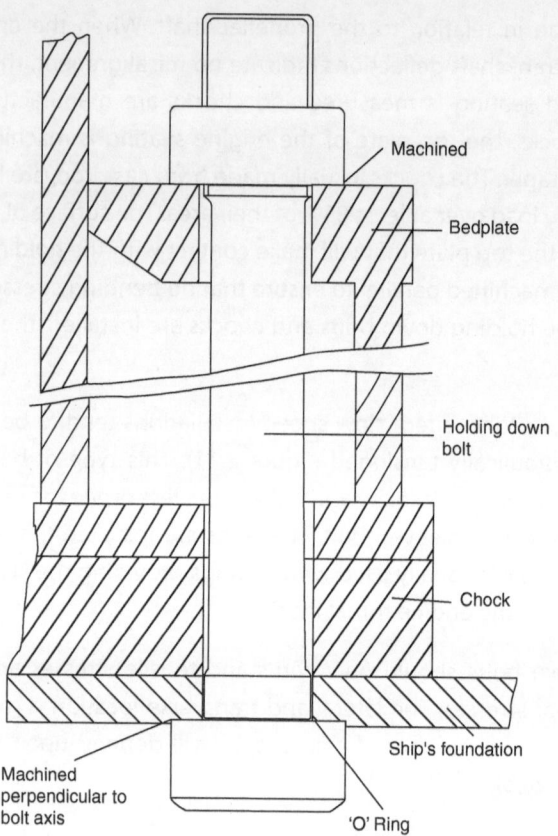

Machined

Bedplate

Holding down
bolt

Chock

Ship's foundation

Machined
perpendicular to
bolt axis

'O' Ring

▲ **Figure 2.11** *Long-sleeved holding down bolt*

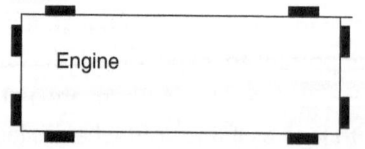

Engine

(a) Short engine: with two sets of side chocks

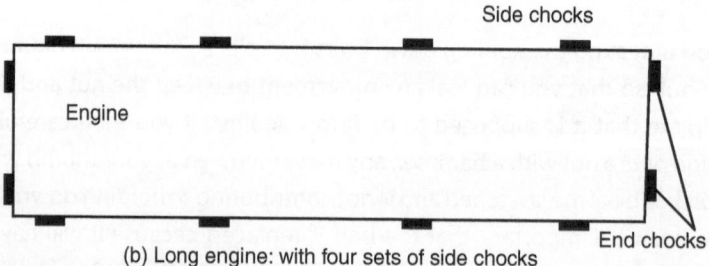

Side chocks

Engine

End chocks

(b) Long engine: with four sets of side chocks

▲ **Figure 2.12** *Side and end chocking*

An alternative to the traditional chocking materials of cast iron or steel is epoxy resin. This material, originally used as an adhesive and protective coating, was developed as a repair technique to enable engines to be realigned without the need for the machining of engine seating and bedplate. It is claimed that the time taken to accomplish such a repair is reduced, thus reducing the overall costs. Although initially developed as a repair technique, the use of epoxy resin chocks has become widespread for new buildings.

Resin chocks do not require machined foundation surfaces, thus reducing the preparation time during fabrication. The engine must be correctly aligned with the propeller shaft without any bedplate distortion before the resin application. This is done in the usual way with the exception that it is set high by about 1/1,000 of the chock thickness to allow for very slight chock compression when the installation is bolted down. The tank top and bedplate seating surfaces must then be thoroughly cleaned with an appropriate solvent to remove all traces of paint, scale and oil.

As resin chocks are poured, it is necessary for 'dams', made from foam strip, to be set to contain the liquid resin. Plugs or the holding down bolts are now inserted. Fitted bolts are sprayed with a releasing agent and ordinary bolts are coated with a silicone grease to prevent the resin from adhering to the metal. The outer sides of the chocks are now blocked off with thin section plate, fashioned as a funnel to facilitate pouring and 15 mm higher than the bedplate to give a slight head to the resin. This is also coated to prevent adhesion. Prior to mixing and pouring of the resin it is prudent to again check the engine alignment and crankshaft deflections.

The resin and activator are mixed thoroughly with the equipment that does not entrain air. The resin is poured directly into the dammed-off sections. Curing will take place in about 18 h if the temperature of the chocking area is maintained at about 20–25°C. The curing time can be up to 48 h if the temperatures are substantially below this. During the chocking operation, it is necessary to take a sample of resin material from each batch for testing purposes.

The advantages claimed for 'pourable' epoxy resin chocks over metal chocks include:

- Quicker and cheaper installation.
- Lower bolt tension by a factor of 4 when compared to metal chocks.
- Elimination of misalignment due to fretting and bolt slackening. As a result of the intimate fit of resin chocks and the high coefficient of friction between resin and steel, the thrust forces are distributed to all chocks and bolts, thus reducing the total stress on fitted bolts by about half (figure 2.13).

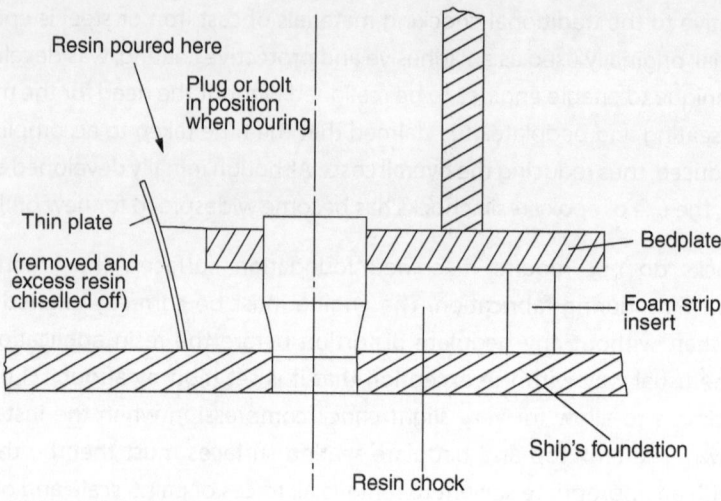

Resin poured here

Plug or bolt
in position
when pouring

Thin plate
(removed and
excess resin
chiselled off)

Bedplate

Foam strip
insert

Ship's foundation

Resin chock

▲ **Figure 2.13** *Poured resin chocks*

Resilient mountings

A possible disadvantage of rigidly mounted engines is the likelihood of noise being transmitted through the ship's structure. This is particularly undesirable on a passenger carrying vessel where low noise and vibration levels are necessary for passenger comfort and consequently manufacturers are now installing diesel engines on resilient mountings.

Diesel engines generate low frequency vibration and high frequency noise, both of which can be transmitted to the hull of the vessel. The adoption of resilient mountings will successfully reduce both noise and vibration. An illustration of this reduction can be seen in figure 2.14.

Figure 2.15 shows how the diesel engine is aligned and rigidly mounted to a fabricated steel subframe. This can be via either solid or resin chocks. The subframe is then resiliently mounted to the ship's structure on standard resilient elements.

In geared engine applications, the engine is again mounted, via solid or resin chocks, to a subframe, which is resiliently mounted to the ship's structure. The engine is then coupled to the reduction gearbox through a highly elastic coupling. It is necessary to limit the total amount of lateral and longitudinal movement of the engine relative to the ship's structure and this is accomplished by stopper devices built into the holding down arrangement.

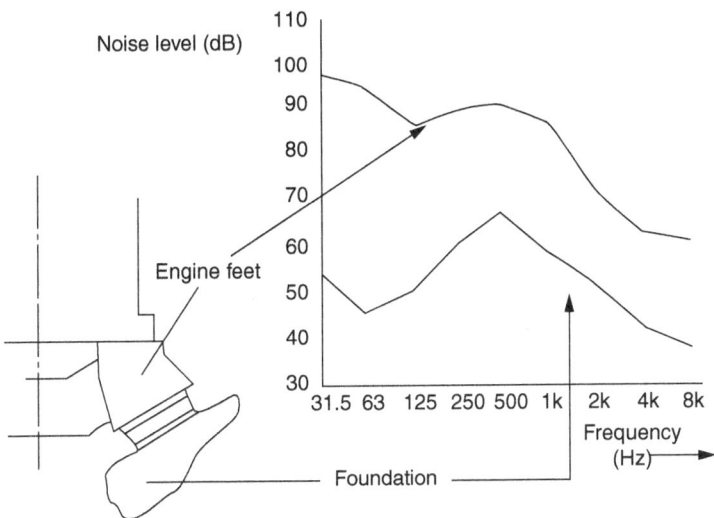

▲ **Figure 2.14** *Reduction in structure-borne noise achieved by resilient mountings*

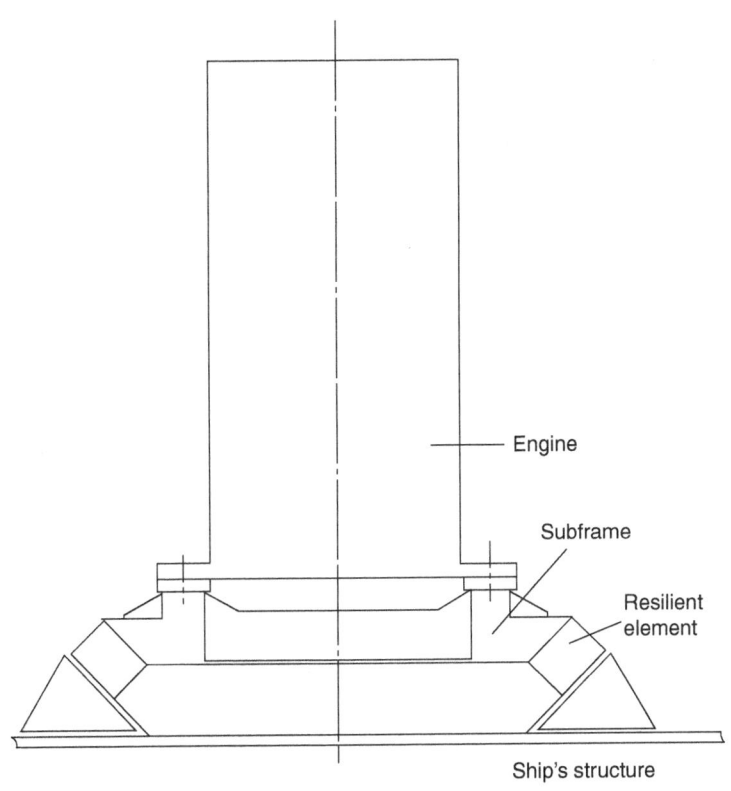

▲ **Figure 2.15** *Subframe-type resilient engine mounting*

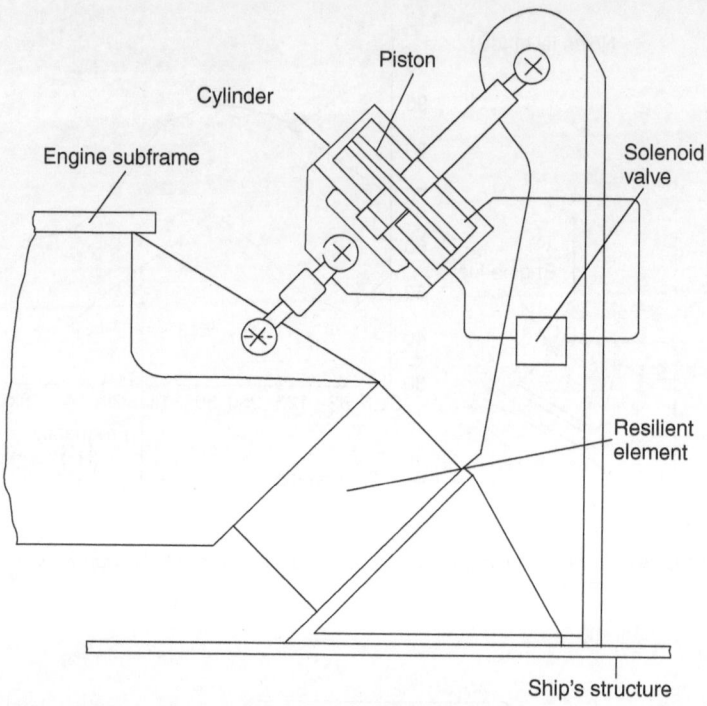

▲ **Figure 2.16** *Hydraulic locking device for engine movement limitation during starting/stopping*

When starting and stopping resiliently mounted diesel engines, large transitory amplitudes of vibration can be encountered. One manufacturer's solution to this problem is to install a hydraulic damper, which is shown in figure 2.16. It has a piston that is able to move within a cylinder and an interconnecting pipework via a shut-off valve that links both sides of the piston. During normal running the connecting valve is open, allowing the piston to displace oil between the upper and lower chambers freely, thus allowing the running vibration to be absorbed by the resilient mounting. During the engine's start and stop procedure, however, this valve will be closed, and the device will then prevent a relative movement between the engine and ship's structure, thus reducing the loading on the mountings.

Crankshafts

The crankshaft is a major structural part of the diesel engine. Despite being subjected to very high and complex stresses, the crankshaft must nonetheless be extremely reliable since not only would the cost of failure be very high, as the whole engine would have to be dismantled, but also the safety of the vessel and personnel could be placed in jeopardy.

Crankshafts must be extremely reliable and virtually maintenance free for the effective life of the engine. If the stresses set up in a crankshaft are examined, the need for extreme reliability will be appreciated. Figure 2.17 shows a crank unit where it could be regarded as a series of simple beams that are then subject to a number of different loads. Figure 2.17a indicates the load generated by the combustion pressure. This is a variable load acting at the centre of the crankpin, which is supported at either end by its two main bearings. If the bearings were flexible, for example spherical or ball shaped, then a simply supported beam equivalent would be the overall characteristic of this part of the structure.

Examining the crank throw in greater detail, figure 2.17b shows how the crankpin itself extends from the centreline of the crankshaft, which could be viewed as a beam with an evenly distributed load placed along its length and which varies with crank position. Each crank web is like a cantilever beam subjected to bending and twisting. Journals would be principally subjected to twisting, but a bending stress must also be present if we refer back to diagram (a).

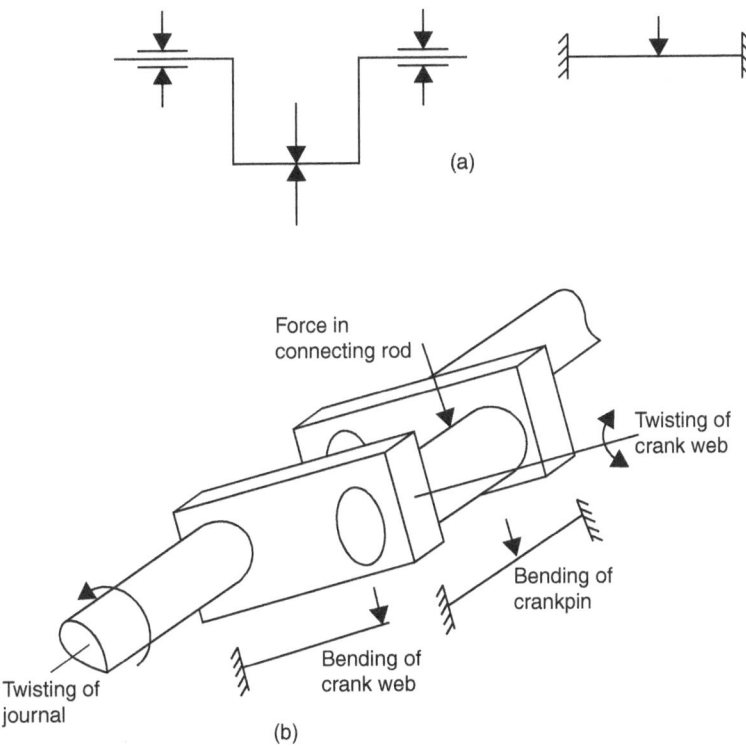

▲ **Figure 2.17** *Stresses in crankshaft*

A further consideration is the twisting or torque induced along the length of the crankshaft. The end of the crankshaft nearest to the flywheel at the output end of the engine will have to transmit all the torque induced along the entire length of the shaft, whereas at the other end, the crankshaft will only have to transmit the torque induced by one and then two of the cylinders in turn. A design feature of the G-type MAN Diesel & Turbo is that considerable weight saving has been realised by making the cylinders closer together at the forward end of the engine. This in turn means that the crankshaft can be shorter, thus saving weight.

Bending causes tensile, compressive and shear stresses and twisting causes shear stress. As the crankshaft is subjected to a series of complex fluctuating stresses, it must be built of a material that will resist the effects of fatigue.

This complex requirement means that the material and the method of manufacture must be chosen carefully. The highest fatigue resistance comes from a forging, which is preferable to casting. This is because metal has a grain structure and, unlike a casting, forging exhibits directional 'grain flow'. The properties of the material in the direction transverse to the grain flow are significantly inferior to those in the direction longitudinal to the grain flow. Under these circumstances, the drop in fatigue strength may be as much as 25–35%, with similar reductions in strength and ductility. Forging methods, therefore, ensure that the principal direction of grain flow is parallel to the major direct stresses imposed on the crankshaft (figure 2.18). Smaller crankshafts are drop forged from one piece of metal, which means that the grain structure runs all along the length of the shaft. Larger crankshafts are just too big to be manufactured in this way.

The materials chosen for forged and cast crankshafts are essentially the same. The composition of the steel will vary depending upon the bearing type chosen. For a

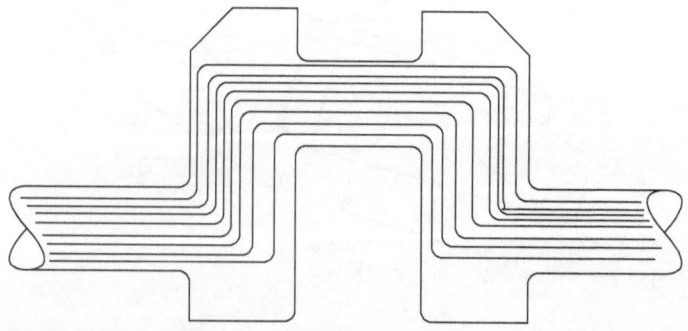

▲ **Figure 2.18** *Direction of grain flow in forged crankshaft*

crankshaft with white metal bearings, a steel of 0.2% carbon may be chosen; this will have an ultimate tensile strength (UTS) of approximately 425–435 MN/m². For higher output applications with harder bearing materials the carbon content is in the range of 0.35–0.4%, which raises the UTS to approximately 700 MN/m². To increase the hardness of the shaft still further, alloying agents such as chromium-molybdenum and nickel are added. For smaller engines, such as automotive applications, the crankshafts are surface hardened and fatigue resistance is increased by nitriding.

There are two broad categories of crankshafts:

1. One-piece construction.
2. 'Built up' from component parts.

One-piece construction

One-piece construction, either cast or forged, is usually restricted to smaller medium- and high-speed engines. Following the casting or forging operation, the component is rough machined to its approximate final dimensions and the oil passages are drilled. The fillet radius and crankpin are then cold rolled to improve the fatigue resistance and to reduce the micro-defects on the surface. Following machining, the crankshaft is then tested for surfaced and subsurfaced defects by using a combination of methods from dye-penetrant testing for surface defects to non-destructive testing for any defects that could be deeper within the structure.

'Built' crankshafts

These crankshafts are too big to be constructed in one piece and there have been three categories of 'built' crankshafts:

1. *Fully built*: this is where the webs are shrunk onto both the journals and crankpins (figure 2.19a).
2. *Semi-built*: the webs and crankpin are forged or cast as one unit and then fitted onto the journals by shrink fitting them in the right place (figure 2.19b) (the most favoured for modern construction).
3. *Welded construction*: a web, journal and crankpin section is forged or cast and the sections are then welded together across the main journal (figure 2.19c). (Although very promising, this method of construction did not catch on and there were only a few built; see the following pages.)

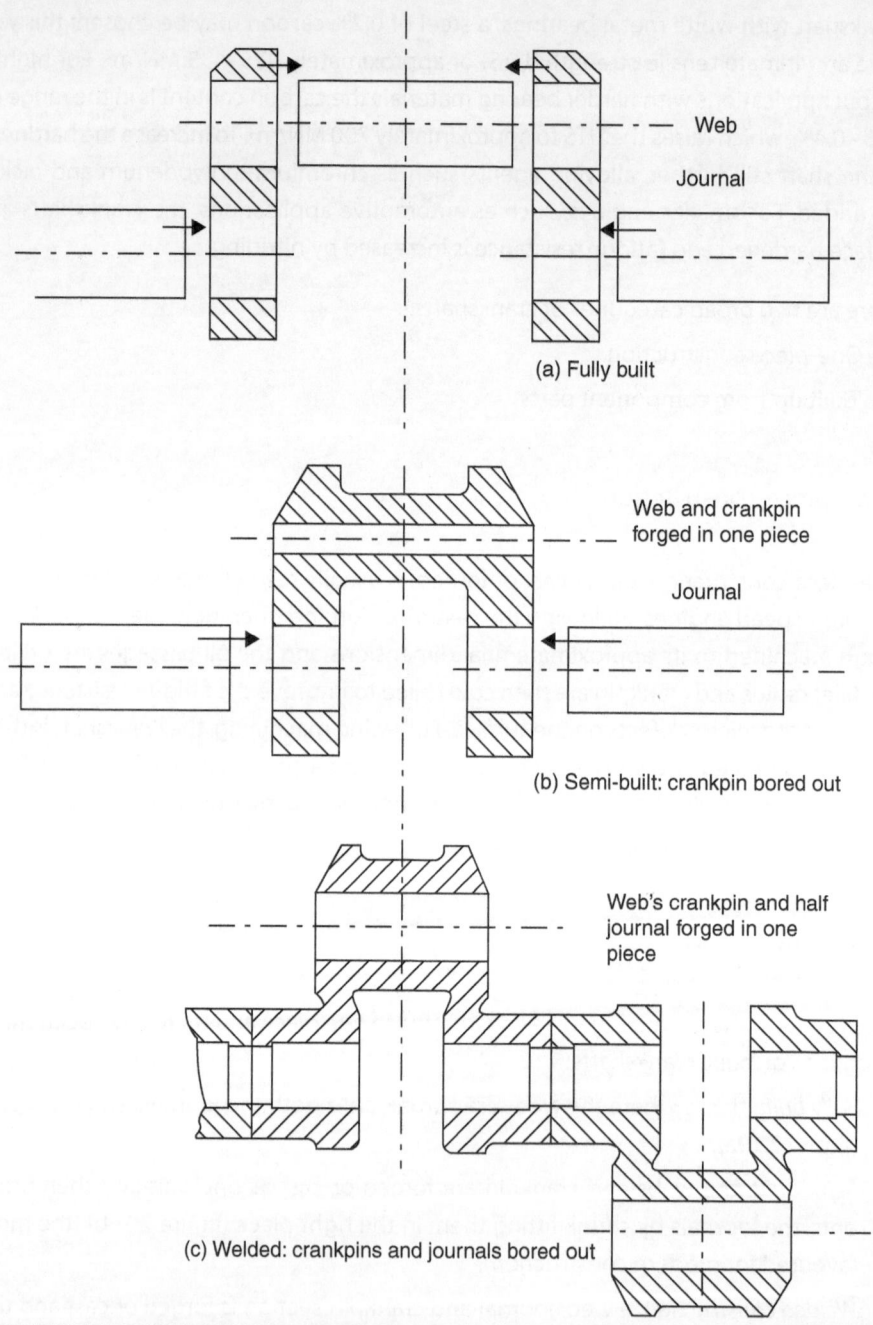

(a) Fully built

Web

Journal

Web and crankpin
forged in one piece

Journal

(b) Semi-built: crankpin bored out

Web's crankpin and half
journal forged in one
piece

(c) Welded: crankpins and journals bored out

▲ **Figure 2.19** *Three types of crankshaft assembly*

Fully and semi-built crankshaft construction

To minimise the risk of distortion, fully and semi-built crankshafts can be assembled in the vertical position. Various jigs are required to ensure the correct crank angles and to provide support for the crankshaft as it grows. The webs are heated to about 400°C and the journals and pins are then inserted into position. Raising the temperature any higher would bring the steel to a critical temperature where the material's characteristics would change and the shape would deform. When the assembly has cooled, the web material adjacent to the journal will be in tension.

The level of stress in the heated components must be well below the limit of proportionality to ensure that, when the components are cooled, the material returns to its original dimensions. If it went past its yield point the metal would not shrink properly, which would reduce the forces holding the web in place on the journal, leading to fretting and probable slippage of the web. To ensure an adequate shrinkage, an allowance of 1/550 to 1/700 of the shaft diameter is usual. Exceeding this allowance would simply increase the working stress in the material without appreciably improving the grip.

When the component parts of the crankshaft have been built up, the journals and pins are machined and the fillet radii are cold rolled (figure 2.20a). The crankshaft is then subjected to thorough surface and subsurface tests using, for example, ultrasound and metal particle techniques. To reduce the weight and the out-of-balance effects of the crankshaft, the crankpins may be bored out hollow (figure 2.19a–c).

The fully built crankshaft has now been replaced by the semi-built type, which displays improved 'grain flow' in the webs and crankpin. They are stiffer and can be shorter than the fully built type, due to a reduction in the thickness of the webs. The largest crankshafts can weigh over 200 tonnes but the machining of solid forged crankshafts requires larger production equipment than is the case for semi-built crankshafts. The safety margin against production error is also greater, as any individually defective cranks and main bearing journals can be scrapped separately, up to the point of assembly. Each crank of a fully built shaft would be substantially heavier than for a semi-built crankshaft, as shrink fitting requires a minimum amount of surrounding material, which is not the case in the crankpin of the semi-built shaft. Fully built crankshafts were common in the past, but today there are no problems in casting, forging and machining the larger sections of the semi-built types and therefore this is the preferred method of producing the large crankshafts on two stroke low-speed engines.

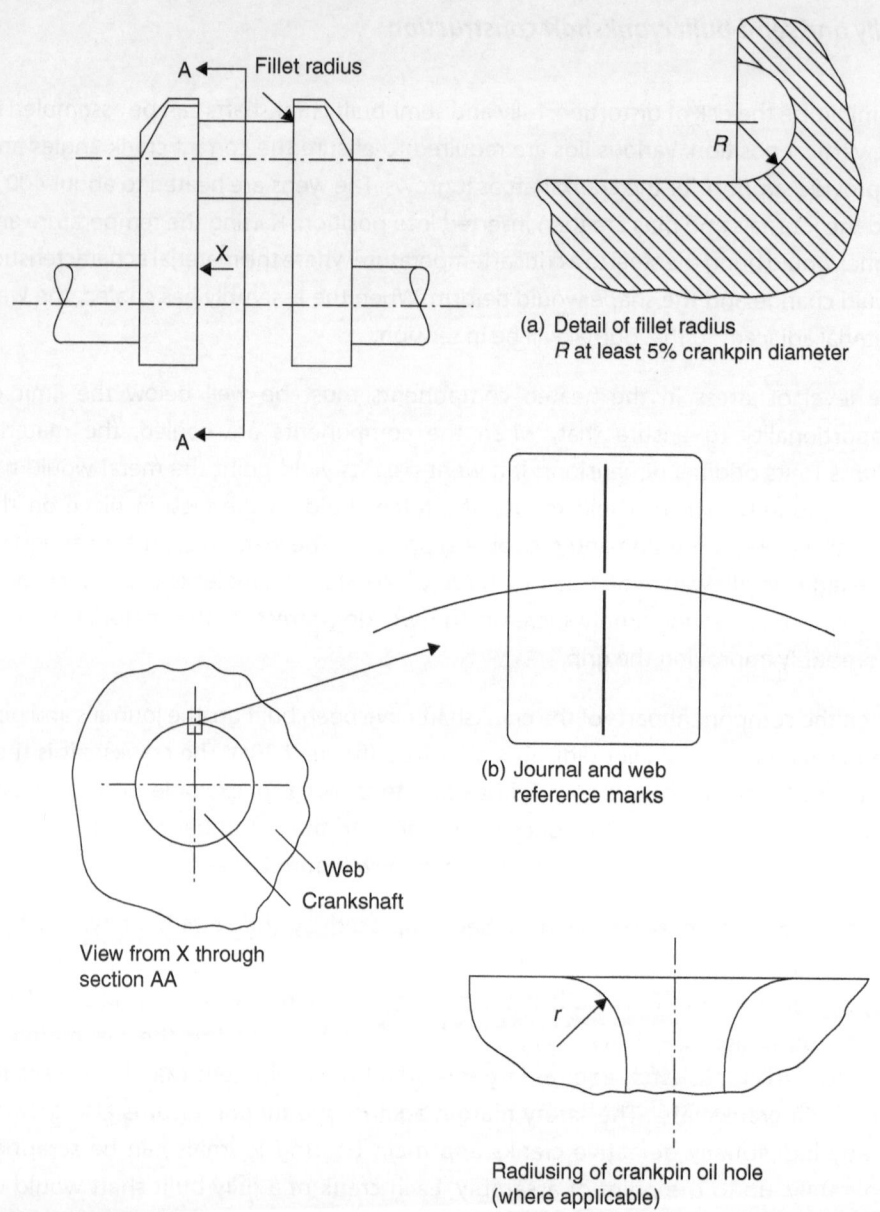

A ← Fillet radius

X

A ←

(a) Detail of fillet radius
    R at least 5% crankpin diameter

Web
Crankshaft

View from X through
section AA

(b) Journal and web
    reference marks

Radiusing of crankpin oil hole
(where applicable)

▲ **Figure 2.20** *Crankshaft construction details*

The effectiveness of the interference fit due to the shrinkage process of production for both the fully built and the semi-built depends upon:

1. The correct amount of shrinkage, which will result in setting up the correct level of stress in the web and journal.

2. The quality of surface finish of the journal and web. Good-quality surface finish will give the maximum contact area between web and journal.

Dowels are not used to locate the shrink since this would introduce a stress concentration that could lead to fatigue cracking. When a crankshaft is built by shrink fitting, reference marks are made to show the correct relative position of web and journal (figure 2.20b). These marks should be inspected during crankcase inspections as slippage could occur under one of the following conditions:

- If the starting air is applied to the cylinders when they contain water or fuel, or when the turning gear is engaged.

- If an attempt is made to start the engine when the propeller is constrained by, for example, ice or any other obstacle such as a log.

- If during operation the propeller strikes a submerged object.

- If the engine comes to a rapid unscheduled stop.

Following these circumstances, a crankshaft inspection must be made and the reference marks checked. Slippage will result in the timing of the engine being altered, which if not corrected will at best result in inefficient operation and possible poor starting and at worst could cause additional stresses that would compound the defect. If the slippage is small, for example up to 15°, then re-timing of the affected cylinders may be considered. If, however, the slippage is such that re-timing may affect the balance of the engine then the original journal and web relative positions must be restored. This is accomplished by heating the affected web while cooling the journal with liquid nitrogen and jacking the crankshaft to its original position. This is a difficult process and would be undertaken by specialist personnel or contractors under controlled conditions.

Welded construction

The development of the large marine crosshead two-stroke engine has resulted in ever higher outputs without an accompanying increase in physical size. This trend is starting to impose difficulties on the traditional shrink fitting of journals and web designs. To transmit the torques required, the traditional shrink fitting method requires that the web is of a minimum width and radial thickness. This will inevitably lead to a larger crankshaft and consequently a larger engine.

Welded construction has been a possible solution in the future; however, the cost of production and the difficulties of quality assurance (weld quality) mean that to date only a few of these have been built. These few have in fact given good service and this method of building a large crankshaft may return if a cost-effective method of production can be found.

There are two methods of assembly:

1. Welding two crank-arms together and then making a crankshaft by welding the crank-arms together (figure 2.21a).

2. Forging a crank-throw complete with half journals and then welding them with others to form the crankshaft (figure 2.21b).

The welding technique chosen is submerged arc narrow gap. This technique is automated and produces a relatively small heat-affected zone, which produces minimal residual stresses and distortion. At the completion of welding, the crankshaft is heated in a furnace to 580°C followed by a slow cooling period. Following heat treatment the crankshaft is tested using ultrasonic and metal particle techniques. If flaws are found, the weld is machined out and re-welded.

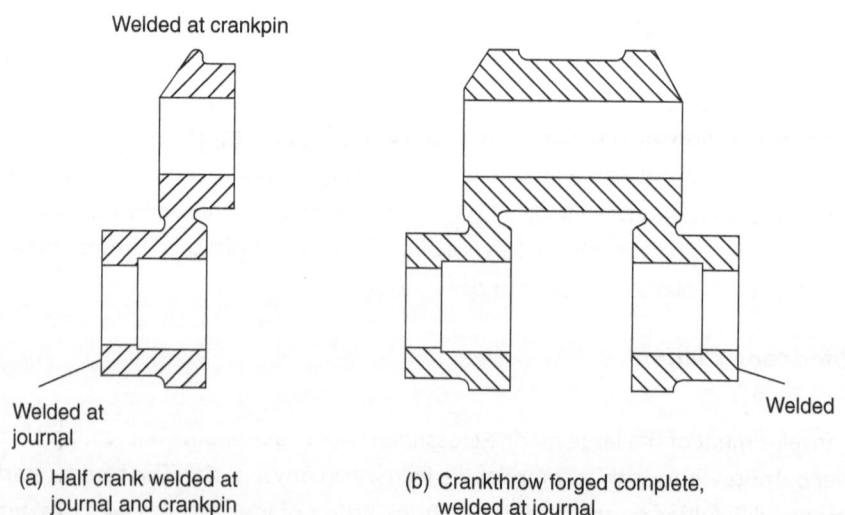

(a) Half crank welded at journal and crankpin

(b) Crankthrow forged complete, welded at journal

▲ **Figure 2.21** *Two options of welded crankshafts*

The advantages claimed for welded crankshafts are:

1. Reduced principle dimensions of the engine.
2. Reduced web thickness results in a considerable reduction in weight.
3. Reduced web thickness allows journal lengths to be increased, resulting in lower specific bearing loads.
4. Freedom to choose large bearing diameters without overlap restrictions.
5. Increased stiffness of crankshaft, resulting in higher natural frequencies of torsional vibration.

Crankshaft defects and their causes

Misalignment

If we assume that alignment was correct at initial assembly then the possible reasons for misalignment are as follows:

1. Worn main bearings.
   (a) Caused by incorrect bearing adjustment, leading to overloading.
   (b) Broken, badly connected or choked lubricating oil supply pipes, causing lubrication starvation, leading to excessive or adverse bearing wear.
   (c) Contaminated lubricating oil, leading to excessive adverse bearing wear.
   (d) Vibration forces.
2. Excessive bending of engine framework.
   (a) This could be caused by incorrect cargo distribution but is unlikely; the more probable cause would be grounding of the vessel and being re-floated in a damaged condition. It is essential that all bearing clearances are checked and crankshaft deflections are taken after such an accident.

Vibration

This can be caused by:

1. incorrect power balance
2. prolonged running at or near critical speeds
3. slipped crank webs on journals
4. light ship conditions leading to impulsive forces from the propeller (eg forcing frequency four times the revs, for a four-bladed propeller)

5. near presence of running machinery

6. excessive wear down of the propeller shaft bearing (this can lead to whipping of the shafting in bad weather conditions)

7. vibration-accentuated stresses, which can be increased to exceed fatigue limits and considerable damage could result. It can lead to fixings and fastenings working loose, for example coupling bolts, bearing bolts, bolts securing balance masses to crank webs and lubricating oil pipes.

Other causes

On modern vessels, with the aid of CAD and other mathematical modelling tools, incorrect manufacture leading to defects is a rare occurrence but not unheard of. In the past, failure has been caused by:

- slag inclusions
- poor control of the heat treatment and machining processes, for example badly radiised oil holes and fillets
- careless use of tools resulting in impact marks on crankpins and journals, which can also lead to failure.

These defects all result in the creation of stress concentrations that, because of the cyclic loading of the crankshaft, can raise the local level of stress in a given component to above the level of the fatigue limit, shown on the *S/N* graph, in figure 1.16 in Chapter 1, resulting in fatigue cracking and ultimately failure. The condition can be exacerbated if the engine is run at or close to the critical speed, which is the rotational speed that causes the crankshaft to vibrate at its natural frequency of torsional vibration.

It is the speed that induces resonance, the consequence of which is to cause the crankshaft to vibrate in the torsional mode with large amplitudes. Stress, being proportional to amplitude, increases and may rise sufficiently to reduce the number of working cycles of the crankshaft before failure occurs.

Bottom end bolts on medium- and high-speed four-stroke diesel engines are subjected to fluctuating stresses and are therefore also exposed to potential fatigue failure. Four-stroke engine bottom end bolts experience large fluctuations of stress during the cycle. This is due to the inertia forces experienced in reversing the direction of the piston over TDC on the exhaust stroke. The forces experienced by bottom end bolts in this situation is high. Reference to the *S/N* graph in Chapter 1 will show that to ensure maximum serviceability, stresses should be commensurate with a level below the fatigue limit. Since:

$$\text{stress} = \frac{\text{load}}{\text{area}}$$

it can be seen that for a given load the stress can only be reduced by increasing the area and therefore increasing the size and weight of the bottom end bolt. Designers opt for a compromise: they design a bolt that will experience a level of stress ABOVE that of the fatigue limit and specify the number of cycles the bolt should remain in service before it is replaced. It is therefore of vital importance that the running hours of four-stroke engines are known in order to monitor the safe working life of bottom end bolts.

In addition to this, designers will specify that bottom end bolts:

- are manufactured to high standards of surface finish
- have rolled threads
- be of the 'wasted' design with generous radii
- have increased diameter at mid shank to reduce vibration
- be tightened accurately to the required level.

As part of the maintenance programme, bolts should be examined for mechanical damage that would cause stress concentration and damaged bolts should be replaced.

Fretting corrosion

This occurs where two surfaces forming part of a machine, which in theory constitute a single unit, undergo slight oscillatory motion of a microscopic nature.

It is believed that the small relative motion causes removal of metal and protective oxide film. The removed metal combines with oxygen to form a metal oxide powder that may be harder than the metal (certainly in the case of ferrous metals), thus increasing the wear. Removed oxide film would be repeatedly replaced, increasing further the amount of damage being done.

Fretting damage increases with load, amplitude of movement and frequency. Hardness of the metal also affects the attack; in general, damage to ferrous surfaces is found to decrease as hardness increases.

Oxygen availability also contributes to the attack; if oxygen level is low, the metal oxides formed may be softer than the parent metal, thus minimising the damage. Moisture tends to decrease the attack.

Bearing corrosion

In the event of fuel oil and lubricating oil combining in the crankcase, weak acids may be released, which can lead to corrosion of copper lead bearings. The lead is removed from the bearing surface so that the shaft runs on nearly pure copper, which raises bearing temperature, causing the lead rises to the surface when it is removed. The process is repeated until failure of the bearing takes place. Scoring of crankshaft pins can then occur. Use of detergent lubricating oil can prevent or minimise this type of corrosion because the detergency properties of the oil hold the small particles in suspension and the alkalinity of the oil will neutralise the acids produced.

Water in the lubricating oil can lead to white metal attack and the formation of a very hard black incrustation of tin oxide. This oxide may cause damage to the journal or crankpin surface by grinding action. Water also combines with any sulphur to form sulphuric acid, which creates a further need for the oil to neutralise the acids.

Bearing clearances and shaft misalignment

The condition of the bearings and their correct position is very important to give feedback to the engineer about the condition of the engine. The new methods of Condition-Based Monitoring/Maintenance (CBM) utilise the monitoring of bearing clearance to give real-time data about the state of the engine. In the past, bearing clearances were checked in a variety of ways:

- A rough check is to observe the discharge of oil, in the warm condition, from the ends of the bearings.
- Feeler gauges can be used, but for some of the bearings they can be difficult to manoeuvre into position in order to obtain readings.
- Clock (or as they are sometimes called, dial) gauges can be very effective and accurate providing the necessary relative movement of the crankshaft webs can be achieved; this can prove to be difficult in short engines that have a stiff crankshaft.
- Finally, the use of lead wire. This required the bearing keeps to be removed, the lead wire inserted and the keeps replaced. The wire would be deformed to a thickness equal to the bearing clearance. It was then just a matter of removing the keeps and measuring the lead wire thickness with a micrometer.

Engine manufacturers have been keeping records of the results produced by the techniques described above. MAN Diesel & Turbo found that as many as 7,000 ships each year have their engine's bearings viewed as part of 'open up' inspections but only 1% of defects are found during those inspections. Not only that, but as part of the growing trend of 'maintenance-induced failures' over 2% of bearing defects are caused by the inspection itself. Proximity sensors have now been developed that will measure the bearing wear on a two-stroke main engine in real time. This has led to the recommendation not to open up main engine bearings if this monitoring equipment is fitted.

The equipment indicates wear in the crank train: main, crankpin and crosshead bearings. The temperature of the main bearing can also be monitored, as can the extent of any water in the lubricating oil and the electrical potential between the propeller shaft and hull, all of which may have an adverse effect on bearing life.

The system is made up of analogue inductive sensors fitted to each cylinder and located on a bracket, which is fixed to the engine frame so that the field effect from the sensor is interfered with each time the bottom of the crosshead guide reaches the bottom of its travel. There are several technologies available to achieve this but the effect on the generated field differs according to the physical position of the base of the crosshead guide. This position obviously takes into account the position of all the components in the crank train including the main, crankpin and crosshead bearings.

The proximity sensor interprets results from the disturbance within the sensor field and transmits an electrical signal proportional to the disturbance. A signal processing unit (SPU) mounted outside the engine processes the signal and sends it to the interface unit mounted in the engine control room, which forms the connection to the main Amot Monitoring System (AMS) and also allows local system access to the engineers in the control room (figure 2.22).

A further link can be made using an Ethernet connection via a PC on the ship's network. The calibrated SPU communicates wear data to the human–machine interface (HMI), which provides a clear graphic display of bearing wear.

If the crankshaft is aligned correctly and is straight in the engine, the main bearing clearances should be zero at the bottom. If misalignment is suspected, due to bearing wear, it might be necessary to check the main bearing clearance. Measurement of this value is usually very difficult as there is not much space to get a set of feeler gauges in place to take the measurement.

Some engines are provided with facilities for obtaining the bottom clearance (if any) of the main bearings. This is with the aid of special feelers and without the need to remove the bearing keep. Another method is to first arrange in the vertical position a clock gauge so that it can record the movement of the crank web adjacent to the main bearing. The main bearing keep is then removed, shims are withdrawn and the keep is

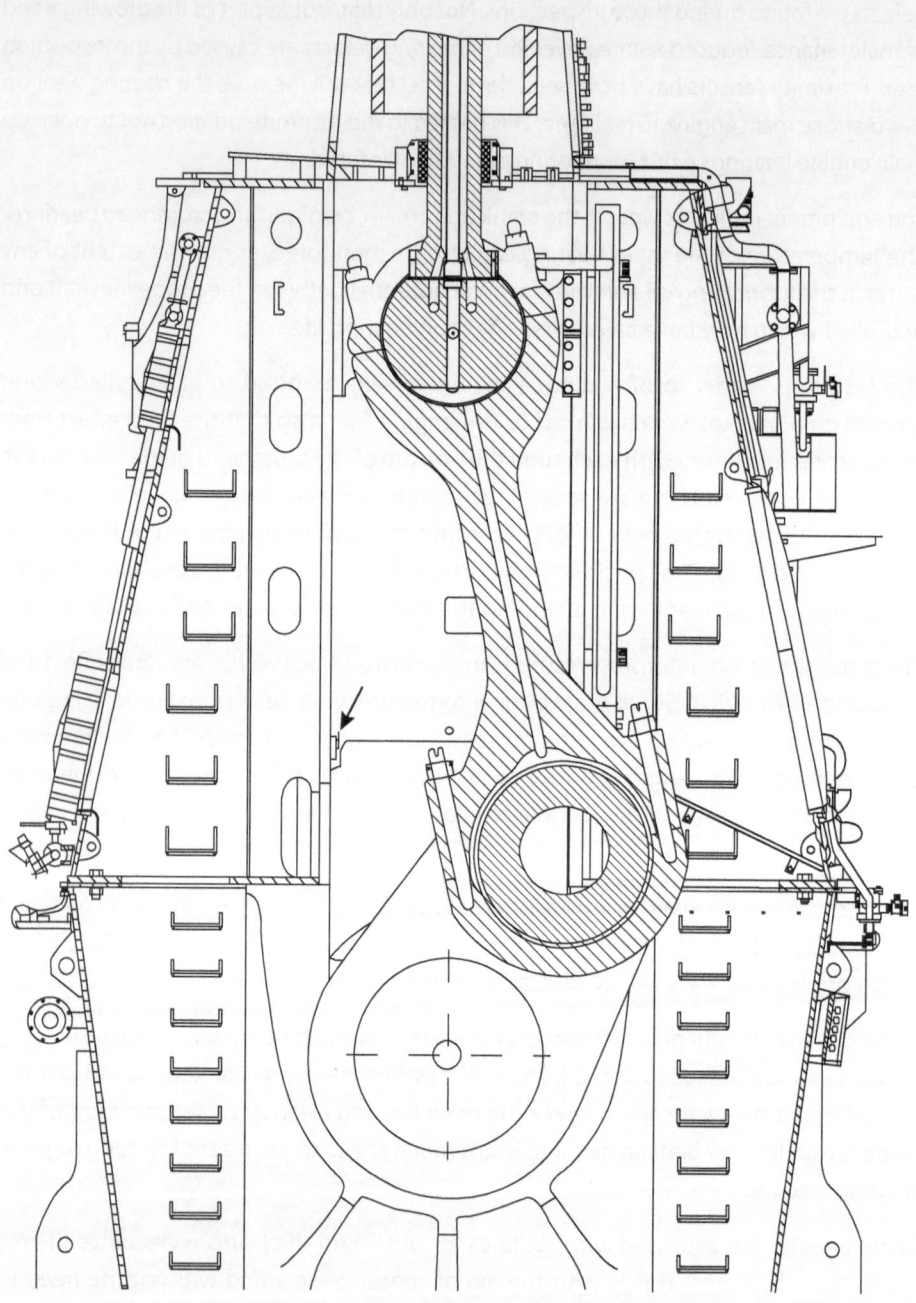

▲ **Figure 2.22** *Position of bearing wear sensor (shown by the arrow)*

replaced and tightened down. The vertical movement of the shaft, if any, is observed on the dial gauge.

Obviously, if the main bearing clearance is not zero at the bottom, the adjacent bearing or bearings are also high by comparison and then the shaft is out of alignment.

Crankshaft alignment can be checked by taking deflections. If a crank throw supported on two main bearings is considered, the vertical deflection of the throw in mid span is dependent upon: shaft diameter, distance between the main bearings, type of main bearing and the central load due to the running gear. A clock gauge arranged horizontally between the crank webs opposite the crankpin and ideally at the circumference of the main journal (see figure 2.23a–c) will give a horizontal deflection, when the crank is rotated through one revolution that is directly proportional to the vertical deflection.

In figure 2.23a, it is assumed that the main bearings are in correct alignment and no central load is acting due to running gear. Then, vertical deflection of the shaft would be small – say zero. With running gear in place and crank at about bottom centre, the webs would close in on the gauge as shown – this is negative deflection. With crank on top centre webs open on the gauge – this is positive deflection. In practice, the gauge must always be set up in the same position between the webs each time, otherwise widely different readings will be obtained for similar conditions. Usually the manufacturers will place dot marks in the web for the ends of the dial gauge to fit into to ensure this requirement takes place.

An alternative is to make a proportional allowance based on distance from crankshaft centre. Obviously, the greater the distance from the crankshaft centre the greater will be the difference in gauge readings between bottom and top centre positions.

Since, due to the connecting rod, it is generally not possible to have the gauge diametrically opposite the crankpin centre when the crank is on bottom centre, an average of two readings would be taken, one on either side during the turning of the crank. Table 2.1 shows some possible results from a six-cylinder diesel engine.

The dial gauge would be set at zero when crank is in, say, port side near bottom position and gauge readings would be taken at port horizontal, top centre, starboard horizontal and starboard side near bottom positions. Say x, p, t, s and y as per figure 2.24, but before taking each reading the turning gear should be reversed to unload the gear teeth, otherwise misleading readings may be obtained.

Any engines still with spherical main bearings will have greater allowances for crankshaft misalignment than those without. Spherical bearings have been used when increased flexibility is required for the crankshaft, as might have been the case for opposed piston engines with large distances between the main bearings.

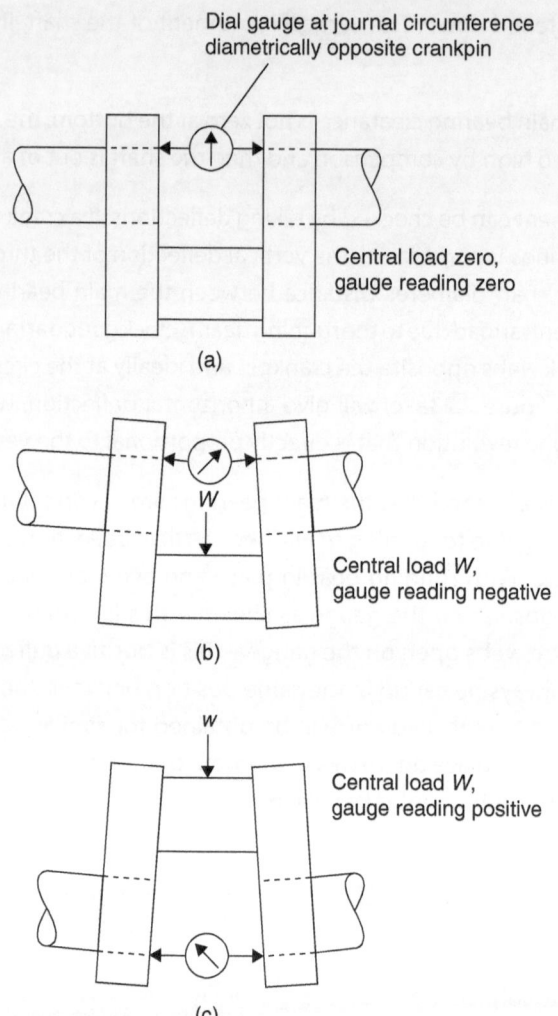

Dial gauge at journal circumference diametrically opposite crankpin

Central load zero, gauge reading zero

(a)

Central load *W*, gauge reading negative

(b)

Central load *W*, gauge reading positive

(c)

▲ **Figure 2.23** *Checking crankshaft alignment*

The vertical misalignment figures shown in figure 2.24 give the reader information that the end main bearing adjacent to No. 1 cylinder and the main bearing between Nos 3 and 4 cylinders are high. Vertical and horizontal misalignments can be checked against the permissible values supplied by the engine builder, often in the form of a graph. If any values exceed or equal maximum permissible values, bearings will have to be adjusted or renewed where required. Indication of incorrect bearing clearances may be given when the engine is running. In the case of medium- or high-speed diesels, load reversal at the bearings generally occurs. With excessive bearing clearances, loud knocking takes place and then white metal usually gets hammered out.

Table 2.1 *Gauge readings in mm/100*

Crank position	Cylinder number					
	1	2	3	4	5	6
X	0	0	0	0	0	0
p	5	2	6	−8	−3	1
t	10	3	12	−14	−8	4
s	5	3	6	−8	−6	3
y	−2	2	−2	0	0	−2
b=(x+y)/2	−1	1	−1	0	0	−1
Vertical misalignment (t–b)	11	2	13	−14	−8	5
Horizontal misalignment (p–s)	0	−1	0	0	3	−2

If bearing lubrication for a unit is from the same source as piston cooling, then a decrease in the amount of cooling oil return may be observed in the sight glass, together with an increase in its temperature. If bearing clearances are too small, overheating and possible seizure may take place. Increased oil mist and vapour at a particular unit may be observed – together with an increase in bearing temperature, which could then lead to a crankcase explosion. Regular checks must be made to ascertain the oxidation rate of the oil. If this increases, high temperatures are encountered and as the oil oxidises (burns) its colour blackens.

The use of wireless technology is due to enter the industry by storm once systems become reliable for use in the marine environment, where the transmission of information must be 100% accurate inside what is essentially a steel box. However, there are some places where the technology is already starting to take hold and one area is with the recording of crankshaft deflections. The job traditionally was very tedious, messy and awkward. Engineers invariably had a 'crankcase' boiler suit that they would wear for jobs such as this and then discard upon completion due to the oil that will have dropped onto them inside the crankcase.

The Bluetooth-enabled measuring device can be placed in position between the marks on the crank webs while they are close to the open crankcase door. The crankshaft can then be turned with the engineer being outside the crankcase. The receiver collects and stores the measured information from the wireless signal sent from the measuring device. The engineer can then download the information to a laptop, also using a wireless connection. Dimensions are still recorded with the crankshaft in the positions indicated in figure 2.24.

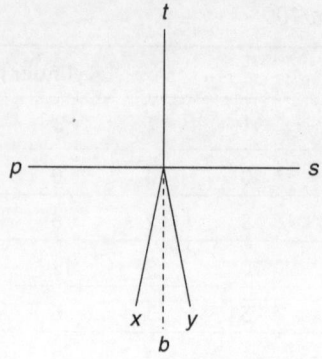

Crank positions for deflection readings

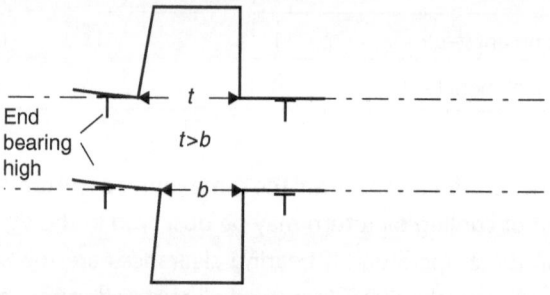

Effect of bearing misalignment

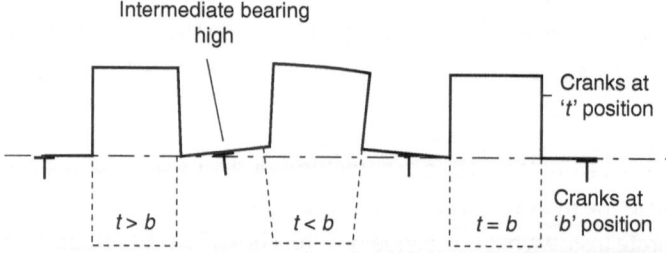

▲ **Figure 2.24** *Crank positions for deflection*

The increased load on the crank throw of the modern crankshafts, see fig 2.17, is causing problems that might not seem probable. The bending moment set up in the arms of the crank exert the twist on the central run of the shaft. The fluctuating forces set up vibrations that are reduced by the vibration dampers, but the resultant overall twist throws out the precise position of the next piston down the line from the one under load.

Additional position measuring or compensating calculations need to be fed into the overall computational algorithms that work out the exact moment to inject the fuel into each cylinder. This has become more of an issue as the accuracy of the control mechanisms grows. Stratified fuel injection, for example, needs very accurate timing to work as it was designed and older fuel systems could not match the function of their modern counterparts.

Choice, Maintenance and Testing of Main Engine Lubricating Oil

Choice of base oil

There is more about the composition of lubricating oil in Volume 8 of the Reeds series. However, it is sufficient here to say that lubricating oil has several roles to play in protecting the moving components of an engine. It reduces the heat stress by cooling the internal components of the engine and it also protects the components by keeping them apart as they rotate or move in the engine. Finally, the oil must keep the components clean and free from the effects of combustion, which includes neutralising any acids that may form as a result of the engine's operation.

The 'trunk type' engine is more susceptible to the fuel and products of combustion finding their way into the crankcase, and contaminating the oil, than is the crosshead type engine. Diesel engine lubricating oil should have a detergent additive; these oils are sometimes called 'heavy duty'. Additives in these oils deter the formation of deposits on the metal components, by keeping substances such as carbon particles in suspension. They also counteract the corrosive effect of sulphur compounds; some of the fuels used may be low in sulphur content and in this case the alkaline additive in the lubricating oil could be less.

The lubrication of large two-stroke low-speed engines is twofold. The crankcase oil will use a straightforward mineral oil, generally with an antioxidant and corrosion inhibitor added because the working cylinder is separate from the crankcase and there is less in the way of contaminants from the product of combustion.

Due to the crankcase being separate from the underside of the cylinder, no oil is thrown onto the cylinder walls and therefore this type of engine has to have cylinder oil lubrication, which is a total loss system and is separate from the crankcase oil. The requirement for modern tonnage to burn very low sulphur fuel (LSF) in specially designated areas means that the engineers will have to change over to the higher cost LSF as the vessel approaches from its deep-sea voyage.

The sulphur content of the fuel has an effect on its lubricity and on its readiness to produce acids when combined with water. Therefore, when the type of fuel is changed over, the grade of cylinder oil needs to be changed as well.

Maintenance

When the main or large auxiliary engine is new, the manufacturer's quality assurance process and careful pre-commissioning should deliver a clean system free from sand, metal, dust, water and other foreign matter. Final checks will be made by both the owner's team and the manufacturer's staff before the engines are run for the first time during the trials.

Checks that are made could include checking the individual parts using a simple hammer test to ensure that any rust flakes, scale and weld spatter are not left inside the engine to work loose when the engine is running and cause damage.

A good flushing oil could be used if necessary, where a clear discharge should be obtained from the outlet pipes before they are reconnected to the lubrication system; filters must also be opened up and cleaned where necessary at this stage. Finally, the flushing operation could be frequently repeated with a new charge of oil of the type to be used in the engine. When the engine is running, continuous filtration and centrifugal purification is essential, with more frequent checking taking place during the first few months of the engine's life.

Oxidation of the oil is one of the major causes of its deterioration; it is caused by high temperatures. This may be due to:

1. Small bearing clearances (hence insufficient cooling).
2. Not continuing to circulate the oil after stopping the engine.

In the case of oil-cooled piston types, piston temperatures could rise as the residual heat soaks into the piston and the static oil within them becomes overheated.

3. Incorrect use of oil pre-heater for the purifier, for example, shutting off oil before the heat or running the unit part full.
4. Metal particles of iron and copper, which can act as catalysts that assist in accelerating oxidation action. Rust and varnish products can behave in a similar fashion.

When warm oil is standing in a tank, water that may be in it can evaporate and condense out upon the upper cooler surfaces of the tank not covered by oil. Rusting could take place and vibration may cause this rust to fall into the oil. Tanks should be given some protective type of coating to avoid rusting.

Oil from the scavenge space and stuffing box drains should not be put into the main oil system and the piston rod gland (commonly known as the stuffing box) and any telescopic pipe glands (older type water-cooled engines only) must be maintained in good condition to prevent entry of water, fuel and air into the oil system.

Regular examination and testing of the main circulating oil is important. Samples should be taken from a pipeline in which the oil is flowing and not from a tank or container in which the oil is stationary, where it could possibly have stagnated and accumulated contaminants that are not from the engine. If this happens, a representative sample of the oil lubrication of the engine will not have been taken and if an adverse analysis is subsequently returned then the wrong corrective action could be taken.

Smelling the oil sample may give an indication of fuel oil contamination, or if an acrid smell is present this could be a sign of heavy oxidation. Dark colour gives an indication of oil deterioration possibly due to oxidation and a black colour denotes the presence of carbon.

Dipping fingers into the oil and rubbing the tips together might detect reduction in oiliness – generally due to fuel contamination – and the presence of abrasive particles. The latter may occur if a filter has been incorrectly assembled, damaged or automatically bypassed. Water vapour can condense on the surfaces of sight glasses, thus giving an indication of water contamination. But various tests are available to detect water in oil, for example immersing a piece of glass in the oil, water finding paper or paste (copper sulphate crystals change colour from white to blue in the presence of water). Plunging a piece of heated metal such as a soldering iron into the oil causes spluttering if water is present.

A check on the amount of sludge being removed from the oil by the purifier is important; an increase would give an indication of oil deterioration. Lacquer formation on bearings and excessive carbon formation in oil-cooled pistons are other indications of oil deterioration.

Oil samples for analysis ashore should be taken about every 1,000–2,000 h (or more often if defects are suspected) and it would be recommended that the oil be changed if one or more of the following limiting values are reached:

1. 5% change in the viscosity from new. Viscosity increases with oxidation and by contamination with heavy fuel, diesel oil can reduce viscosity.
2. 0.5% contamination of the oil.
3. 0.5% emulsification of the oil; this is also an indication of water content. Fresh water is generally permissible up to 0.2%, but sea water is dangerous.
4. 1.0% Conradson carbon value. This is a measure of the carbon residue left after evaporation and pyrolysis of an oil sample and is intended to provide notice of incomplete combustion of fuel oil.

5. 0.01 mg KOH/g total acid number (TAN). The TAN is the total inorganic and organic acid content of the oil. Sulphuric acid from engine cylinders and chlorides from sea water give the inorganic and oxidation produces the weak organic acids. The overall number gives an indication of the overall quality of the oil.

Determining the quality of lubricating oil in an engine is the basis of the modern CBM schemes, which are discussed in more detail in Chapter 12 of Volume 8 of the Reeds series. However, regular oil sample analysis will produce a string of results from which trends can be seen and the internal condition of an engine can be determined. This internal condition can then be used as a basis for determining the maintenance required.

Lubrication Systems

Diesel engine bearings are kept in good condition for the working life of the engine by effective lubrication. The designs of the lubrication system can be the make or break of the success of the engine. The objective for good bearing lubrication is to create the correct environment for 'hydrodynamic' lubrication to occur. This is where a film of oil is set up to lift the journal away from the bearing surface. In this case the two surfaces never touch and the only friction is that which is within the structure of the oil (figure 2.25).

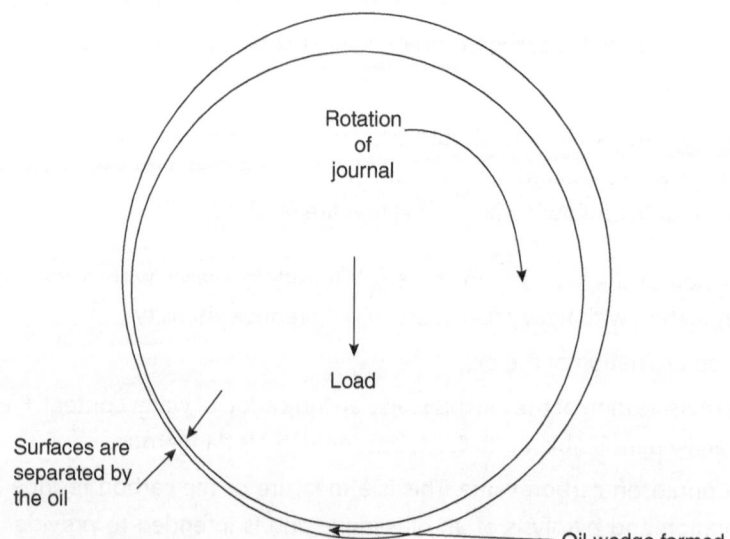

▲ **Figure 2.25** *Hydrodynamic lubrication set up by moving surfaces*

Hydrodynamic lubrication is a continual operating process all the time a shaft journal is rotating in a bearing. The oil forms a wedge between the two components and the journal travels in its circular motion on the wedge of oil. Relating to a crankshaft main bearing, as the load on the shaft grows due to the forces of combustion, the oil wedge is subject to more stress and becomes thinner. Eventually the oil film would break down and the journal would come into contact with the bearing surface. Similarly if the viscosity of the oil was low – possibly due to contamination or by using the wrong type of oil – then again the oil film will break down.

The crosshead bearing, on the other hand, is a different proposition. This bearing carries out an oscillating motion; therefore as soon as it starts moving in one direction the bearing stops and reverses direction. This means that it does not have sufficient time to set up efficient hydrodynamic lubrication and as a consequence the effective lubrication of this bearing has been difficult. The 'two–pillar' design, which until recently has been a feature of these bearings, does not help due to the reduced surface area.

Engine lubrication systems for the bearings and guides, etc should be simple and effective. Considering the lubrication of a bottom end bearing, various alternative routes are available to channel the oil to the appropriate places and the objective would be to choose a route that will be the most reliable, least expensive and least complicated.

Oil could be supplied to the main bearing and by means of holes drilled in the crankshaft the oil could then be sent to the bottom end bearing. This method may be simple and satisfactory for small engines but with a large diesel it presents machining problems, which would also enhance the stress involved. In one large type of diesel the journals and crankpins were drilled axially and radially, but to avoid drilling through the crank web and the *shrinkage surfaces* the oil was conveyed from the journal to the crankpin by pipes. A common arrangement that used to be adopted with engines having oil-cooled pistons is to supply the bottom end bearing with oil that is led down a central hole in the connecting rod from the top end bearing (figure 2.26).

On older engines, some of which could still be in service, a telescopic pipe system was used along with a swinging arm; the disadvantage of the latter is that it has three glands, whereas the telescopic has only one. However, it is more direct and is less expensive, especially if it saved a bearing from failure.

With the majority of bearings, as outlined earlier, the main objective is to provide an efficient hydrodynamic film of lubricant. The factors assisting hydrodynamic lubrication are as follows:

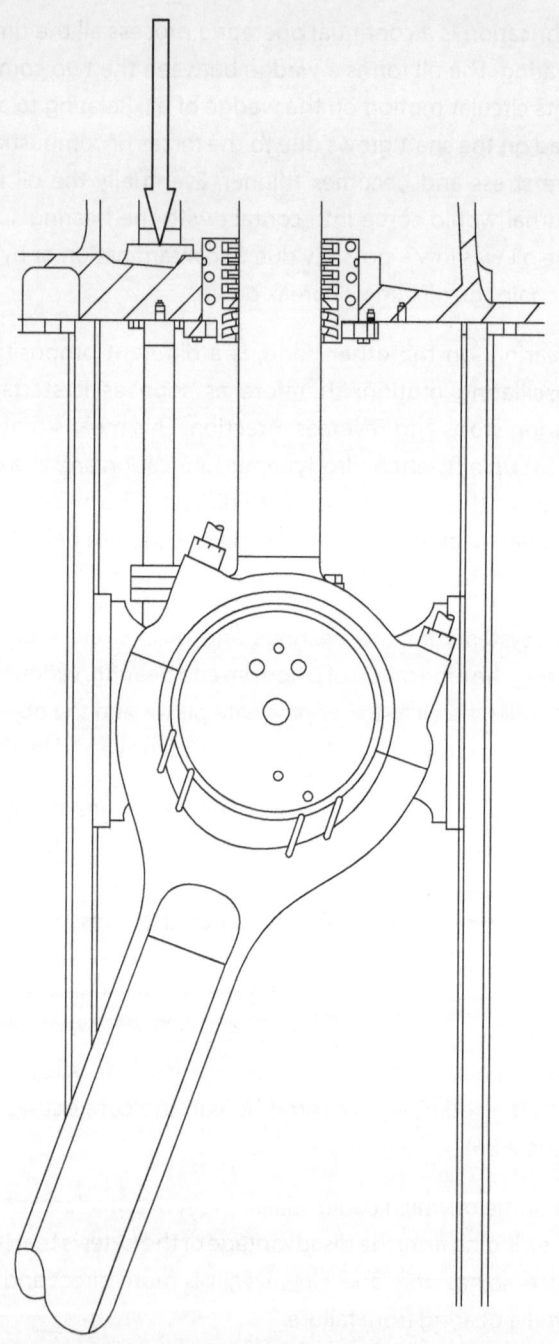

1. *Viscosity:* If the oil viscosity is increased there is less likelihood of oil film breakdown. However, too high a viscosity increases viscous drag and power loss.

2. *Speed:* Increasing the relative speed between the lubricated surfaces pumps oil into the clearance space more rapidly and helps promote hydrodynamic lubrication.

3. *Pressure:* Increasing the load on the bearing increases the pressure (load/area) on the oil, causing a breakdown in the oil film. Increasing the load at the engine design stage can be offset by increasing the area of the bearing surface, which could be done by making the pin diameter larger – this will also increase relative speed. Manufacturers have now redesigned the crosshead bearing on their latest engines, to increase the surface area available to take the load from the forces of combustion (figure 2.27). Previous designs had much lower surface areas, as can be seen in figure 2.28.

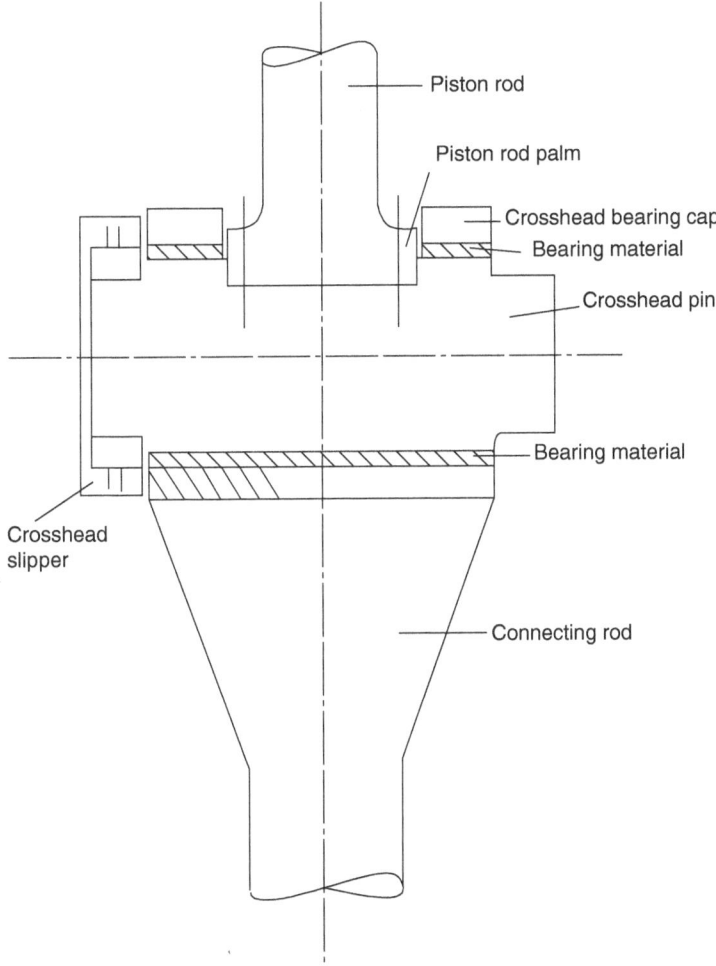

▲ **Figure 2.27** *One-piece lower bearing crosshead design. Sketch shows LHS slipper*

4. *Clearance:* If bearing clearance is too great then the inertia forces lead to 'bearing knock'. This impulsive loading results in pressure above normal and breakdown of the hydrodynamic layer. Figure 2.29 illustrates these points graphically for a rotating journal type of bearing.

Referring to the graph in figure 2.29, hydrodynamic lubrication should exist in the main, bottom end and guide bearings. The top end and crosshead bearings will have a variable condition, for example when the unit is at TDC the relative velocity between crosshead guide and the bearing surface is zero and the pressure from combustion is building. The swing of the connecting rod is, however, building the relative speed of the crosshead bearing surface across the crankpin journal as the forces of combustion are taking effect. Reflecting upon this point will enable the student to appreciate the

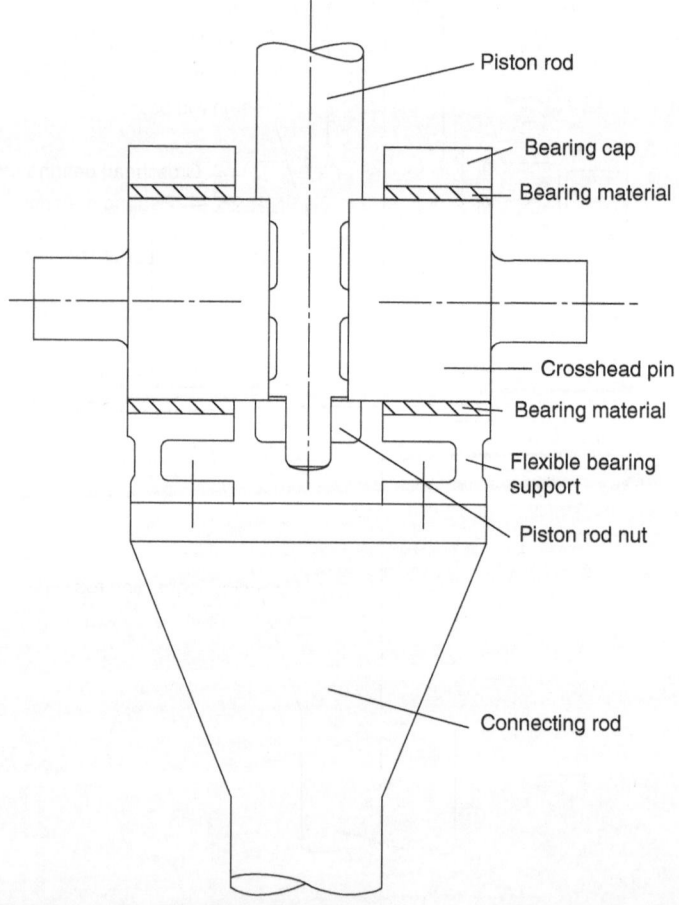

▲ **Figure 2.28** *Crosshead with flexible bearing supports*

complications involved not only in effective crosshead design in the large two-stroke engine but also with 'small end' design in the four-stroke engine.

Methods of improving top end bearing lubrication are as follows:

1. Reduction in the load on top end exerted by inertia forces – only required with four-stroke engines, which are usually medium- or high-speed diesels, although at the time of writing Akasaka Diesels were still producing a slow-speed four-stroke engine.

2. Use as large a surface area as possible, that is, the complete underside of the crosshead pin (figure 2.27).

3. Avoid large axial variation of bearing pressure by more flexible seating and design (figure 2.28).

4. Increase oil supply pressure – on the modern engines this is achieved by using a pump external to the engine to supply oil pressure directly to the top end bearing, which tends to keep the crosshead pin 'floating' at all times as the crosshead bearing rocks to and fro.

5. An increase in oil supply pressure to the crosshead bearing can also be accomplished by the oil inlet to the engine being at the position of the crosshead, enabling the full pressure of the lubrication oil system to be placed on the crosshead bearing.

Current design practice for manufacturers to accomplish successful bearing lubrication is split between the two systems: they either introduce a separate lubricating oil feed to the crosshead bearing (figure 2.30) via a booster pump or divert the oil inside the engine with the correct volume going to the crosshead and the cylinder oil cooling.

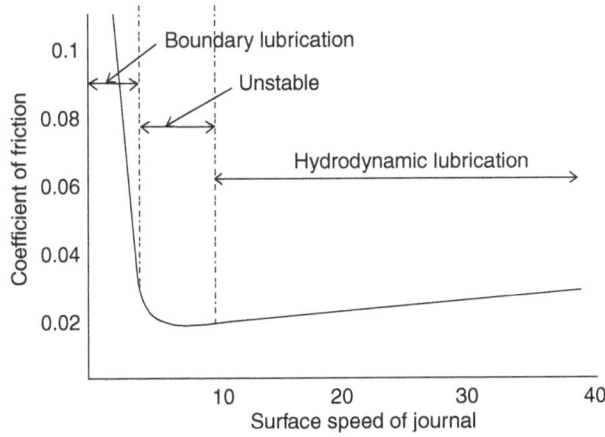

▲ **Figure 2.29** *Relationship between coefficient of friction and surface speed*

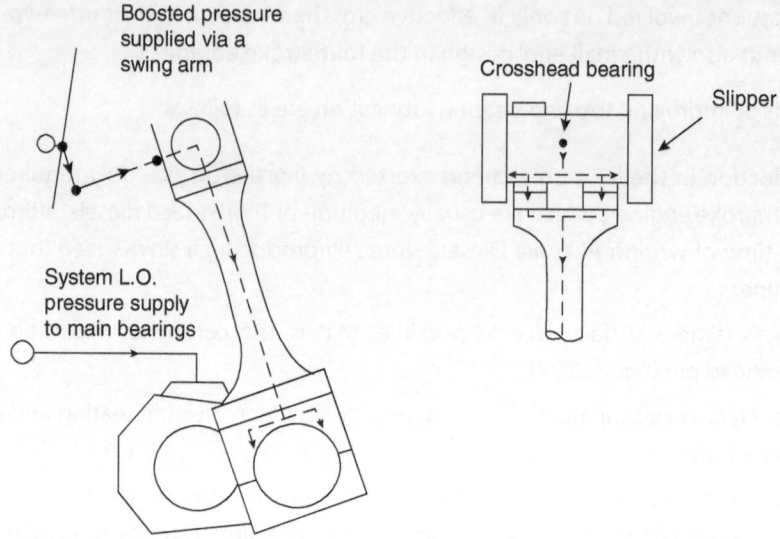

▲ **Figure 2.30** *Lubrication system for main bearings*

Cylinders and Pistons

Cylinders

Figure 2.31 shows a section through a typical cylinder liner from a large two-stroke low-speed engine. Modern liners are manufactured from good-quality alloyed cast iron and must satisfy the conflicting requirements of being thick and strong enough to withstand the high pressures and temperatures that occur during combustion and thin enough to allow good heat transfer.

This conflict is reconciled by the use of bore cooling. Figure 2.32 illustrates that by boring the upper part of the liner at an angle to the longitudinal axis, the bore at mid-point is close to the surface of the liner. The close proximity of the liner surface to the cooling water results in effective heat transfer. By using this technique of bore cooling, good heat transfer is accompanied by high overall strength.

Maintaining the correct surface temperatures in the vicinity of the combustion space by good heat transfer does, however, cause the risk of low temperature corrosion or cracking occurring in the lower portions of the liner. The solution to this problem is to either insulate the cooling water spaces that are at risk, or utilise a load-controlled cylinder cooling system to maintain optimum cylinder liner temperature (figure 7.14 in Chapter 7).

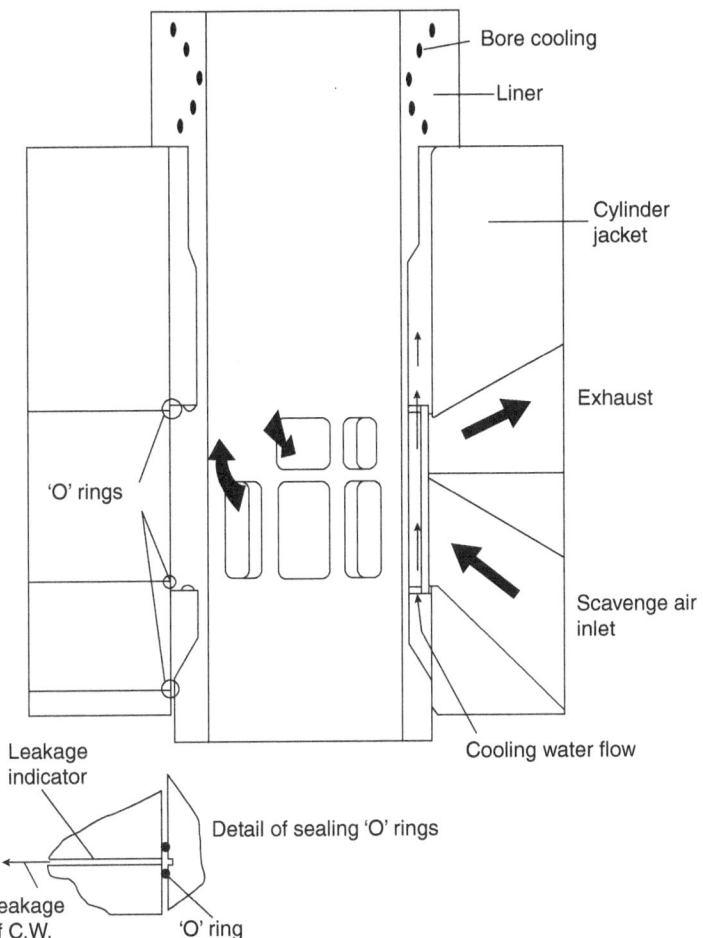

Bore cooling

Liner

Cylinder jacket

Exhaust

Scavenge air inlet

'O' rings

Leakage indicator

Cooling water flow

Detail of sealing 'O' rings

Leakage of C.W.

'O' ring

▲ **Figure 2.31** *Two-stroke cylinder liner*

Longitudinal expansion of the liner takes place through the lower cooling jacket. The sealing of the cooling water is accomplished by silicone rubber 'O' rings installed in grooves machined in the liner, which slide over the jacket as the liner expands and contracts (figure 2.31). The 'O' rings are in groups of two, the space between them being open to the atmosphere via an inspection hole. Leakage of water past an 'O' ring will be seen at the inspection hole, which not only alerts the engineers but also prevents leakage into the scavenge space. Great care must be exercised to ensure that the 'O' rings are not damaged when refitting a cylinder liner into the cylinder jacket. The good practice of keeping the engine cooling water at the normal operating temperature when the engine is shut down will mean that the liner expansion/contraction is kept as small as possible.

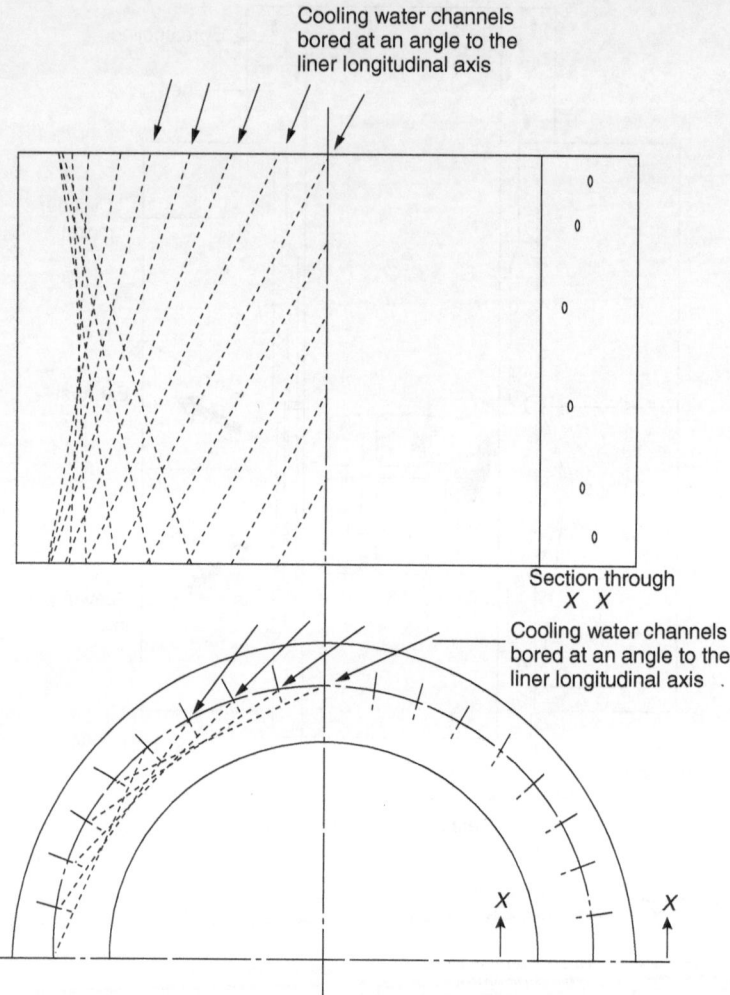

Cooling water channels
bored at an angle to the
liner longitudinal axis

Section through
X X

Cooling water channels
bored at an angle to the
liner longitudinal axis .

▲ **Figure 2.32** *Diagrammatic view of cylinder liner bore cooling*

Modern materials have recently enabled significant advances in cylinder liner design, with the liner walls becoming much thinner but still giving the same strength as before. The cylinder liners in MAN Diesel & Turbo two-stroke engines are made from alloyed cast iron and are placed in the engine monoblock by sitting on a flange, which is designed about two-thirds of the way down the liner. The low position of the flange is a departure from the more traditional designs, still retained by Wärtsilä, where the resting flange is at the top of the liner, as shown in figure 2.31.

The top of the liner is fitted with a light cooling jacket and the liner extends up above the cylinder frame more than in other designs. This means that the liner is not supported by its neck, which is also the area that is subjected to the full firing loads of combustion.

Cylinder liner lubrication

Friction between the cylinder liner and the piston rings is a major loss of power in the drive line of two-stroke diesel engines. Therefore, cylinder lubrication is very important on a four-stroke engine; the liner is 'splash' lubricated from the oil that is leaving the surfaces of the big end bearing as it rotates during operation and a large volume of the oil will also be returned to the crankcase by the action of the 'scrapper' ring on the piston. Some four-stroke engines are provided with additional means of liner lubrication either from the lower part of the piston or from a mechanism outside the cylinder liner that supplies oil through holes in the lower section of the liner. Either way the oil used is the same oil as the main engine lubricating oil. Some of the oil will find its way past the piston rings and will be burned. On average, a large well-maintained trunk type four-stroke engine will consume about 0.3–0.5 g/kWh of lubricating oil.

The modern large two-stroke engine, however, has the liner separated from the crankcase by a diaphragm and the piston rod gland. Therefore, the cylinder liner must be lubricated by injecting cylinder lubricating oil onto the surface of the liner as the piston moves past the holes.

This system is a 'total loss' lubricating system, meaning that when the oil is injected to lubricate the cylinder and having completed its task it is either burned or it comes out in the scavenge space. This system of having the two lubricating systems separate means that a different cylinder oil can be used to the main lubricating oil for the engine.

As the lubricating oil is 'lost' it also means that this operation is costly and therefore as technology has improved so has the research into lowering consumption by studying the path of the cylinder oil through the engine. The problem has recently been compounded by the notification of the introduction of the use of Low Sulphur Fuel (LSF) (below 1.5% sulphur). Some manufacturers are recommending that two types of cylinder oil are used, one to be used with the high sulphur fuel (HSF) (between 1.5% and 4.5% sulphur) and another with the LSF. At least one major oil company has now produced a cylinder oil that can be used with both LSF and HSF.

The reason for two types of fuel is that at the time of writing the countries that are signatories to the International Maritime Organization (IMO) have agreed that there will be areas around the world, available to ships, that are to be classed as Emission Control Areas (ECA). When a vessel sails into those areas, the emissions (mostly nitrogen oxides NOx and sulphur oxides SOx) from flue gases must be below certain levels.

Currently the answer to this is to switch from common heavy fuel oil to the LSF as the vessel approaches the designated area. The experience of ship operators has suggested that the difficulty is not so much a question of switching fuels but more about ensuring that the correct lubricant is being used with a particular grade of fuel.

Engine manufacturer guidelines state that lubricants that are suitable for use with high sulphur fuel (HSF) are not suitable for use with LSF and, technically, a different base number (BN) lubricant should be used. A low BN, typically 50 or 40, corresponds with LSF and a high BN, typically 70, with HSF. This means that the ship's staff will also be required to switch lubricants when they switch from one fuel to another.

Basicity of Oils

The basicity of the oil is at the heart of the chemical structure relating to the oil's ability to react with and neutralise acids. In chemistry, the Acid-Base relationship is studied in a number of compounds. Water, for example, is interesting as it can be a base or an acid. It is referred to as amphoteric and can be used as a universal 'comparator' to determine the relative strength of a group of bases or acids.

The measure that will be of interest to the marine engineer will be the Total Base Number (TBN). The higher base numbers will indicate a higher ability of an oil to react with and neutralise the acids that are being formed as a result of the combustion of fuels.

Additives can also be used to boost the base number and the higher base numbers are associated with the cylinder oils that are used in two-stroke engines.

The base number of the lubricant is calculated in accordance with the procedures set out in the standard ASTM D-2896.

The use of lower basicity cylinder lubricants, used with low sulphur fuels within an Emission Control Area (ECA), runs directly counter to the lubrication requirements for slow steaming or other conditions outside ECAs (where different conditions or sulphur content of the fuel exists), which conversely requires owners and operators to run different specific lubricants under the differing circumstances or conditions.

Oil majors have worked to introduce a lubricant that not only has the high basicity, and detergency, required for slow steaming but also the low BN characteristics needed for a lubricant being used with LSF. The principal objectives of cylinder lubrication are as follows:

1. To separate sliding surfaces with an unbroken oil film.
2. To form an effective seal between piston rings and cylinder liner surface to prevent blow-past of gases.

3. To neutralise corrosive combustion products and thus protect cylinder liner, piston and rings from corrosive attack.

4. To soften deposits and thus prevent wear due to abrasion.

5. To remove deposits to prevent seizure of piston rings and keep the engine clean.

6. To cool hot surfaces without burning.

In practice, some oil burning will take place; if excessive, this would be indicated by blue smoke and increased oil consumption. As the oil burns, it should leave as little and as soft a deposit as possible. Over-lubrication is now avoided with the use of computer-controlled systems such as the MAN Diesel & Turbo 'Alpha Adaptive Cylinder-oil Control' (Alpha ACC) or the Wärtsilä RTA Pulse Lubrication System (PLS). The average oil consumption for this type of system would be 0.7–0.9 g/kWh.

When the engine is new, cylinder lubrication rate should normally be greater than when the engine becomes run in. Reasons for this initial increased lubrication are as follows:

1. Liner surface unevenness will cause localised high temperatures, which in turn will cause increased oxidation of the oil and reduce its lubrication properties.

2. Sealing of the rough surfaces is more difficult.

3. Worn metal needs to be washed away.

The actual amount of lubricating oil to be delivered into a cylinder per unit time depends upon stroke, bore and speed of engine, engine load, cylinder temperature, type of engine, position of cylinder lubricators and type of fuel being burned.

Position of the cylinder lubricators for injection of oil has always been a topic of discussion. The following points are of importance:

1. They must not be situated too near the ports; oil can be scraped over the edge of ports and blown away.

2. They should not be situated too near the high temperature zone or the oil will burn easily.

3. There must be sufficient points to ensure as even and as complete a coverage as possible.

The modern systems will deliver the oil to the point of use within 8–10 ms and therefore timed injection is possible. The objective is to inject the cylinder lubricating oil into the piston ring pack when the piston rings pass the quill level. Some systems deliver a measured amount of oil with every revolution, while others deliver a larger dose every fifth or sixth revolution.

Once the lubricator pump has delivered the cylinder oil to the quills in the cylinder liner, it gets picked up by the piston ring pack as they pass the openings in the liner wall. The key to success is for an even distribution of the oil on the cylinder liner's running surface and to keep the oil on the cylinder wall replenished to provide sufficient additives to neutralise the acid that will be forming and to keep the liner clean.

An example of the amount of oil required could be shown by the operation of the Wärtsilä 84T engine. This engine has a bore of 840 mm and a stroke of 3,150 mm, therefore the surface area of each cylinder is 8.3 m^2. There are eight quills separated around the liner 900 mm from the top. The lubricator delivers 310 mm^3 to each time it is activated, which, depending upon the speed and load, could be every two to five revolutions of the engine. Therefore, 8.3 m^2 is lubricated by 8×310 mm^3, which then has to be evenly distributed across, up and down the liner.

Vertical distribution of the cylinder oil is mainly performed by the piston rings during their travel up and down the cylinder. The cylinder oil is injected into the piston ring pack during the piston's upward stroke; factors such as oil viscosity, feed rate and the volume of oil per injection are calculated by the control software. The correct oil viscosity is important to encourage the spreadability of the cylinder oil. The applied feed rate and volume of oil injected for each operation of the lubricator are key factors in the critical balance between under- and over-lubrication.

There was some experimentation with the new PSL system before the 'zigzag' groove principle was adopted for all new engines and retrofitted to a high number of large-bore engine cylinder liners. This system measures the temperature in two diametrically opposite positions near the running surface in the upper part of each cylinder liner. It then filters and interprets the development of the temperatures, and in case the temperature level escalates, a 'high friction alarm' is generated.

Under-lubrication could lead to corrosion, accumulated contamination from unburned fuel and combustion residues and, in the worst case, metal-to-metal contact, known as 'scuffing'. Over-lubrication can lead to a number of problems, including:

- the loss of unused oil in the scavenge ports
- piston rings being prevented from moving (rotating) in their grooves by 'hydraulic lock'
- 'chemical bore polish'
- 'mechanical bore polish'.

Cylinder liner wear

Cylinder liner wear can be divided into:

- Abrasive wear
- Corrosive wear.

Abrasive wear

This occurs when abrasive particles enter the combustion space with the scavenge air or as a result of poor quality or contaminated fuel. Maintenance of the engine's air filtration system is very important for the continued good health of the engine's internal components.

The quality of bunker fuel will be discussed in more detail in Chapter 3 of this volume. However, it is appropriate to say at this point that because marine engines burn residual fuel oil it is very easy to also have within the fuel components that are left over from the refining process. Catalytic cracking of oil during the refining process uses very hard particles of aluminium and silicon that come about due to the catalytic cracking process in the refinery. They are in a form of complex alumino-silicates and can vary in size and hardness. Bunker specification should mean that these are not in the final product but all too often they get through the quality assurance processes. It is important to have samples taken at the vessel's bunker station while bunkering is taking place. If there is any damage to the engine that could be because of catalytic 'cat' fines (left over from the oil refining process), then evidence will be required and without the samples a successful outcome to any insurance claim might be difficult.

Corrosive wear

Overall liner wear is reducing as understanding about the lubrication process improves. Corrosive wear is still a more common cause of cylinder liner wear and comes from burning heavy fuel containing significant amounts of sulphur. As the fuel burns, the sulphur combines with oxygen to produce oxides of sulphur, which further form sulphuric acid on contact with water. To minimise the formation of acids it is important that cylinder liner temperatures are maintained above the dew point (figure 2.33).

Good operational practice is the key to keeping cylinder liner wear as low as possible. It is very important that ships' engineers understand how to operate the engine correctly. This includes:

- Correct quantity and grade of cylinder lubrication.
- Correctly fitted piston rings.

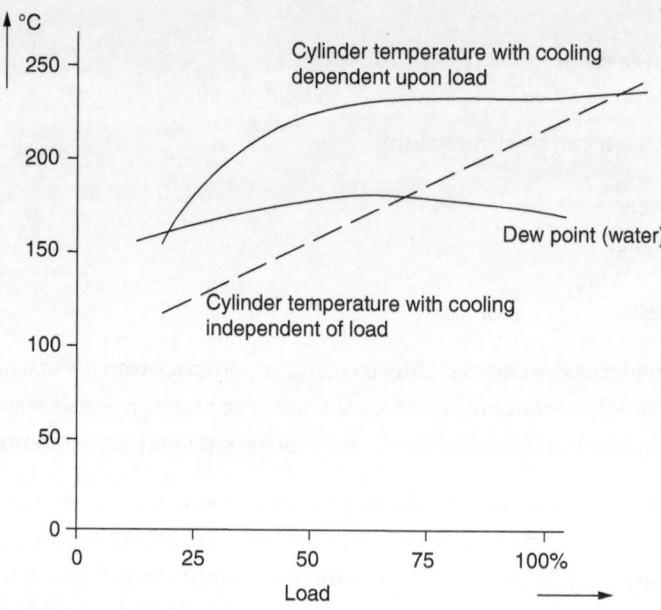

▲ **Figure 2.33** *Temperature of cylinder liner surface throughout. Engine load range*

- Correct warming through prior to starting.
- Well-maintained and timed fuel injectors.
- Well-managed fuel storage and purification plant.
- Correct cooling water and lubricating oil temperatures.
- Correct scavenge air temperatures.
- Engine load changes carried out gradually.
- Well-maintained equipment.

The deterioration of fuel quality that has taken place over the years coupled with the increased pressures and temperatures that occur during the combustion process have resulted in liners and piston rings operating under very severe conditions. Despite these adverse operating conditions, cylinder liner wear rates have been reduced with large two-stroke manufacturers claiming 0.03 mm/1,000 h and medium-speed four-stroke engine manufacturers claiming wear rates of 0.02 mm/1,000 h when operating on heavy fuel, and with the recent trial of a stepped piston and liner arrangement MAN Diesel have reported a wear rate as low as 0.0045 mm/1,000 h. These advancements are helping to extend the time between overhaul (TBO) and manufacturers are working towards running main engines from dry dock to dry dock without the need of a major overhaul where the piston has to be removed.

These wear rates have been achieved as a result of a number of factors, such as:

- The development of highly alkaline lubricating oils to neutralise the acids formed during combustion.
- The development of load-dependent temperature control of cooling water, which maintains the cylinder liner temperature at an optimum level (figure 7.14 in Chapter 7) [cooling water section].
- The use of good-quality alloyed cast iron with sufficient hard phase content for cylinder liners.
- Careful design of piston ring profiles to maximise lubricating oil film thickness.
- Improvements in lubricating oil distribution across cylinder liner surface. This includes multi-level injection in two-stroke engines and forced piston skirt lubrication in four-stroke engines (figure 2.37).
- Improved separation of condensate from scavenge air.

Cylinder liner wear profile

Figure 2.34 shows the wear profile of both a two-stroke and four-stroke engine cylinder liner. It can be seen that the greatest wear occurs in the upper part of the liner adjacent to the firing zone. This is due to:

- The high temperatures and pressures that occur at this point.
- Because the piston reverses direction at this point, hydrodynamic lubrication is not established.
- Acids formed during combustion attacking the liner material.

Slow steaming

During different economic climates, ships will need to operate in different ways, for example a high-speed liner service between two ports might be the correct strategy during a rising or buoyant economic climate. However, it might not be correct if the trading markets fall. When the changes in economic fortune come about during the lifetime of a vessel, owners have difficult choices to make. If the engine is flexible and is able to operate efficiently at different speeds and power outputs then it will prove much more useful to the owner due to its flexibility.

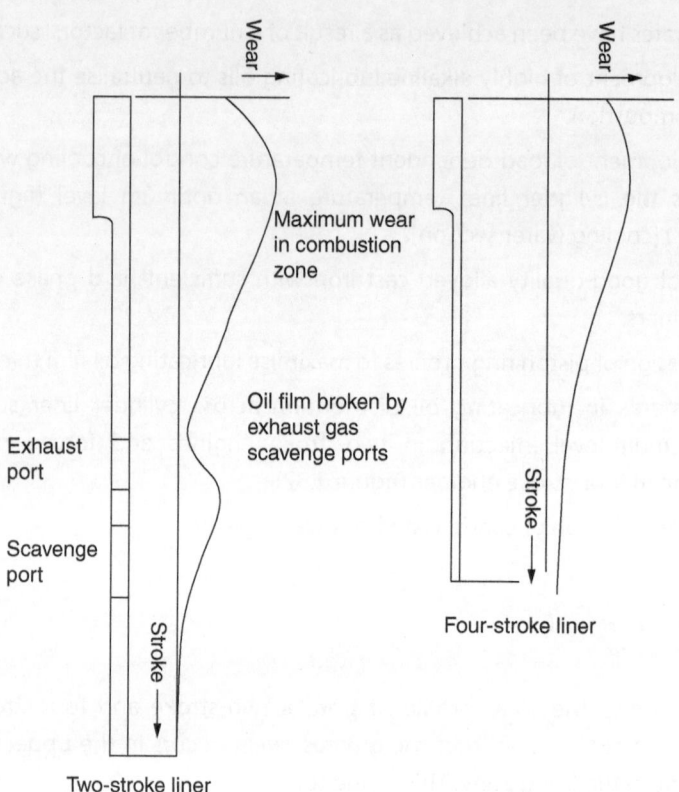

▲ Figure 2.34 *Liner wear profile*

Engine manufacturers are working hard to increase the working envelope of their engines so that they will operate efficiently at full design power/speed output and also at reduced power/speed combinations. In the past this has proved to be very difficult to achieve as engine components, such as fuel injectors, had to be changed when slow steaming. With the introduction of electronic control and a better understanding of processes such as cylinder liner lubrication, engines can run much more efficiently and under varying conditions.

However, the best check to ensure that the engine is running at its best is by regular inspection of the piston, piston rings and combustion chamber by viewing them through the scavenge ports when the engine is shut down and in port. Judgements can then be made about whether to open up a unit for further inspection or if everything is operating correctly.

Cloverleafing

Despite the close control of cylinder surface temperatures, acids are still formed, which must be neutralised by the cylinder lubricating oil. This requires that the correct quantity and TBN grade of oil is injected into the cylinder. As soon as the oil enters the cylinder it starts neutralising the acids, becoming less alkaline as it does so. If the TBN of the oil is too low then its alkalinity may be depleted before it has completely covered the liner surface. Further contact with the acids may lead to the oil itself becoming acidic. This will lead to the phenomenon known as 'cloverleafing' in which high corrosive wear occurs on the liner between the oil injection points (figure 2.35). Severe cloverleafing can result in gas blow-by past the piston rings and ultimate failure of the liner.

Micro-seizure

This is due to irregularities in the liner and piston rings coming into contact during operation as a result of a breakdown of lubrication due to an insufficient quantity of lubricating oil, insufficient viscosity or excessive loading. This results in instantaneous seizure and tearing. In appearance, micro-seizure resembles abrasive wear since the characteristic marks run axially on the liner. Micro-seizure may not always be destructive, indeed it often occurs during a running-in period. It becomes destructive if it is persistent and as a result of inadequate lubrication.

Pistons and Rings

Pistons

Pistons must be strong enough to withstand the very high firing pressures that are common today, be able to dissipate sufficient heat to maintain the correct piston crown temperatures and withstand the stresses imposed by friction. Pistons are manufactured from cast steel, forged steel and cast iron, although all of these materials have limitations. Cast iron is weak in tension, especially at elevated temperatures. It does, however, have high compressive strength, which enables it to resist the hammering that occurs at

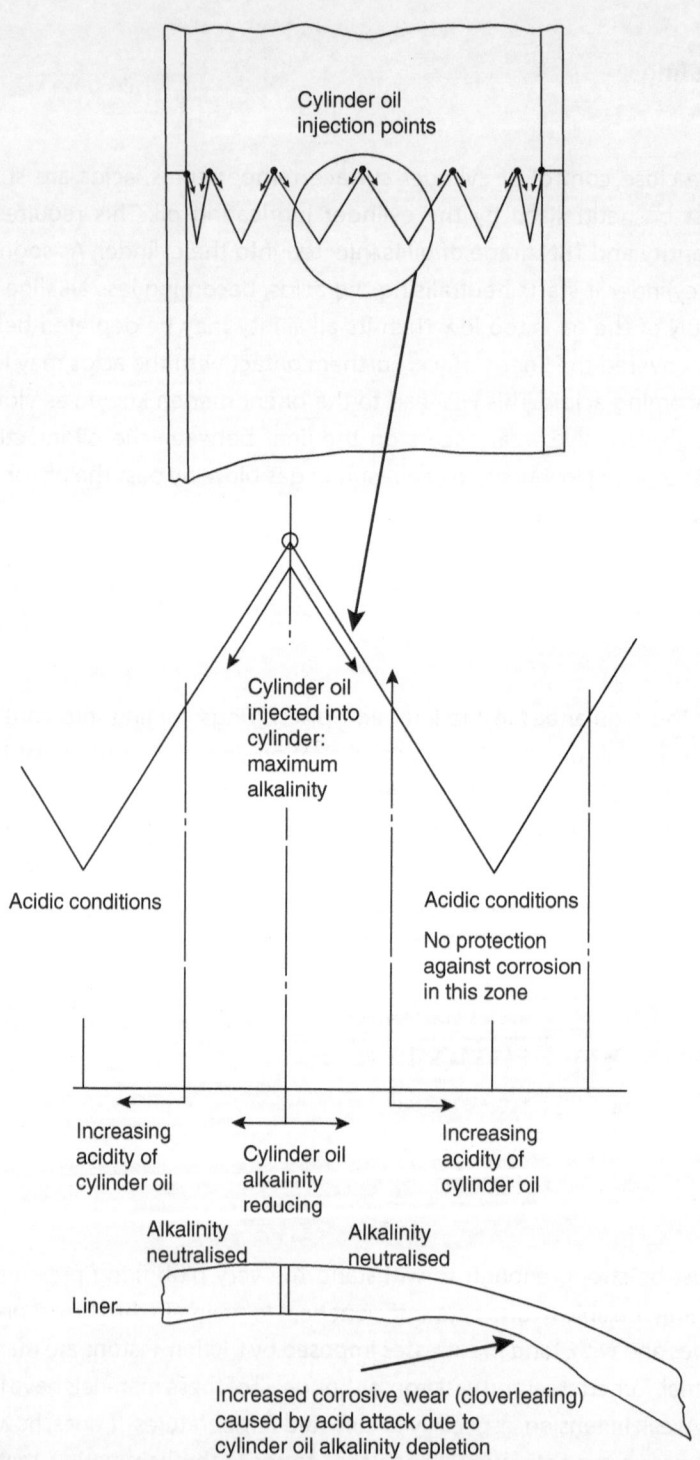

▲ **Figure 2.35** *Increased corrosive wear of cylinder liner (cloverleafing)*

the ring grooves. As a result of its graphite content, cast iron performs well when exposed to rubbing. This makes it a suitable material for piston skirts. Cast steel resists heat stresses better than cast iron but is difficult to ensure that the molten material flows to the extremities of intricate moulds. Cast steel also requires extensive heat treatment to relieve casting stresses. Forged steel is a suitable material because the directional grain flow exhibited as a result of forging produces a strong tough component. Forged steel is prone to high wear at the ring grooves and also requires a greater degree of machining, which tends to increase the production cost.

Modern pistons are composite components, made from materials that exhibit suitable properties for the different parts of the structure.

- Piston crowns that are highly stressed mechanically and thermally are made from cast or forged steel. Cast iron inserts are fitted to the ring grooves to resist wear.
- Piston skirts are made from cast iron, which has superior rubbing properties to either cast or forged steel. To reduce weight and reduce inertia loads, aluminium is used in some medium-speed four-stroke applications.

The design trend with marine diesel engines has moved away from using water as a cooling medium for the piston. The hazard of introducing water into the scavenge space or the crankcase has led manufacturers away from this practice, although the designs are still included in Flag State examinations because the seagoing engineer may find an example still in service and will need to be prepared to look after the system while on an ocean-going passage. Figure 2.36, however, shows a piston for a large Sulzer (now part of the Wärtsilä Corporation) engine. The crown is of forged steel and combines strength with good heat transfer.

Strength is achieved by using an overall thick section piston crown, which is then bore cooled. Intensive cooling is achieved by the cocktail shaker effect of the water. With air present in the piston (this comes from the telescopic system, it being necessary to provide a cushion and prevent water hammer) together with water, the inertia effect coupled with the bore cooling leads to very effective cooling as the piston goes over TDC.

Figure 2.37 shows an oil-cooled piston typical of the design of the large two-stroke engines in service today. The piston crown of this piston is also manufactured from forged steel but in this case the section is relatively fine, strength being achieved by the 'strong back' principle, which supports the piston crown from inside. Bore cooling is employed by some designs of large-bore two-stroke engines. In this design the bores act as nozzles through which the oil flows radially, spraying onto the underside of the piston crown, before flowing to the drain (similar to the water-cooled design).

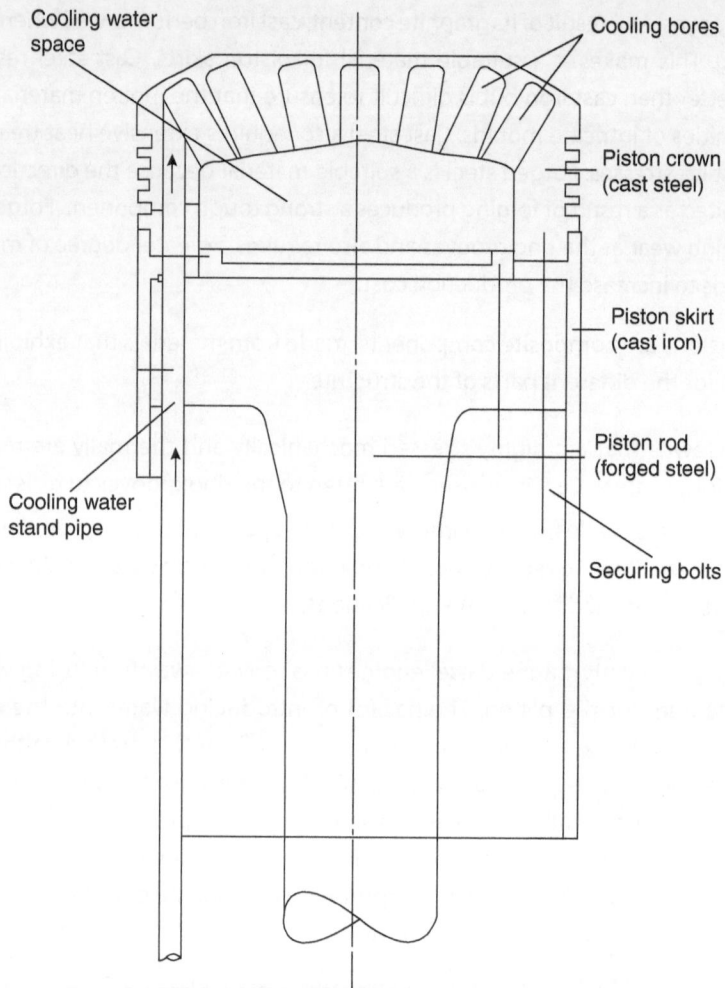

▲ Figure 2.36 *Water-cooled piston with bore cooling*

The latest large-bore two-stroke designs (figure 2.38) have a series of branches extending from the top of the piston rod to direct the oil onto the underneath of the piston crown, enhancing the cooling effect of the oil on the piston. The oil is fed up the middle of the piston rod and flows back through channels in the outer part of the piston rod.

Figure 2.39 shows a piston from a medium-speed four-stroke engine. This type of engine design transmits the combustion forces directly through the gudgeon pin and onto the connecting rod. The piston crown is of forged or cast steel while the skirt is of nodular cast iron. Cooling is effected by oil flowing from the connecting rod into the piston crown then flowing radially outwards to effectively cool the piston. In Wärtsilä engines some of this oil is then taken out through four nozzles that feed the oil distribution

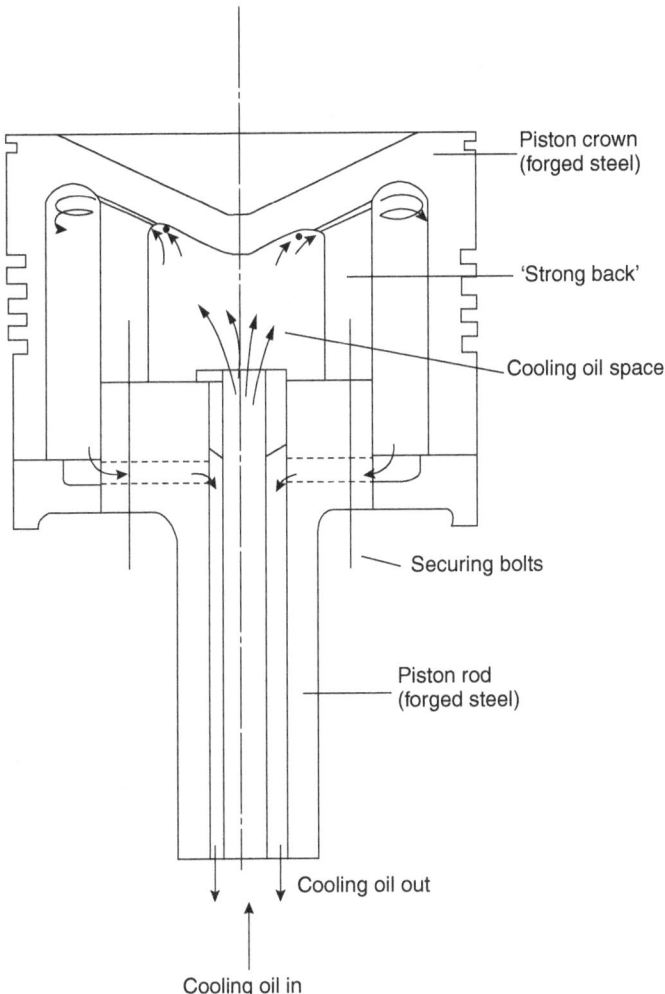

Piston crown
(forged steel)

'Strong back'

Cooling oil space

Securing bolts

Piston rod
(forged steel)

Cooling oil out

Cooling oil in

▲ **Figure 2.37** *Oil-cooled piston from small-bore two-stroke engine*

groove in the piston skirt. The manufacturers claim that this design, which they have patented, provides an even oil film formation that reduces liner wear.

The choice of water or oil for piston cooling

The choice for the latest designs of marine main propulsion engines is not an issue as all the manufacturers have opted for oil-cooled designs. However, it is expected that the modern marine engineer will still have an appreciation of the advantages and disadvantages between the oil- and water-cooled designs.

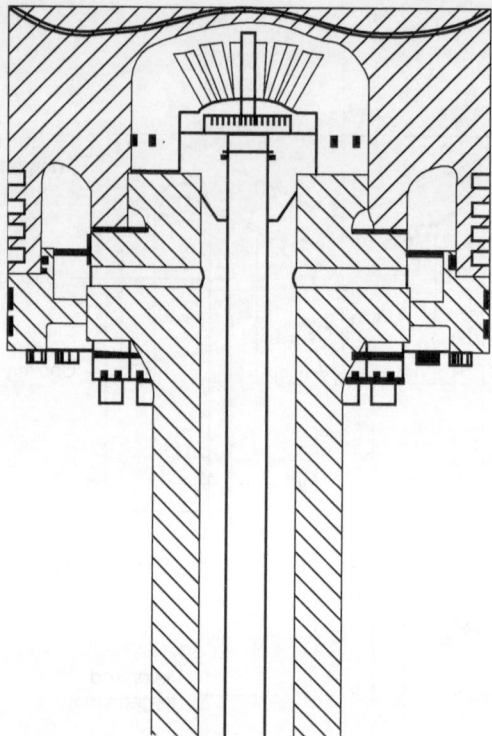

▲ **Figure 2.38** *Piston from the MAN 'M' series large-bore engines*

Distilled water, kept free from impurities and in the correct alkaline state, has some advantages over oil:

- It is relatively cheap and plentiful.
- Internal surfaces are kept free from deposits.
- Water-cooled pistons can be operated at higher temperatures.
- Water has a specific heat capacity nearly twice that of oil. (This means that, for the same mass flow rate, water is able to carry away nearly twice as much heat as oil, so lower mass flow rates can be specified.)

The disadvantages of water are that:

- Leakage into the crankcase will result in serious contamination of the lubricating oil.
- Additional pumps and coolers are required.
- Telescopic pipes and glands are required to transport the water to and from the piston.

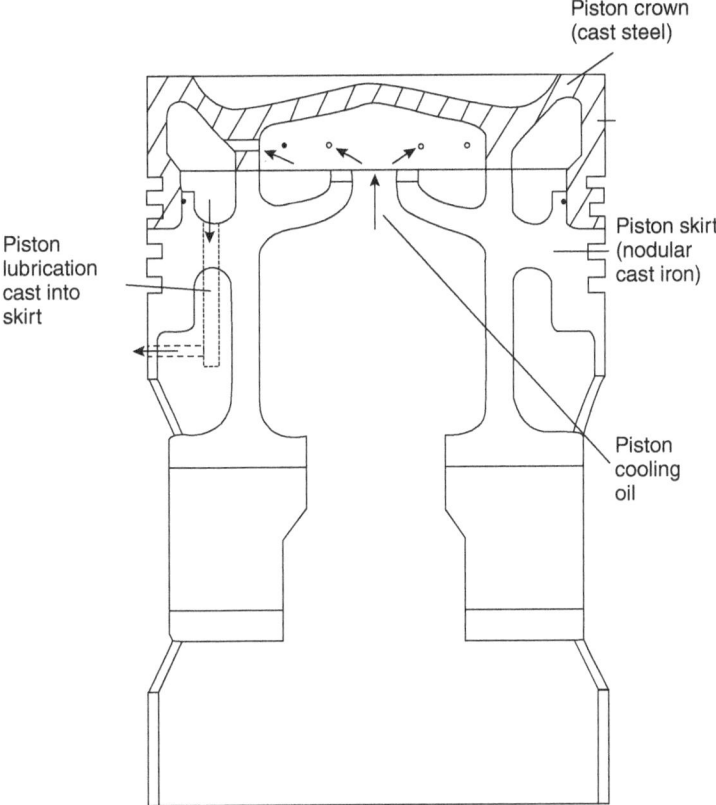

Piston crown
(cast steel)

Piston
lubrication
cast into
skirt

Piston skirt
(nodular
cast iron)

Piston
cooling
oil

▲ **Figure 2.39** *Composite piston suitable for a high-output medium-speed four-stroke diesel engine*

Oil is used extensively in modern engines and in all of the new building. The advantages claimed for this medium are:

- Simplified supply of the oil to the piston is achieved.
- Leakage into the crankcase does not present contamination problems.
- The lower thermal conductivity results in a less steep temperature gradient over the piston crown.

The disadvantages of oil are:

- The temperatures must be kept relatively low in order to limit oxidation of the oil.
- If overheating occurs there is a possibility that carbon deposits could form on internal surfaces and there is a danger that carbon particles could enter the lubricating oil system. This could have an insulating effect and make the cooling process less efficient.

Failure of pistons due to thermal loads

When a piston crown is subjected to high thermal load, the material on the gas side attempts to expand but is partly prevented from doing so by the cooler metal under and around it. This leads to compressive stresses within the piston cross section, in addition to the stresses imposed mechanically due to the variation in cylinder pressures.

At very high temperatures the metal can creep to relieve this compressive stress and when the piston cools a residual tensile stress is set up, hence residual thermal stress. If this stress is sufficiently great, cracking of the piston crown may result.

At normal working temperatures the piston and cylinder liner surfaces should be parallel. Since there is a temperature gradient from the top to the bottom of the piston, allowance must be made during manufacture for the top cold clearance to be less than the bottom. The temperature gradient is generally non-linear and thermal distortions produce tensile stresses on the inner wall of the piston; gas forces tend to bulge the piston wall out, thereby reducing the tensile stress. This variable tensile stress at very high thermal loads could lead to cracks propagating through from the inside of the piston to the piston ring grooves. Modern engine designers use complex mathematical models to evaluate these stresses before the engine is manufactured. However, information from the running test engines is used to update the algorithms in the computer models. For this reason, MAN Diesel have their four-cylinder 500-mm bore test engine at their research centre in Copenhagen and Wärtsilä have the RTX5 test engine in Trieste, Italy (figure 2.40).

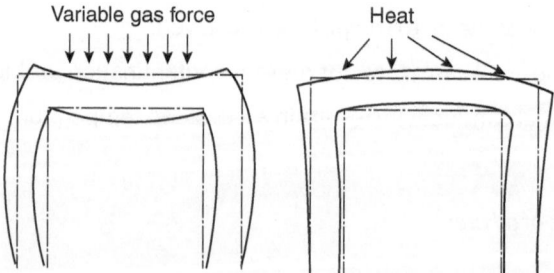

▲ **Figure 2.40** *Effect of gas and heat*

Piston rings

Properties required of a piston ring are as follows:

1. Good mechanical strength, it must not break easily.
2. High resistance to wear and corrosion.
3. Self-lubricating.
4. Great resistance to high temperatures.
5. Must at all times retain its tension to give a good gas seal.
6. Be compatible with cylinder liner material.

The above properties are the ideal and therefore difficult to achieve in practice. Materials that are used to obtain as many of the desired properties as possible are as follows:

1. Ordinary grey cast iron, in order that it may have good wear resistance and self-lubricating property, must have a large amount of graphite in its structure. This, however, reduces its strength.
2. Alloyed cast iron, elements and combinations of elements that are alloyed with the iron to give finer grained structure and good graphite formation are: molybdenum, nickel and copper or vanadium and copper.
3. Spheroidal graphitic iron, very good wear resistance, not as self-lubricating as the ordinary grey cast iron. These rings are usually given a protective coating, for example chromed or aluminised etc to improve running-in.

It is possible to improve the properties by treatment. In the case of the cast irons with suitable composition, they can be heat treated by quenching, tempering or austempering. This gives strength and hardness without affecting the graphite.

Piston rings are often contoured to assist in the establishment of a hydrodynamic lubricating oil film and so reduce liner wear (figure 2.41). It is common practice for manufacturers to specify a ring pack in which the first and second compression rings, subjected to higher temperatures and pressures, differ from the lower rings. It is important when installing new rings that the manufacturer's recommendations are followed since ring failure may result if incorrect rings are fitted.

In addition to compression rings, four-stroke medium-speed engines also employ oil control or oil scraper rings (figure 2.42). Unlike compression rings, which help promote

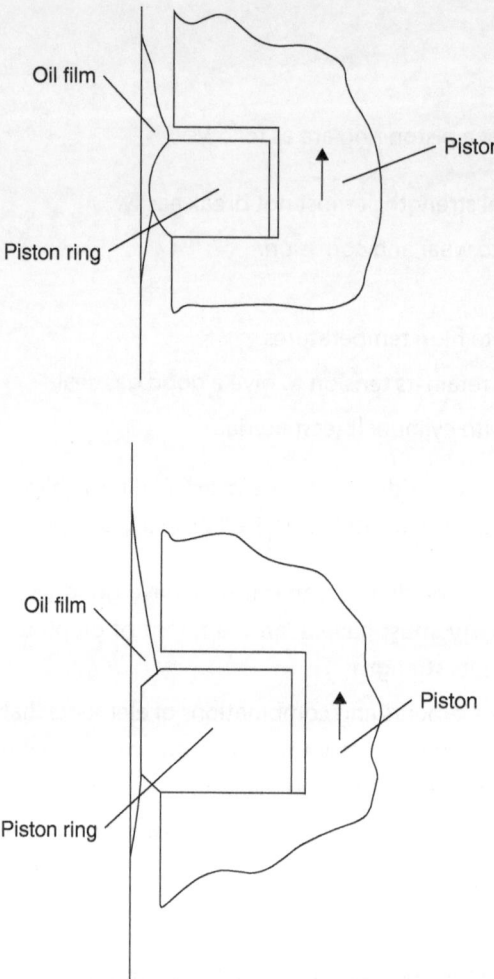

▲ **Figure 2.41** *Two-stroke engine piston ring profile*

the formation of an oil film, oil scraper rings scrape the oil from the cylinder liner and return it to the sump. Many designs of oil scraper rings can only be fitted in one direction and care must be exercised when installing these rings. Without these rings, lubricating oil in the upper cylinder would be burned during combustion, resulting in extremely high oil consumption. As the oil scraper rings wear, their effectiveness in returning the oil to the sump reduces with high oil consumption as the consequence. Oil scraper ring wear may be the limiting factor when deciding cylinder overhaul periods for medium-speed four-stroke engines.

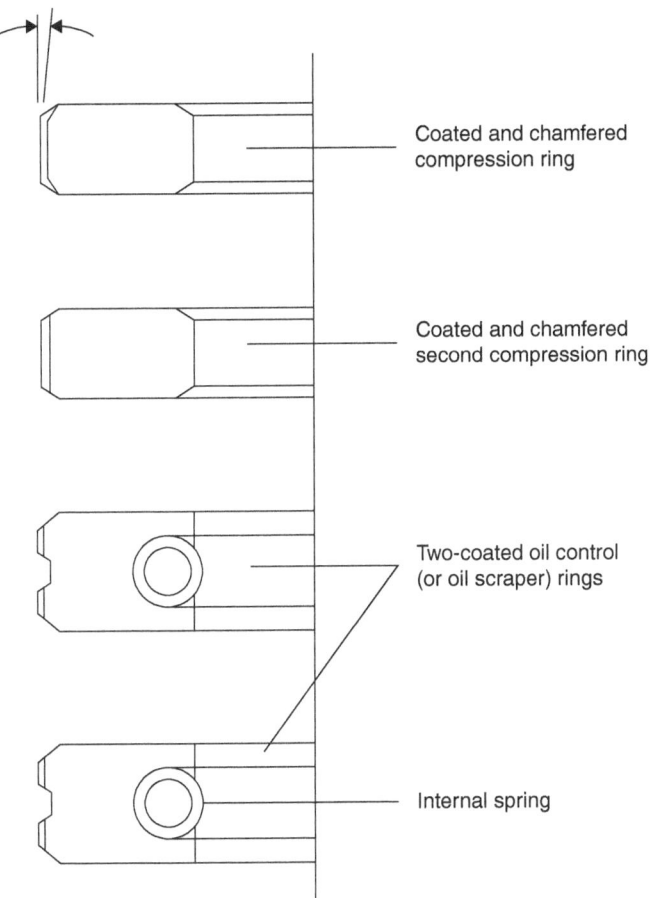

▲ **Figure 2.42** *Four-stroke piston rings*

Manufacture

1. Statically cast in sand moulds to produce either a drum from which a number of piston rings would be manufactured or an individual ring.

2. Centrifugally cast to produce a fine-grained non-porous drum of cast iron from which a number of piston rings will be machined. The statically cast rings, either drum or single casting, may be made out of round. The out-of-round blanks are machined in a special lathe that maintains the out of roundness. Rings manufactured in this way are expensive but ideal.

Most piston rings are made from circular cast blanks machined to a circular section on their inner and outer diameters. In order that the rings may exert radial pressure when fitted into the cylinders, they are split in tension. Tensioning is done by cold deformation of the inner surface by hammering or rolling. The finished ring would be capable of exerting a radial pressure from 2 to 3 bar and have a Brinel hardness from 1,600 to 2,300 (SI units). Large diesel engine cylinder liners have a hardness range similar to the above.

Piston ring defects and their causes

1. Incorrectly fitted rings. If they are too tight in the grooves the rings could seize, causing overheating, excessive wear, increased blow-past, etc. If they are too slack in the grooves, angular working about a circumferential axis could cause ring breakage and piston groove damage. If the butt clearance is too great, excessive blow-past will occur.

2. Fouling due to deposits on the ring sides and their inner diameters; this could lead to rings sticking, breakage, increased blow-past and scuffing.

3. Corrosion of the piston rings can occur due to attack from corrosive elements in the fuel ash deposits.

4. If the ring-bearing surfaces are in poor condition or in any way damaged (this could occur during installation), scoring of the cylinder liner may take place; if the ring has sharp edges it will inhibit the formation of a good oil film between the surfaces.

Due to uneven cylinder liner wear, the piston ring diameter changes during each stroke; this leads to ring and groove wear on the horizontal surfaces. This effect obviously increases as differential cylinder liner wear increases. Oscillation of the piston rings takes place in the cycle about a circumferential axis approximately through the centre of the ring section, and if the inner edges are not chamfered they can dig into the piston groove lands. Keeping the vertical clearance to a working minimum will reduce the oscillatory effect. If the cylinder liner has become worn at the top of the piston travel, there could be a ridge from which the piston ring hits at the top of its travel. This ridge can be chamfered during unit overhaul to reduce the damage to the top piston ring.

When considering piston rings, perhaps the most destructive force at work is hammering. This is caused by relative axial movement between piston and ring as a result of gas loading and inertia when the piston changes direction at BDC. The hammering results in enlargement of the piston ring groove and may result in ring breakage. Cast steel, forged steel or aluminium pistons usually have ring groove landing surfaces protected to minimise the effects of hammering. This can be either:

• Flame hardening – top and bottom on upper grooves.
• Chromium plating – top and bottom on upper grooves.
• The fitting of cast iron inserts.

In two-stroke engines, piston rings have to pass ports in the cylinder wall. Each time they do, movement of segments of the rings into the ports can take place. This would be more pronounced if the piston ring butts are passing the ports. It is possible for the butts to catch the port edge and bend the ring. In order to avoid or minimise this possibility, piston rings may be pegged to prevent their rotation or they may be specially shaped.

Inspection of pistons, rings and cylinders

Withdrawal of pistons, their examination, overhaul or renewal, together with the cleaning and gauging of the cylinder liner, is a regular feature of maintenance procedures, frequency of which depends upon numerous factors, such as: piston size; material and method of cooling; engine speed of rotation; type of engine, two- or four-stroke; fuel and type of cylinder lubricant used. The manufacturers of modern engines have the design aim of extending the time between overhauls. Modern research and development techniques have led to a much better understanding of cylinder liner lubrication and modern materials are lasting much better than in the past. Therefore, the student must be careful about answering a question about servicing intervals because although the manufacturers might still recommend servicing upon operational hours, it will be the responsibility of the ship's engineering staff to ensure that the engine is running efficiently.

With high-speed four-stroke diesel engines, as a general statement, the running time between piston overhauls used to be greater than that for large slow-speed two-stroke engines. This was attributed to the fact that the engine is usually unidirectional, hence reduced numbers of stops and starts with their attendant wear and large fluctuations of thermal conditions. Small-bore engines are easier to cool, cylinder volume is proportional to the square of the cylinder diameter, hence increasing the diameter gives greatly increased cylinder content and high thermal capacity. Thus overhaul time across all types of engines can now vary between about 10,000 and 32,000 h. In fact, manufacturers look to extend the time between major overhauls as being the same as the time from one dry dock to the next, meaning that the main engines do not have to be opened up in-between docking periods.

The modern two-stroke engine can have its pistons and cylinder liners inspected without having to remove the piston. After scavenge spaces have been cleaned of inflammable oil sludge and carbon deposits, each piston can in turn be placed at its lowest position. The cylinder liner surfaces can then be examined with the aid of a light introduced into the cylinder through the scavenge ports. The cylinder liner surfaces should have a mirror-like finish. However, black dry areas at the top of the liner indicate blow-past of combustion gases. Dull, vertically striped areas indicate breakdown of oil film and hardened metal surface (this is caused by metal seizure on a micro-scale leading to intense heating).

After inspection of a cylinder the piston can be raised in steps in order to examine both the piston and the rings. Heavy carbon deposits on piston crown and burning away of metal would indicate incorrect fuel burning and poor cooling. Piston rings should be free in the grooves, have a well-oiled appearance, be unbroken and worn smooth and bright on the outer surface. If they are too worn then sharp burrs can form on the edges that enable them to act as scraper rings, preventing good oil film formation. A dull surface of the piston ring will indicate that it is most likely broken somewhere around its circumference and it is no longer functioning as a piston ring.

Cylinder covers – large two-stroke engines

On the older loop-scavenged two-stroke engines, the cylinder covers tended to be a relatively simple symmetrical design to avoid the problems of differential expansion and the consequent stresses. Early Sulzer (now Wärtsilä) designs consisted of two pieces with a cast iron main component and a central cast steel insert containing the valves. In this design, cooling water is introduced into the cylinder cover through nozzles that ensure the water flows tangentially, thus minimising impingement on internal surfaces and reducing the possibility of erosion. More modern engines, which operate at higher temperatures and pressures, have one-piece forged steel cylinder covers. This design employs bore cooling, which allows the cooling water to pass very close to the combustion chamber, effectively maintaining safe surface temperatures irrespective of operation (figure 2.43).

MAN Diesel & Turbo, Mitsubishi and the Wärtsilä Corporation are currently the only designers of large two-stroke diesel engines that are used for marine propulsion. These designers do allow licensees to carry out the manufacturing process and some licensees are allowed to incorporate their own name into the engine model's name. However, all the basic designs have become standardised around the uniflow scavenging system. This system started life with opposed piston engines where the exhaust piston was arranged to uncover the exhaust ports. Approximately one-third of the engine's power was transmitted through the exhaust piston and eccentrics arranged on the crankshaft on either side of the unit's main journal. Figure 2.44 shows the arrangement of the latest Wärtsilä engine, showing the cooling channels and the position of one of the three injectors.

The modern designs have removed the exhaust piston and associated running gear and have arranged for the exhaust gases to be removed through a poppet valve situated at the centre of the cylinder cover. On the early designs the valve was operated by a traditional system, which was a mechanical push rod operated from a cam on the camshaft. The valve was closed and kept in position with a mechanical spring.

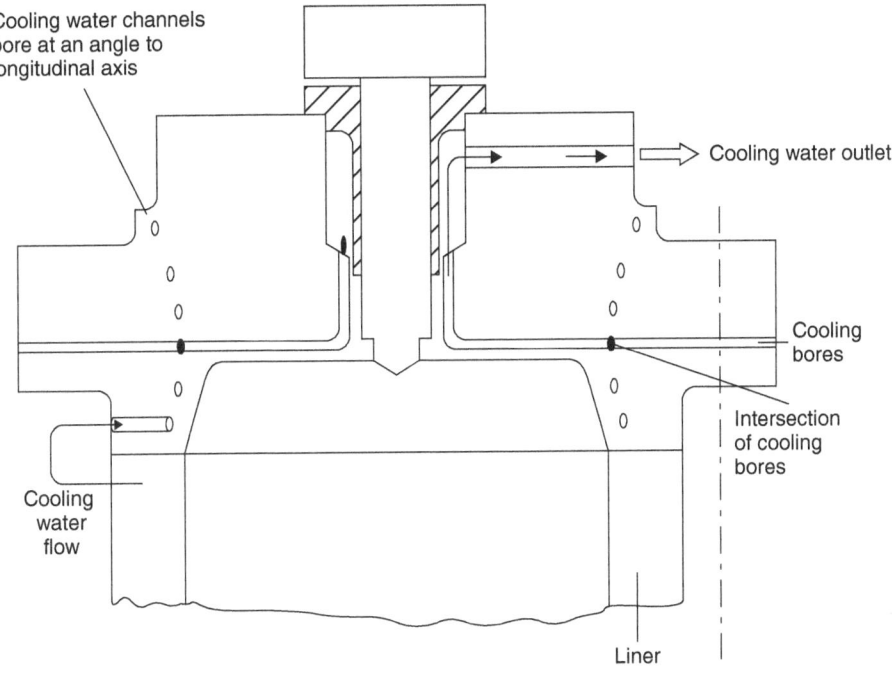

Cooling water channels bore at an angle to longitudinal axis

Cooling water outlet

Cooling bores

Intersection of cooling bores

Cooling water flow

Liner

▲ **Figure 2.43** *One-piece cylinder cover*

The next step in the design was to replace the mechanical push rod with a hydraulic actuator to open the valve, which was closed again by using the mechanical spring. Later versions on the current models show that the mechanical spring has been replaced with a compressed air return spring (figure 2.45). The cylinder cover is manufactured from forged steel and cooling is accomplished through radial cooling bores close to the combustion chamber surface. The exhaust valve cage and seat ring are also bore cooled.

The advantages claimed for the latest configuration of hydraulically operated valves are:

- There is no transverse thrust from hydraulic actuators. Thrust is purely axial, resulting in less guide wear.
- Controlled landing speed, from the air return spring, ensures minimum stress on valve and seat.
- Valve rotation, caused by impellers fixed to the valve spindle, ensures well-balanced thermal and mechanical stress and uniform valve seating.

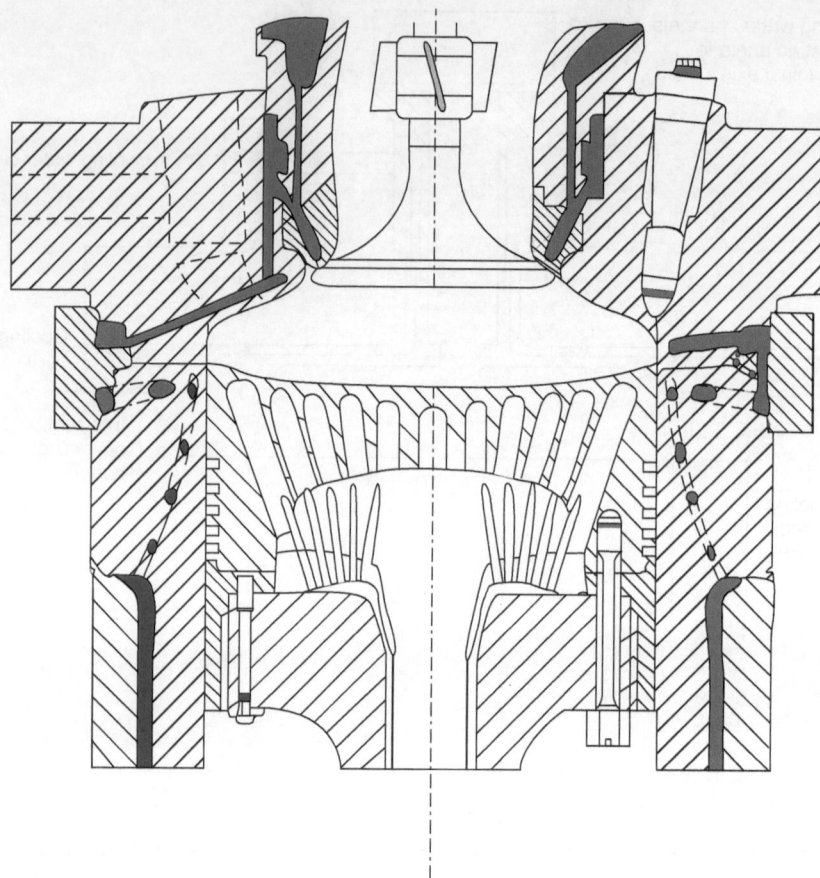

▲ **Figure 2.44** *Latest arrangement of exhaust valves*

The extensive cooling of the valve cage and seating ring results in relatively low exhaust valve seat temperatures which, coupled with the choice of the nickel–chromium–cobalt alloy (Nimonic) for the one-piece valve, increases reliability and the intervals between overhauls even when operating on heavy fuel (figure 2.46).

Medium-speed four-stroke engines

The breathing of an engine is a description given to its ability to get the air into and out of its combustion spaces so that the air can be mixed with sufficient quantities of fuel to give efficient combustion. Efficiency gains in engine design in recent years have been attributed to changes made to the inlet air and exhaust systems.

One of the advances is with the development of four valves for each cylinder. Having the four valves gives a larger opening for the gases to pass when entering and leaving the combustion chamber. Due to the number of openings required for the valves, four-stroke cylinder heads are of a complex shape and for this reason spheroidal graphite cast iron is a suitable material for the manufacturer, since it is relatively easy to cast.

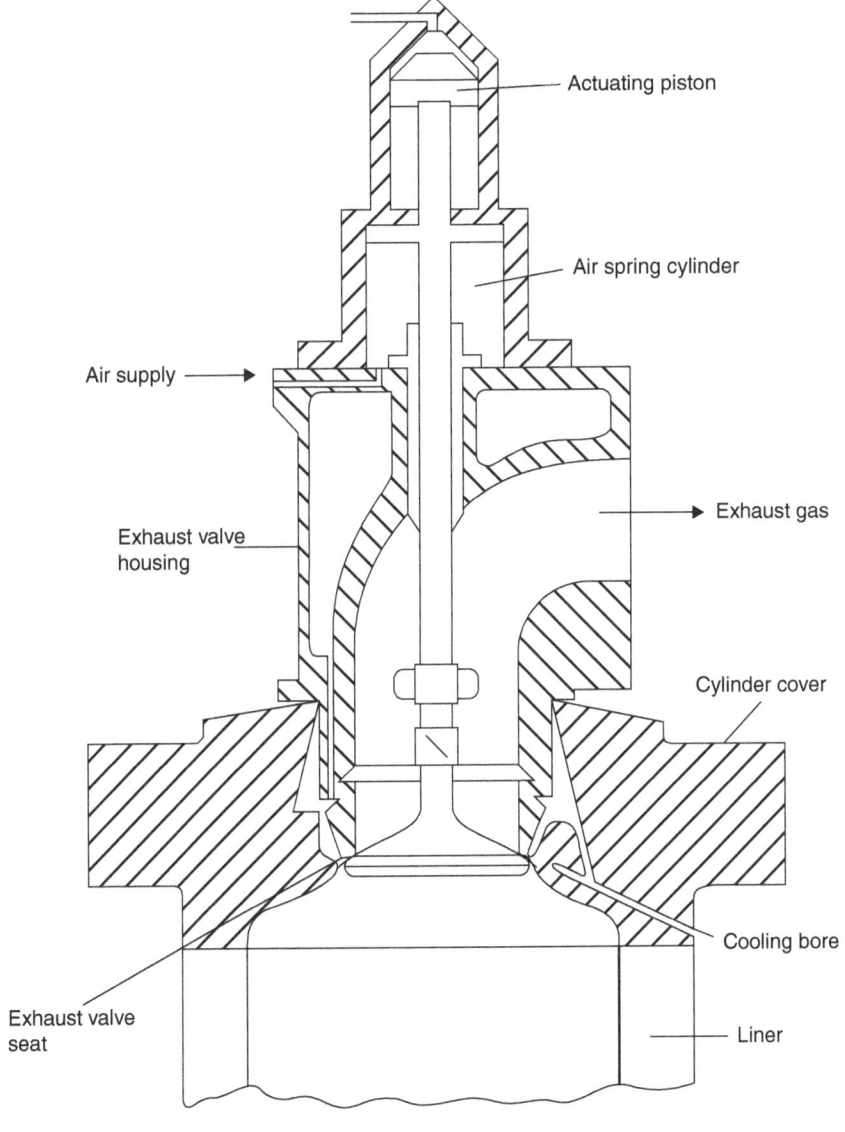

▲ **Figure 2.45** *Hydraulically activated central exhaust valve for large slow-speed two-stroke diesel engine*

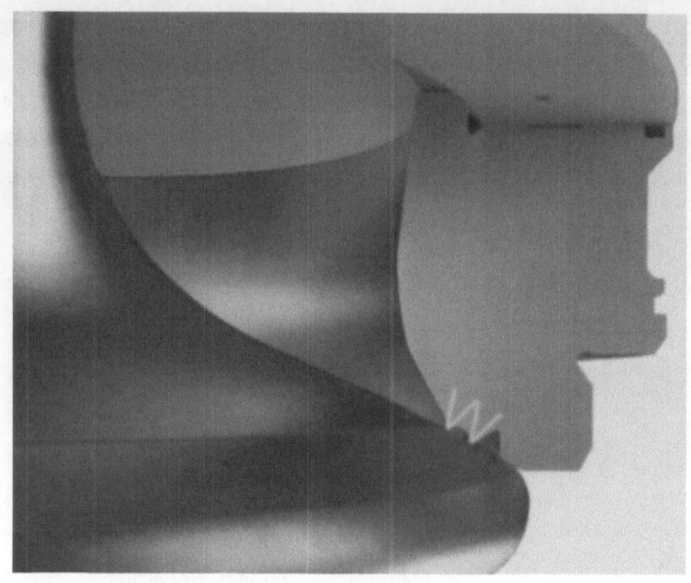

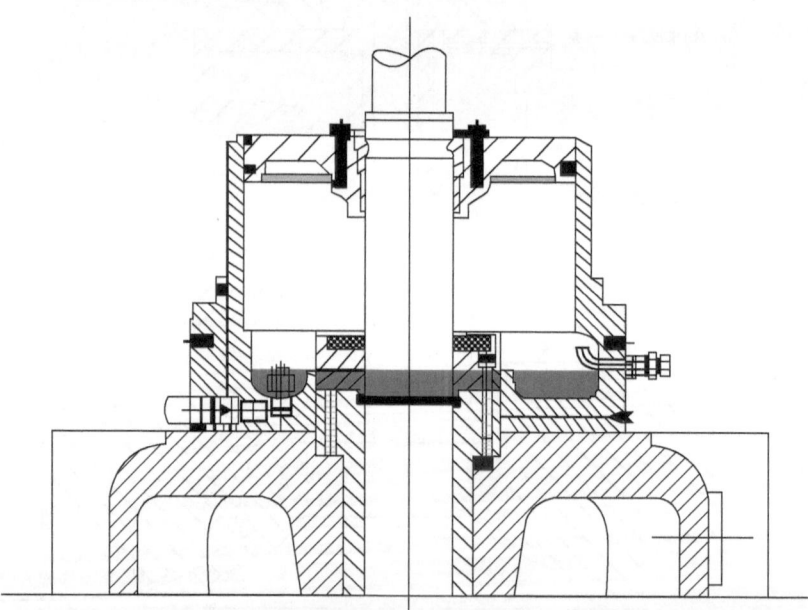

▲ **Figure 2.46** *'W' seat technology (MAN two-stroke engine exhaust valves) and exhaust valve air spring – MAN Diesel*

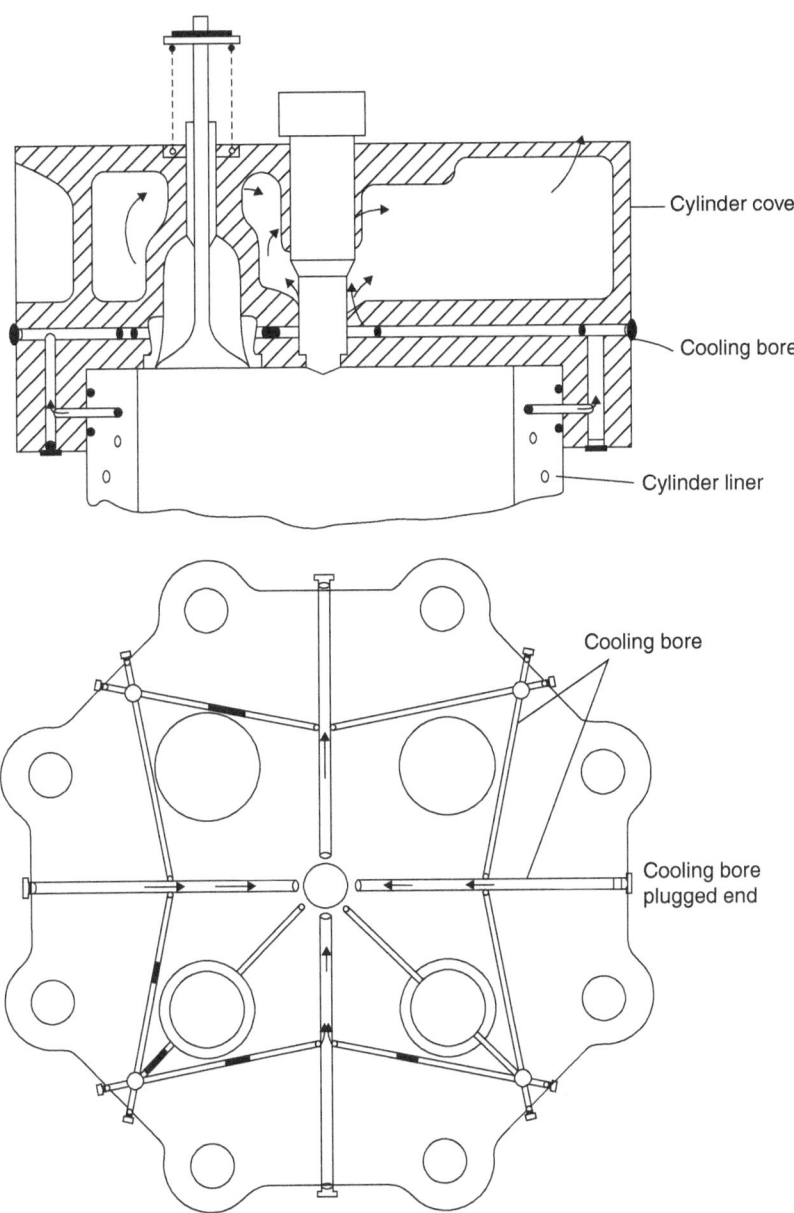

▲ **Figure 2.47** *Four-stroke diesel engine cylinder cover with bore cooling*

A modern cylinder head for an engine operating on heavy fuel is shown in figure 2.47. Such a cylinder head should have:

- A small and even thermal and mechanical deformation with correspondingly low stress levels.
- Low and uniform temperature distribution at the exhaust valves and valve seat.
- Good valve seating due to effective valve rotation and low levels of distortion (important for optimum heat transfer from exhaust valve seats).
- Exhaust valves made from a material that provides resistance to high temperature corrosion.

Effective cooling of the exhaust valve and seat must be accomplished if reliable operation is to be achieved.

To maintain low surface temperatures in the combustion space and at the valve seat, bore cooling is employed. The bore-cooling passages are shown in figure 2.47. Four valves are usually employed on four-stroke engines. This configuration allows the designer to maximise the cross-sectional area (CSA) of the inlet and exhaust ports and so improve the flow through the cylinder. This arrangement results in more complicated valve actuation since the two exhaust and two inlet valves must each be operated by one push rod. Examples of the various ways that the valves are actuated can be seen in figure 2.48. All of the designs shown control the valves together, and it is important that, following maintenance, adjustments are made correctly. Clearance is allowed between the valve stem and the rocker arm when the engine is cold. As the engine attains normal running temperature, this clearance is taken up by expansion. If adjustments leave too little clearance then it is likely that the valve will be prevented from closing correctly by the valve gear, resulting in gas leakage, burning and deteriorating performance. Conversely, too great a clearance may result in reduced valve lift and duration of opening, mechanical noise and reduced performance levels.

Internal inspection via the scavenge space

With the large two-stroke engine, this inspection process can be completed by a person entering the scavenge space and physically surveying the condition of each combustion space in the engine.

The condition of the space itself can be viewed and therefore the general efficiency of each unit can be assessed. Defective parts that can't normally be identified can also be viewed through the scavenge ports within the cylinder.

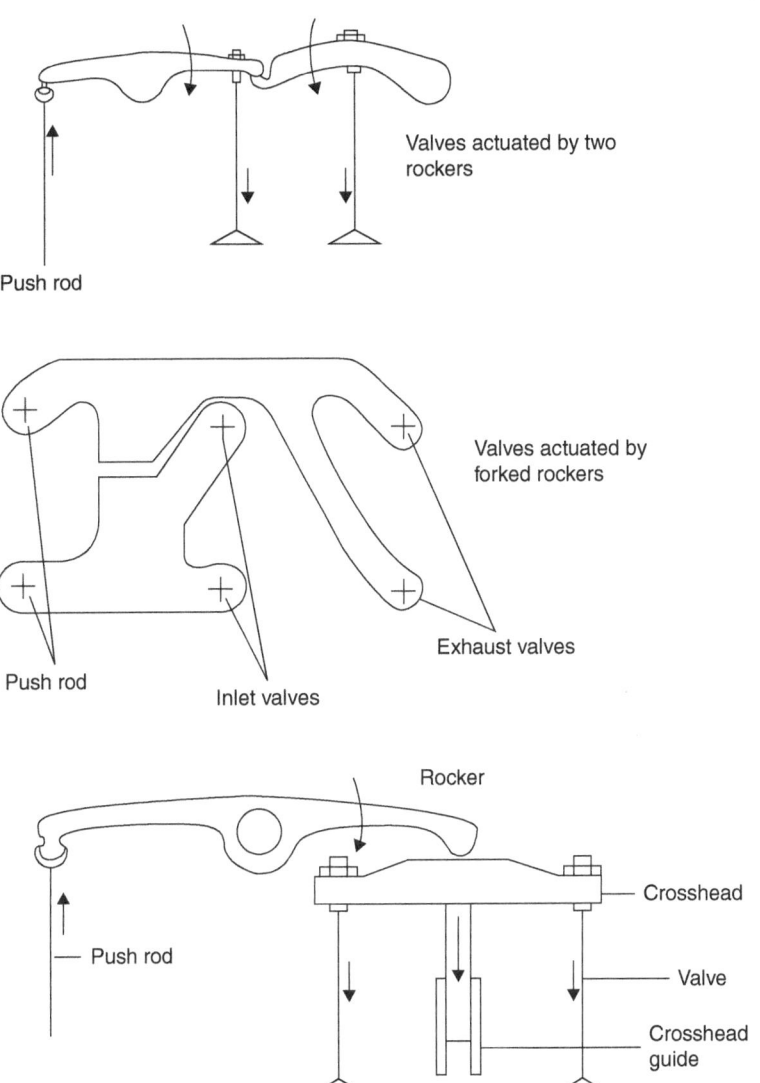

Valves actuated by two rockers

Push rod

Valves actuated by forked rockers

Exhaust valves

Push rod

Inlet valves

Rocker

Crosshead

Push rod

Valve

Crosshead guide

▲ **Figure 2.48** *Alternative methods of valve actuation*

If the engine is turned over with the turning gear then each piston and the associated piston rings can be viewed to ensure that they are working correctly.

The effectiveness of the cylinder lubrication can be assessed by inspecting the condition of the cylinder liner. Any scoring or scuff marks would then require further investigation and the root cause identified.

3

FUEL MANAGEMENT

Introduction

Optimum performance of a diesel engine is drastically affected by the quality of the fuel. In fact, the quality can be so poor that the engine stops working. If this is in the middle of an ocean passage, then the ship and its crew/passengers could be placed in grave danger.

This chapter describes the mechanics of extracting the maximum heat energy from the fuel that is taken onboard and delivered to the combustion chamber of the engine. However, for that process to work well the fuel must be delivered safely, of the correct quality and quantity, and managed carefully before it is used.

Bunkering fuel oil carefully is the first step in getting the highest quality fuel to the engine. Samples must be taken at each refueling stop. Ideally these can be sent to the laboratory for testing and the results returned before the fuel is required. If this is not possible and problems occur when the fuel is used, then the samples sent to the laboratory will be helpful in identifying the root cause of any problem associated with the quality of the fuel.

IMO's marine pollution regulations (MARPOL) require seafarers to ensure that fuel is not spilled from the ship during any operation that involves moving fuel. Therefore, all of these operations must be recorded in the Oil Record Book. A record of the time, quantity, action taken and position of the vessel must be recorded. In addition, when receiving the fuel onboard, the ship's staff must take precautions to ensure that no oil is spilled into the water. The important precautions are shown in Fig 3.20. Students will be able to learn the drawing in Fig 3.20 and should find it useful in answering a related question during their flag state examinations.

Definitions and Principles

Diesel engine combustion

Combustion is the rapid oxidation of a material (fuel) that results in the production of heat energy and to enable combustion the air/fuel ratio must be correct. This ratio is different for different materials but is about 14.6:1 for a marine fuel. The challenge for engine designers is the very short timescale that is available for the combustion to take place, before the cycle must be repeated to keep the engine running.

Figure 3.1 shows a close up view of the fuel droplet. The actual droplet, in the middle of the picture, is too rich to burn, despite there being sufficient heat present, but it does evaporate into the vapour cloud that surrounds the droplet. The fuel vapour then mixes with the oxygen in the air outside of the fuel droplet, and as the fuel mixes with oxygen in the correct air/fuel ratio is achieved and then the combustion starts to happen.

Students will be able to see that the smaller the droplets the faster this process will happen, releasing the heat energy required to run the engine. Smaller droplets are achieved by increasing the fuel injection pressure, which is why common rail engines are a step in the right direction of producing a more efficiently running engine.

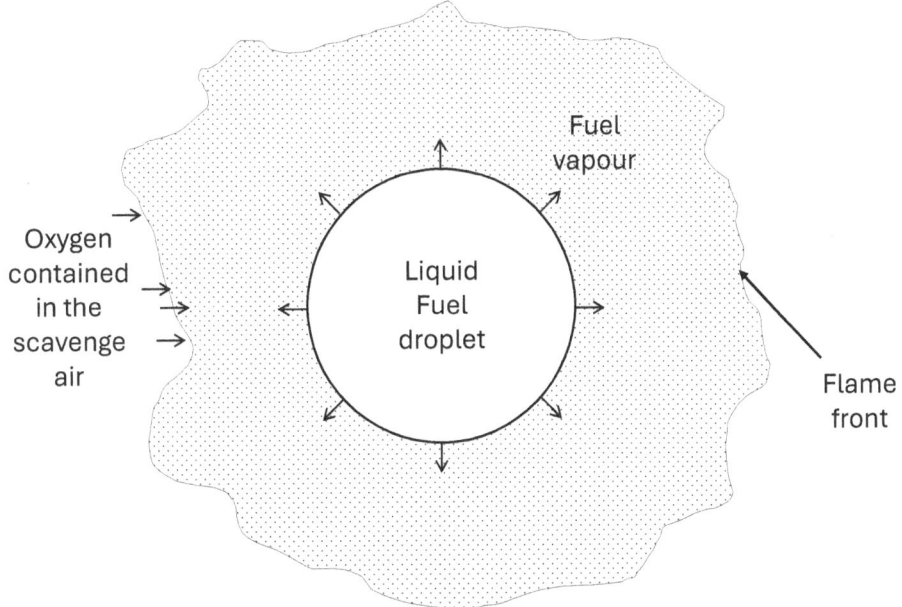

▲ Figure 3.1

Atomisation

The breakup of fuel into very small particles to ensure an intimate mixing of air and fuel oil is known as atomisation. The objective of atomisation is to reduce the size of the fuel particles as much as possible, which increases the 'surface area' to come into contact with the air in the correct ratio for combustion. This is achieved by increasing the injection pressure as high as possible.

The surface area/volume ratio of fuel-oil in droplets increases as the droplets' diameter decreases (figure 3.2). The effect of this is that a smaller droplet can present a greater percentage of its molecules in contact with the available air than can a larger droplet. The smaller the fuel-oil droplets can be made, the more effective is the atomisation, resulting in more rapid and complete combustion and maximum heat release from the fuel.

Turbulence

This is a swirl effect of the air charge as it enters and moves about in the cylinder, which in combination with atomised fuel spray gives an intimate mixing of the air and fuel, both of which helps good combustion. Turbulence is 'designed into' the engine by attention to the arrangement of:

- the inlet system
- the liner and piston
- air pressure and temperature gradients.

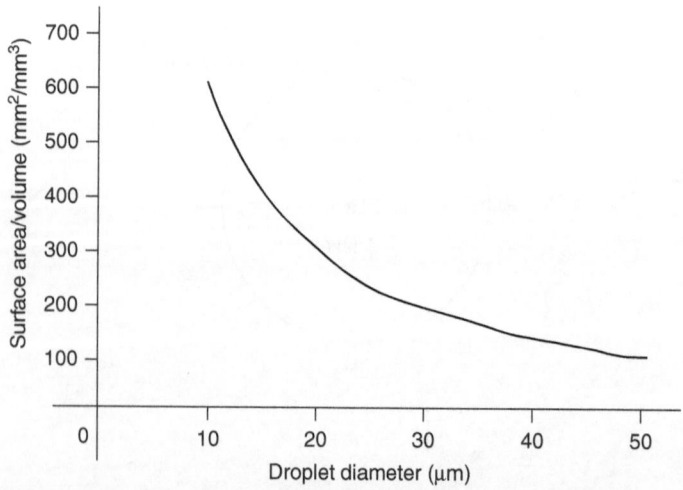

▲ **Figure 3.2** *The relationship between fuel droplet size and surface area/volume ratio*

Penetration

This is the ability of the fuel spray droplets to spread across the cylinder combustion space, allowing maximum utilisation of the available volume within the space allocated for the start of the combustion. Engine manufacturers complete hours of research studying the relationship between fuel injection timing, temperature and pressure, nozzle size and direction, air temperature and fuel quality, the aim being to retain a 'jet' of fuel for exactly the correct time before spreading out by breaking up into droplets.

Once the design has been optimised for the engine concerned, it will be the job of the ship's staff to keep the equipment in correct working order. Good watchkeeping will be the main way to look after the equipment. Keeping the fuel and scavenge temperatures and pressures correct, to the manufacturer's guidelines, will mean that the fuel injection will also be as it was designed to be.

Ensuring that the fuel is of the correct specification is also very important, not only to keep the exhaust emissions to the correct values but also to stop any premature wear of the fuel injection equipment, which will alter the delicate balance described above leading to poor combustion, increased fuel consumption and increased maintenance costs in bringing the system back up to design specification.

Ignition delay

The process of atomisation is to achieve as small a fuel droplet size as possible and penetration has the objective of moving the fuel to each part of the combustion space to achieve an even atomisation. The reason for having very small droplets is that, at the micro level, a drop of pure fuel is too dense to burn instantly. There is therefore a time delay while the fuel atomises and the outer surface of the fuel droplet absorbs heat, evaporates and mixes with the oxygen to form a flammable mixture. The time interval from the start of injection until the start of ignition is called the ignition delay. Ignition delay will be affected by the:

- level of atomisation achieved
- grade of fuel being burned
- quality of the fuel being burned
- conditions in the combustion space (scavenge temperature, jacket cooling temperature and the clean condition of the space).

If the ignition delay is prolonged for any reason then unburned fuel could build in the cylinder space and when ignition does occur the result could be more violent than the engine was designed for and would result in the characteristic noise known as 'diesel knock' (see section below).

Impingement

Excess velocity of fuel spray will result in the fuel droplets making contact with metallic engine parts and resulting in flame impingement. Engine manufacturers have found that this impingement is responsible for a considerable amount of the particulate matter or smoke produced by an engine. If fuel injectors are allowed to become worn then the injection 'spray pattern' would not be correct and this could lead to some of the fuel not being atomised early enough in the combustion space before it reaches the cylinder wall or piston crown.

Diesel knock

The diesel engine is designed to operate on a continuous cycle with every component playing its part and operating at the exact moment in the cycle. The 'timing' of this process is critical with the burning of the fuel at the correct moment probably being the most important of all the processes. When the piston is just into its journey travelling down the cylinder is when it needs the energy boost from the pressure built up by the heat released from the combustion process. The piston will then be assisted in its motion. However, if the piston is hit by a force trying to push it down the cylinder when it is still travelling up the cylinder, then the two will be in opposition and the result will be a 'knocking' sound called diesel knock.

Sprayer nozzle

This is an arrangement at the fuel valve tip to direct fuel in the proper direction with the correct velocity. If the sprayer holes are too short the direction can be indefinite and if too long impingement can occur. If the hole diameters are too small fuel blockage (and impingement) can take place; alternatively, too large diameters would not allow proper atomisation. In practice, each manufacturer has a specific design taking into account method of injection, pressure, pumps, etc. Even with a particular engine, different nozzles may be specified for different applications.

For example, vessels with engines that still have mechanical fuel injection control, and are engaged in slow steaming for reasons of economy, may be supplied with fuel valve sprayer nozzles with smaller holes of differing geometry than engines at higher powers. This measure improves the atomising and penetration performance of fuel valves at part load due to the restoration of fuel velocity through the nozzle. The 'slow steaming' injectors give the engine improved economy at part-load operation. The nozzles must be changed to the original size prior to operating at maximum power. As a generalisation the sprayer hole length:diameter ratio will be about 4:1, maximum pressure drop ratio about 12:1 and fuel velocity through the hole about 250 m/s. This system is still not ideal for complete combustion control at part-load operation. This is one of the primary reasons for the development of electronic control of the combustion process. See the section later in this chapter (page 128).

Viscosity

This may be defined as a fluid's resistance to flow due to the molecular friction that is present within its structure. The internal friction can be changed by heating the fluid where the higher temperature reduces internal friction and thus will also reduce the viscosity.

Pre-heating

The correct atomisation of fuel will depend upon the viscosity at the point of injection. The demands upon modern engines to burn a variety of fuel types is becoming more important. The ability to burn residual fuel in marine diesel engines is still important and therefore to ensure optimum fuel injection it is important that the correct pre-heating is carried out. If the temperature of the fuel is too low then the viscosity will be high, resulting in higher injection pressure and reduced atomising performance, resulting in ignition delay, excessive penetration and possible impingement on internal surfaces.

If the fuel temperature is too high then the viscosity will be low, thus reducing penetration and causing deposits to be left on the nozzle tip, affecting atomisation. The relationship between viscosity and temperature is shown in figure 3.3. Careful control of fuel temperature is required to ensure that the fuel viscosity at the engine fuel rail is inside the range specified by the engine manufacturers. It is modern practice to utilise viscosity controllers that ensure correct fuel viscosity by the careful control of fuel temperature. Despite the manufacturers of the viscosity controller specifying that it is fitted in close proximity to the engine, builders do not always do so. It is important, therefore, that the viscosity controller is adjusted so that any cooling of the fuel that takes place between the heater and the engine does not allow the fuel to move out of the optimum viscosity range for injection. This effect will be exacerbated when burning high-viscosity fuels, such as residual fuel oil, that require considerable heating. Effective trace heating and insulation is an important feature of a well-designed fuel system.

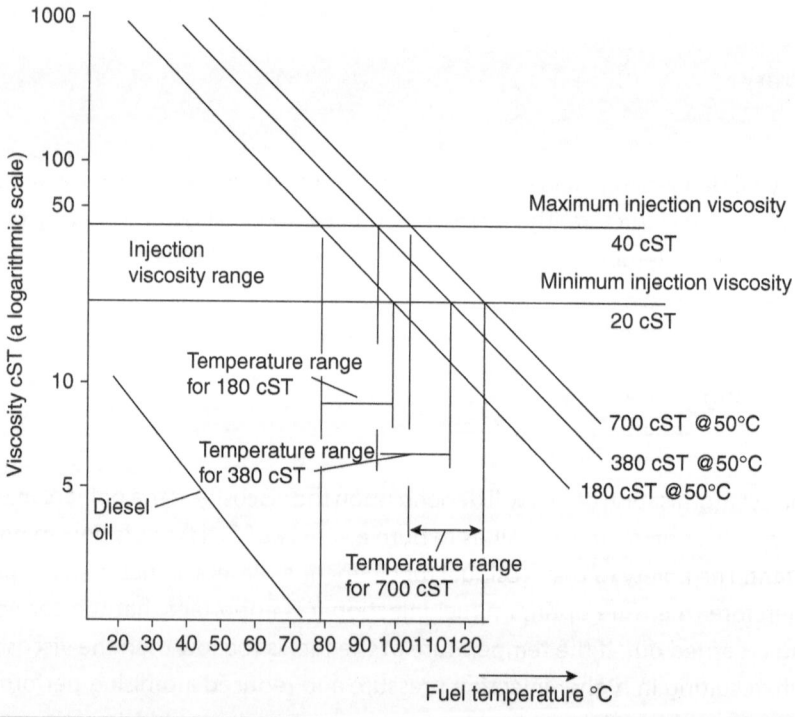

▲ **Figure 3.3a** *Viscosity temperature chart for marine fuels*

The fuel injection system is vitally important to the efficient operation of the engine. It must:

- supply an accurately measured amount of fuel to each cylinder, regardless of load.
- supply the fuel at the correct time for all loads with rapid opening and closing of the fuel valve.
- inject the fuel at a controlled rate.
- atomise and distribute the fuel in the cylinder.

Heat release characteristics of residual fuel oil

The quality of fuel is a problem associated with the use of residual fuels and recently with LSF, which will be explained in more detail later in this chapter. However, the general rate of heat release (ROHR) for residual fuel is affected by the quality of the refining process and any other types of fuel that are used as part of that process. For example, if there is any gas oil left over from the process then this will start to burn first and may affect the overall efficiency of the combustion process.

Figure 3.5b shows that there could be three distinct phases to the heat release process. The naphtha ignites first, starting the process, a middle section where gas oil is ignited is next, followed by the main event where the residual fuel starts to burn. The quality of the bunker fuel could well have an effect on the running of the engine, especially as ships are becoming more reliant on one fuel to power the main engine and the generators, one of which could be a variable, slow-speed, two-stroke engine and the other a constant, high-speed, four-stroke engine.

If the rate of rise in the pressure is too great, then the resultant shock could cause damage to the engine, which results in the characteristic 'knocking' sound from the engine.

However, poor fuel quality could result in poor performance from the engine. The reason for this is often due to an increase in the 'ignition delay'. If the heat from combustion is not released at the correct time then the piston is already travelling down the cylinder, increasing the volume, which in turn will reduce the overall pressure achieved for that cycle.

Students will be able to see the intended production of power within the diesel engine from Chapter 1. The information about measuring the pressure rise in the cylinder in relation to the position of the piston (see indicator diagrams, figure 1.9) assumes that a reliable fuel is being used and it burns with the design features that match the operation of the engine.

The fuel can, however, perform in one of the ways indicated in the graphs below: graph A (ECN=29) shows the way that the fuel is expected to burn; graph B shows the effect of a poorer quality fuel (ECN=18) and graph C shows the heat release characteristic of a poor quality fuel (ECN=8).

If residual fuel oil can be produced and sold to be used in ships' main engines, then that must be a reasonable expectation that the engine will be able to operate using the fuel that has been loaded.

This being the case means that the fuel must be sold to a 'standard' where it would then match the operating envelope of the engine. The first standard developed was the Calculated Carbon Aromaticity Index (CCAI) number, which now appears in the ISO 8217 (from the 2010 edition onwards) specification criteria for marine fuel oils. This number is derived from a combination of the density and the viscosity of a given fuel to show an assessment of the combustion quality and the ignition characteristics of the fuel.

However, the blend of modern residual is becoming more complex and under some cases this standard does not fully describe the ignition quality of residual fuel oils having the similar CCAI numbers but with a different density/viscosity mix.

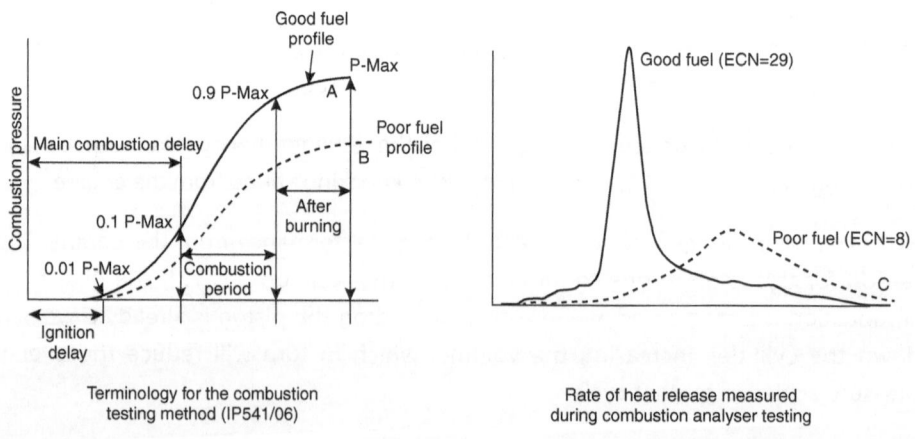

Terminology for the combustion
testing method (IP541/06)

Rate of heat release measured
during combustion analyser testing

▲ **Figure 3.3b** *Ignition delay and heat release*

This could lead to problems where the staff have, following on-board tests, found the bunker fuel within specification when in fact it is not. Therefore, an index called the Estimated Cetane Number (ECN) has been developed.

The ECN is calculated from the time it takes the fuel to start its main combustion. This is called the main combustion delay (MCD). Engine design, condition and load are all variables that can influence the performance of the fuel.

Injection

To achieve effective combustion, the fuel must be atomised and then distributed throughout the combustion space. It is the function of the fuel valve to accomplish this. Most fuel valves on the current generation of marine diesel engines are still of the hydraulic type, the cross section of which is shown in figure 3.4. The opening and closing of this type of valve is controlled by the fuel pressure delivered by the fuel pump. The fuel pressure acts on the needle in the lower chamber and when the force is sufficient to overcome the force of the spring keeping the valve shut, the needle lifts.

Full lift occurs quickly as an extra area of the needle is exposed to the fuel pressure after initial lift, placing additional force against the closing force of the spring. The full action of lift is limited by the needle shoulder, which halts against a thrust face on the injector's body (see figure 3.5). The injector lift pressure varies with the different designs but may be about 140 bar on average with some designs reaching as much as 250 bar.

A fuel valve lift diagram for such an injector is given in figure 1.12 in Chapter 1. Figure 3.4 shows that by removal of the spring cap, the valve lift indicator needle can be removed, reassembled or adjusted. The particular design as sketched is not cooled by itself but is enclosed in an injector holder in the cylinder head that will be kept at the necessary design temperature by the engine's cooling systems. The fuel valve will require a seal, which in the case of the one shown in figure 3.4 is intended to be a face-to-face seal with the cylinder head's pocket. Some injectors are designed to have a copper ring between the fuel valve nut on the lower end and the bottom of the cylinder head fuel valve pocket. Care must be taken here as in the past bad practice has led to additional copper rings being added or thicker ones being used than the system was designed to have. The exact design system must be used because adding a copper ring will change the height of the nozzle tip and thus will change the point of injection relative to the rest of the combustion space, which will have a dramatic effect on the engine's performance due to poor combustion.

Coolant is circulated in the annular space between the injector holder and the holder itself. Direct cooling of the fuel valve as an alternative to this is easily arranged. Coolant connections on the main block would supply and return through drillings similar to that shown for fuel. The choice of oil or water for cooling depends on the engine and

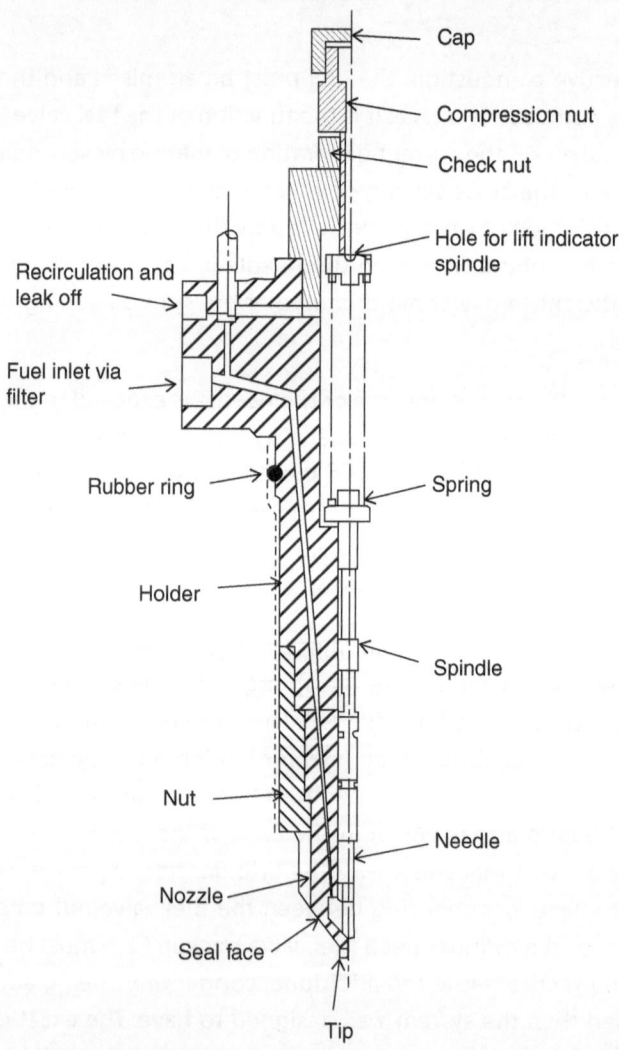

▲ **Figure 3.4** *Fuel valve injector (hydraulic)*

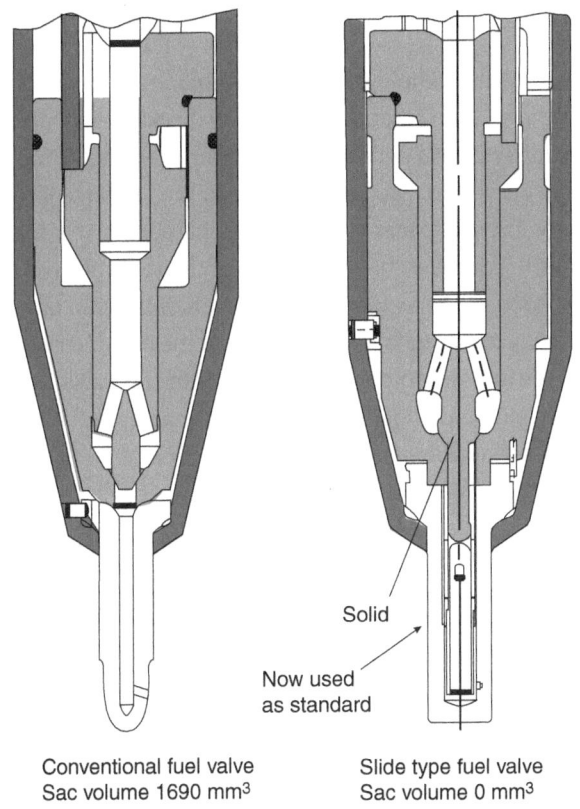

Solid

Now used
as standard

Conventional fuel valve	Slide type fuel valve
Sac volume 1690 mm^3	Sac volume 0 mm^3

▲ **Figure 3.5** *MAN fuel slide valve*

valve design and is also affected by the type of fuel. With hot boiler oil it is necessary to cool right to the injector tip so as to attempt to keep metal temperatures below 200°C. Hydraulic fuel valves usually have a lift of about 1 mm and the action is almost instantaneous.

Fuel injection systems

- Mechanical fuel injection
- Mechanical common rail
- Electronically controlled (common rail) fuel injection (EFI).

Manufacturers are using a number of variations on the basic themes named above as increasingly strict legislation starts to drive the quest for efficiency as well as the experimentation in different types of fuel.

Mechanical fuel injection

This is the traditional system employed in modern marine diesel engines. It is the most commonly used one in existing engines but it is very quickly being replaced by the newer electronic fuel injection systems. Fuel is supplied to the high-pressure fuel pump from a low-pressure fuel delivery pump and associated pipework. The fuel flows into the high-pressure pump where it fills the internal spaces in the delivery chamber. As the pump is operated by a cam on the camshaft, the delivery chamber becomes closed off and the fuel pressure starts to rise dramatically following a few degrees of rotation of the cam operating the plunger. Fuel is delivered directly to spring-loaded injectors via the pump's delivery valve and high-pressure pipework. The fuel injectors are opened up by the hydraulic action of the fuel after the high-pressure fuel pump plunger movement has generated sufficient fuel pressure to overcome the spring pressure keeping the fuel valve closed. The action is designed to be rapid as an aid to the timing and atomisation of the fuel ready for combustion.

Mechanical common rail

Although this is now an obsolete system, it is worth a mention here to show the student that good systems that fell out of favour in the past could make a return to efficiency, especially with the advancement of material science and electronics. This very early system had fuel pumps to deliver fuel to a pressure main and various cylinder valves were opened to the main that allowed fuel injection to the appropriate cylinder. The system required either mechanically operated fuel valves (eg older Doxford engines) or mechanically operated timing valves (eg newer Doxford engines), allowing connection between rail and hydraulic injector at the correct injection timing.

Electronically controlled (common rail) fuel injection (EFI)

Modern engine designers are coming under increasing pressure to build engines that are considerably more fuel efficient than they were just a few years ago. The most important strategy to achieve this is to use sophisticated and very close combustion control, which cannot be achieved by the use of mechanical systems such as described in the first section here – mechanical fuel injection systems.

The fuel is supplied – via a low-pressure delivery pump and associated pipework – to a high-pressure pump and piping system running the entire length of the engine. The electrically operated fuel valves (injectors) are connected to the high-pressure 'rail' pipework. When combustion within a given cylinder is required, an electrical signal from the electronic control unit operates the correct fuel valve.

The important aspect about the EFI system is that the start, duration, quantity and end of fuel delivery are all completely controllable to a very fine degree and they are parameters that can be modified independently of each other and independent of any other consideration such as engine speed or ambient temperature. The more complex systems can also provide a pre-injection and a post-injection sequence, which give an added advantage as described over the next few pages.

Note: Many aspects of fuels are covered in Chapter 2 of Volume 8 of the Reeds series and revision of oil tests, as well as basic definitions relating to specific gravity, Conradson carbon residue, cetane number, etc, is strongly advised.

Indicator diagrams

The development of EFI and the associated equipment means that some of the same control mechanisms can be used to give detailed feedback about the equipment's performance to the engineers. This improved level of information will enable the knowledgeable engineers to carry out performance analysis and fault finding at a level not achievable in the past. The heart of this diagnostic process will be the combustion indicator diagram.

Details have been given of some typical indicator diagrams showing engine faults in Chapter 1 and some fuel injection faults have been outlined including late and early injection, shown on the draw card, fuel valve lift diagrams, as well as related details such as compression cards. Two further typical faults are illustrated in figure 3.6.

Afterburning is generally associated with poor quality fuels (see section on 'heat release') and will be characterised as shown in figure 3.6b by:

- an increased peak pressure
- loss of power
- increased cylinder exhaust temperature
- possible discolouration of exhaust gases.

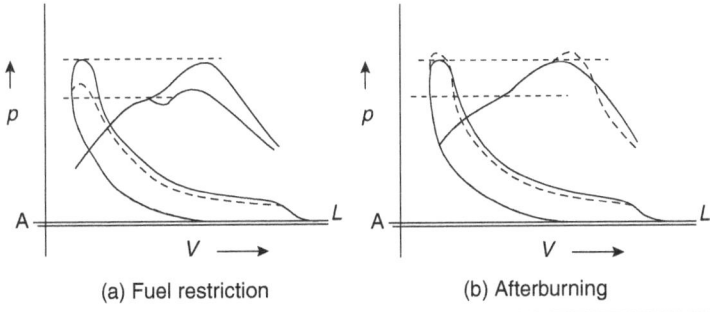

(a) Fuel restriction   (b) Afterburning

▲ **Figure 3.6** *Effects of defective fuel injection*

The effects of fuel restriction on the combustion process can be seen in figure 3.5a. This may be due to:

- blocked fuel filters
- injectors
- incorrect viscosity
- poor fuel quality.

These will all result in a loss of power and reduced maximum pressure.

Fuel Pumps

General

The physical energy demand of fuel injection is substantial. Typical requirements include delivery of about 100 ml of fuel in 1/30 s at 750 bar so as to atomise an area of 40 m. A peak energy input can reach 230 kW. A short injection period at high pressure, arranged to give the desired firing pressure, at the right time and in the right direction is very important. Pilot injection and phased injection of charge is now not a problem with modern engines that have electronic combustion control systems.

Quantity control

Traditionally, diesel engine fuel delivery, injection and atomisation has been carried out by using totally mechanical systems with the majority operating by varying the amount of fuel injected per stroke, which is controlled by varying the effective plunger stroke of the high-pressure fuel pump (see figure 3.10). Mechanical systems achieve this by:

1. Varying the beginning of delivery
2. Varying the end of delivery
3. Varying the beginning and the end of delivery.

Currently, the most popular method of mechanical fuel injection on larger marine engines is by using a single high-pressure fuel pump for each cylinder. Regulation of the quantity of fuel is matched to the load of the engine by a governor operating various linkages that control the output from the pump.

The pump is a single piston moving inside a barrel or cylinder. As the piston is at the bottom of the stroke, the space above fills with fuel. As the plunger rises, the inlet port is closed and the fuel is delivered from the delivery valve at the top of the pump. The control method is to change the effective length of the pump's delivery stroke by altering the end of delivery. This is still an important method of diesel engine control and students will need to be familiar with the constant stroke and helical groove system, which has a constant beginning of injection, described later. This method is more suited to constant speed engines which would require a fixed start of injection and the amount of fuel required would increase as the load increased.

They are regularly fitted to auxiliary engines and give fuel injection early in the cycle at light load, which not only gives higher efficiency but also leads to higher firing pressures. They have also been used with large direct drive engines such as the older Burmeister & Wain (now MAN Diesel & Turbo). However, the large two-stroke engine is a variable speed engine and therefore requires a variable time for the start of injection.

Therefore, the valve-type pump was favoured for large engines, which had a constant end of delivery and regulation of the start of injection accomplished by varying the suction valve closure. Part-load performance of these engines with later injection is always a compromise between economy and firing pressure. With turbocharged engines, the disadvantage of the constant end pump control is more noticeable as reduced firing pressure and efficiency is more marked at part loads due to reduced turbocharger delivery and pressure.

Owing to the limitations of varying the fuel quantity delivered by only varying the beginning of delivery, Sulzer (now owned by the Wärtsilä Corporation) redesigned their fuel pumps to include a suction valve and spill valve. Initially the spill valve was the only controlled valve, resulting in constant beginning with variable end of delivery. Later, however, in the interests of fuel economy and in common with other manufacturers, both valves were controlled to give a controllable beginning and end of delivery.

Injection characteristics

Figure 3.7 illustrates some features of fuel injection based on research carried out on a mechanically fuel injected slow-speed engine. The diagram shows a generic fuel valve injector lift diagram with a total lift of about 1.3 mm and injection period at full load approximately 6° before to 22° after TDC. High firing pressures achieved at full load

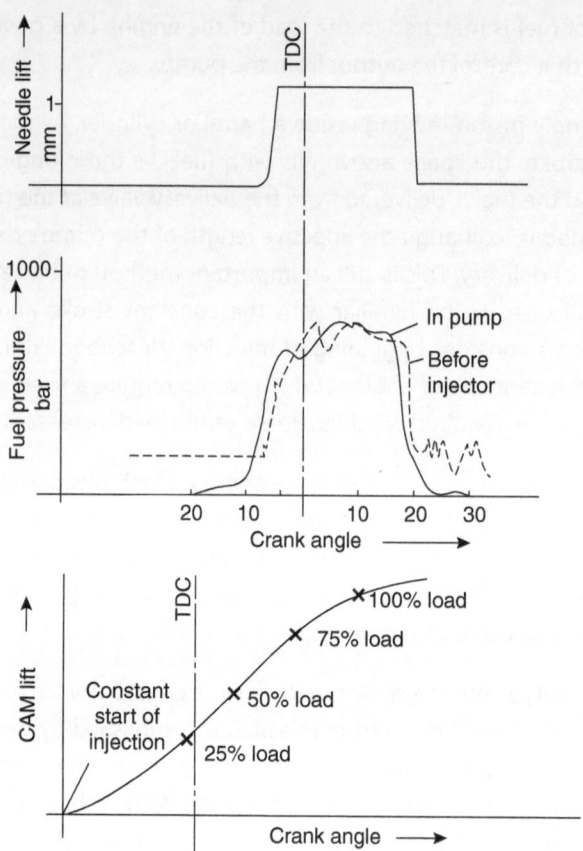

▲ **Figure 3.7** *Injection characteristics*

in the high-powered turbocharged range of engines can be reduced when using the constant beginning of injection method.

The ideal injection profile (for a fuel with a constant ignition characteristic) corresponds to a rectangle with almost constant fuel pressure before and during injection. The practical curves illustrated show almost constant pressure at a reasonable maximum (750 bar). However, fuel injection pumps can be a problem, especially when running on residual fuel oil and also when running at the same load over a prolonged period. Impurities in the fuel cause wear on the barrels and plunger of individual fuel pumps. This may happen to pumps in different amounts. Therefore, one pump might wear more than another and pumps may also wear at different points on their load operating range. This means that as an engine increases in load and all the fuel pumps are operated in the same relationship, not all the pumps will be delivering fuel in the same way due to the different wear characteristics of each pump.

Individual fuel pumps can be adjusted in relation to the other pumps on the engine; however, this must be done with great care. A balanced engine is one where each of the cylinders is taking its fair share of the load. The problem is that one of the pumps could be pumping low at one part of the range but just a small movement further it will be pumping the full amount. Therefore, any slight adjustment might cause a pump to move from under-pumping to pumping more than it should. In many cases only a small adjustment is possible and the best cause of action is to change the pump.

Variable injection timing

The previous section dealt with variable fuel delivery having a constant beginning of injection. This system is no longer acceptable in modern engines because the part-load and low-load performance does not meet the needs of current legislation relating to engine emissions. Also, in the quest for better fuel economy it is modern practice to now vary both the beginning and end of injection. With a constant start of injection system, the maximum firing pressure of the engine will fall almost linearly as the power of the engine is reduced. The brake mean effective pressure (BMEP) of the engine, however, reduces at a slower rate. Since the thermal efficiency of the engine varies as the ratio of P_{max}/P_{MEP} a reduction of firing pressure will result in a reduction of thermal efficiency of the engine.

In order that the thermal efficiency and hence the specific fuel consumption can be maintained at optimum, it is therefore necessary to maintain maximum firing pressures as the engine load is reduced. This is accomplished by advancing the timing of the fuel injection, and the start of combustion, as the engine load is reduced (see figure 3.8). The advancement of the injection timing continues until about 65–70%; thereafter the injection is retarded (figure 3.9).

The ability to retard the fuel injection is extremely important, especially when used to control the emissions from the engine. Retarding the fuel injection delays the heat release and controls the highest peak temperatures in the combustion process. The ability to advance and retard the fuel injection process will reduce the 'diesel knock' at low engine loads that is sometimes experienced in engines without VIT. The inclusion of a 'knock' sensor will provide feedback to the engine control unit, which will then be able to vary the timing accordingly in real time while the engine is in service.

As we shall see later, variable valve control will allow the designers to further change the combustion process to the requirements of local regulation, fuel type and operational conditions.

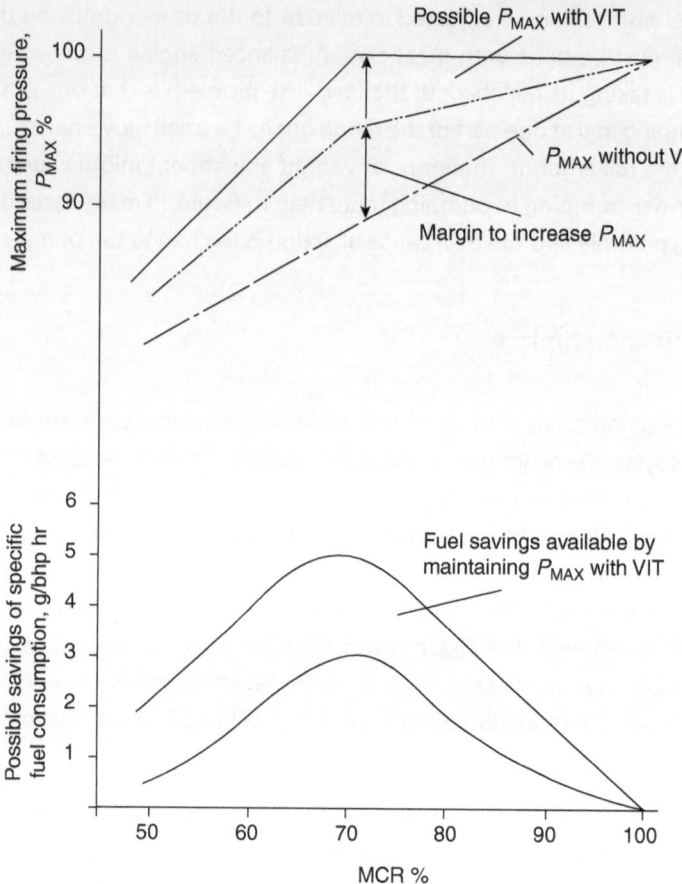

▲ **Figure 3.8** *Fuel savings available by utilising variable injection timing (VIT)*

Scroll-type high-pressure fuel pump

Despite the move towards common rail fuel injection, the scroll-type high-pressure fuel pump is still very important to the industry. Therefore, students should study the principle of operation described here as the examiners will be very interested to find out all you know about the operation of this type of pump. Figure 3.9 shows a scroll-type fuel pump set at approximately 75% load.

The sketch numbered 1 in Figure 3.10 shows the plunger moving down. The pressure in the barrel falls and as the suction and spill ports open to the fuel rail, the fuel flows into the barrel. In sketch 2, the plunger is moving upwards. The fuel is displaced from the barrel through the spill and suction ports. This displacement will continue until the plunger completely covers both ports.

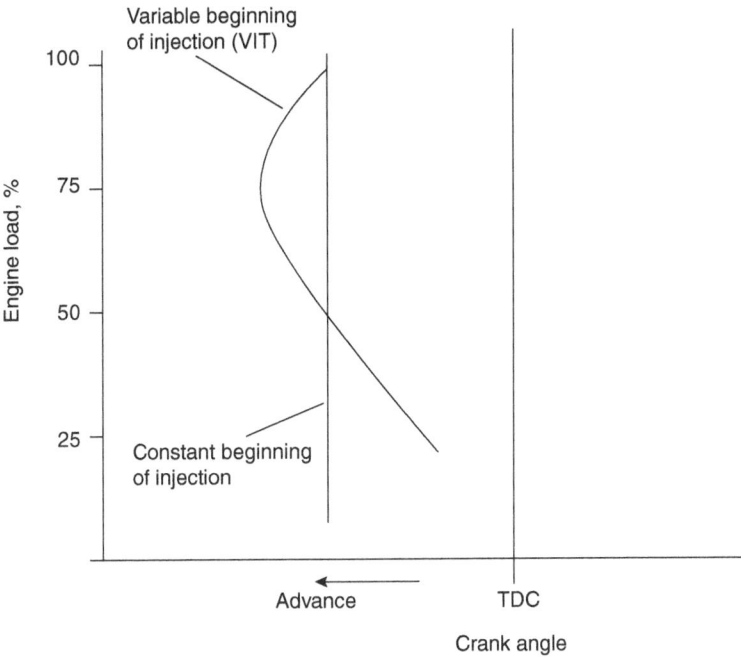

▲ **Figure 3.9** *Injection timing variation with engine load*

Sketch 3 shows the plunger continuing to move upwards and just covering the spill and suction ports. This is the effective beginning of delivery and any further upward movement of the plunger will pressurise the fuel that has already filled the high-pressure pipe between the pump and the fuel valve, and it will open the fuel valve injecting fuel to the engine.

In sketch 4, the plunger continues to move upwards. Injection continues until the point when the lower helical edge of the groove on the plunger uncovers the spill port. The high pressure in the barrel is immediately connected to the low pressure of the fuel suction. There is no longer sufficient pressure to keep the fuel valve needle open and injection ceases.

Sketch 5 shows the plunger turned by means of the rack and pinion control arm so that the longitudinal groove of the plunger is aligned with the spill port. In this position the plunger is unable to deliver any fuel since the spill port does not close during the cycle.

Sketch 6 shows a sectional plan view to illustrate how the plunger can be rotated in order to vary the effective height of the helix relative to the spill port. The plunger base is slotted into a control sleeve rotated by a quadrant and rack bar, which are both under the control of the engine's governor.

With both valve and helical spill designs there can be problems of fuel cavitation due to very high velocities. Velocities near 200 m/s can create low-pressure vapour bubbles

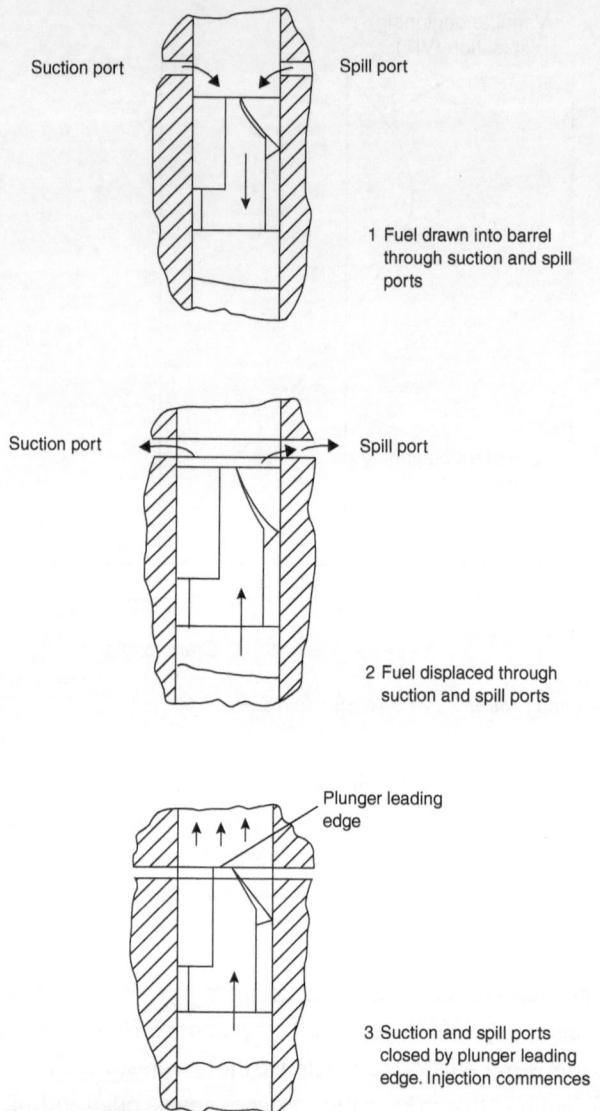

Suction port   Spill port

1 Fuel drawn into barrel
  through suction and spill
  ports

Suction port   Spill port

2 Fuel displaced through
  suction and spill ports

Plunger leading
edge

3 Suction and spill ports
  closed by plunger leading
  edge. Injection commences

▲ **Figure 3.10** *Scroll-type fuel pump operation*

if pressure drops below the vapour pressure. These bubbles can subsequently collapse during pressure changes, which results in shock waves and erosion attack as well as possible fatigue failure. A spring-loaded piston and orifice design can absorb and damp out fluctuation. Many manufacturers utilise a form of the above pump.

The adjustment of the start of injection timing is carried out on Bosch-type fuel pumps by varying the relative height of plunger and suction/spill ports in the barrel (figure 3.11). This may be accomplished in a number of ways:

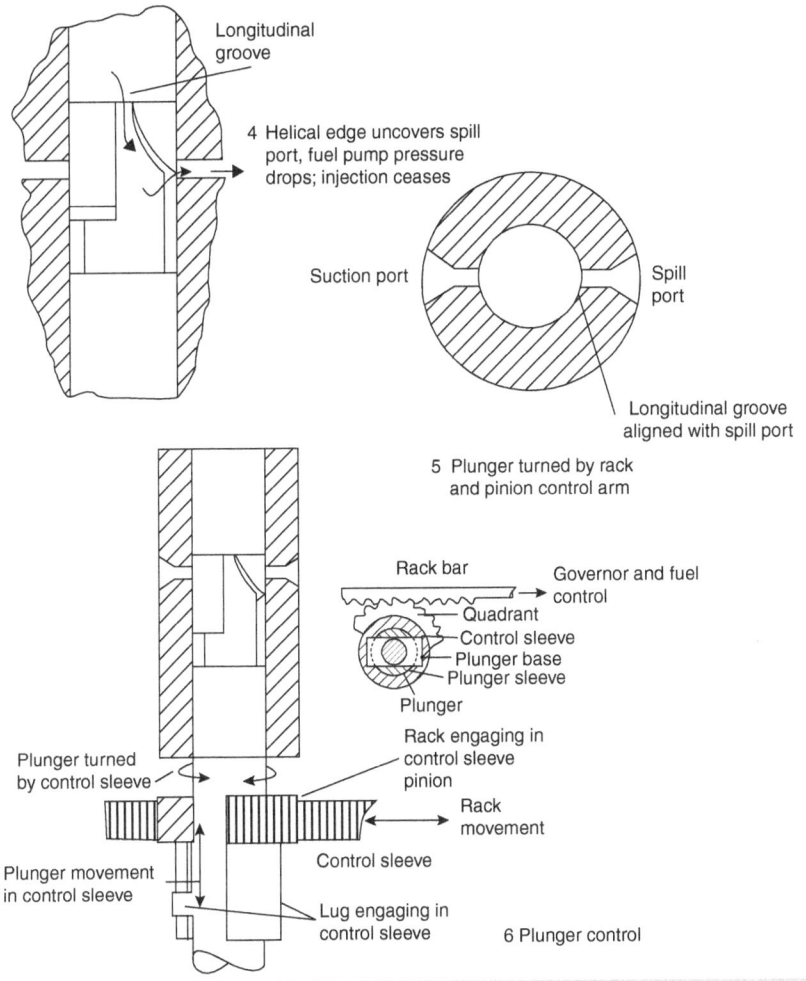

Longitudinal groove

4 Helical edge uncovers spill port, fuel pump pressure drops; injection ceases

Suction port

Spill port

Longitudinal groove aligned with spill port

5 Plunger turned by rack and pinion control arm

Rack bar

Governor and fuel control

Quadrant

Control sleeve

Plunger base

Plunger sleeve

Plunger

Rack engaging in control sleeve pinion

Plunger turned by control sleeve

Rack movement

Plunger movement in control sleeve

Control sleeve

Lug engaging in control sleeve

6 Plunger control

▲ **Figure 3.10** *continued*

1. Adjusting the height of the plunger relative to the barrel (figure 3.11a)
2. Adjusting the height of the barrel relative to the plunger (figure 3.11b).

Adjustment of plunger height can be accomplished in some installations by adjusting the cam follower. Lowering the plunger has the effect of retarding the injection. Raising the plunger advances the injection.

In earlier B&W designs, adjustment of injection timing was carried out by raising the fuel pump barrel in relation to the plunger. In this design the fuel pump top flange has an external threaded portion projecting towards the barrel (figure 3.12). This thread

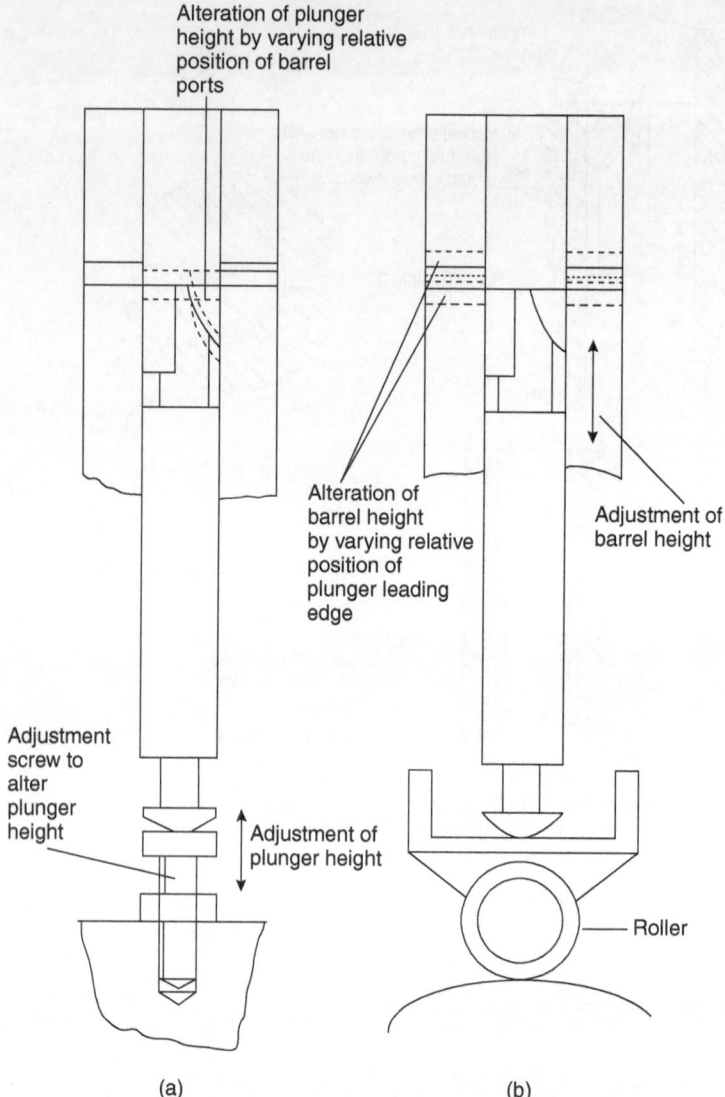

Alteration of plunger
height by varying relative
position of barrel
ports

Alteration of
barrel height
by varying relative
position of
plunger leading
edge

Adjustment of
barrel height

Adjustment
screw to
alter
plunger
height

Adjustment of
plunger height

Roller

(a)   (b)

▲ **Figure 3.11** *Adjustment of fuel injection timing by varying relative position of plunger and barrel*

matches with the external thread of the adjusting ring. The adjusting ring has external gear teeth cut on its upper part, which are engaged by the adjusting pinion. To adjust the injection timing the pinch bolts are released and the adjusting pinion is turned to either raise or lower the barrel. Lowering the barrel will advance the injection timing while raising the barrel will retard the injection timing.

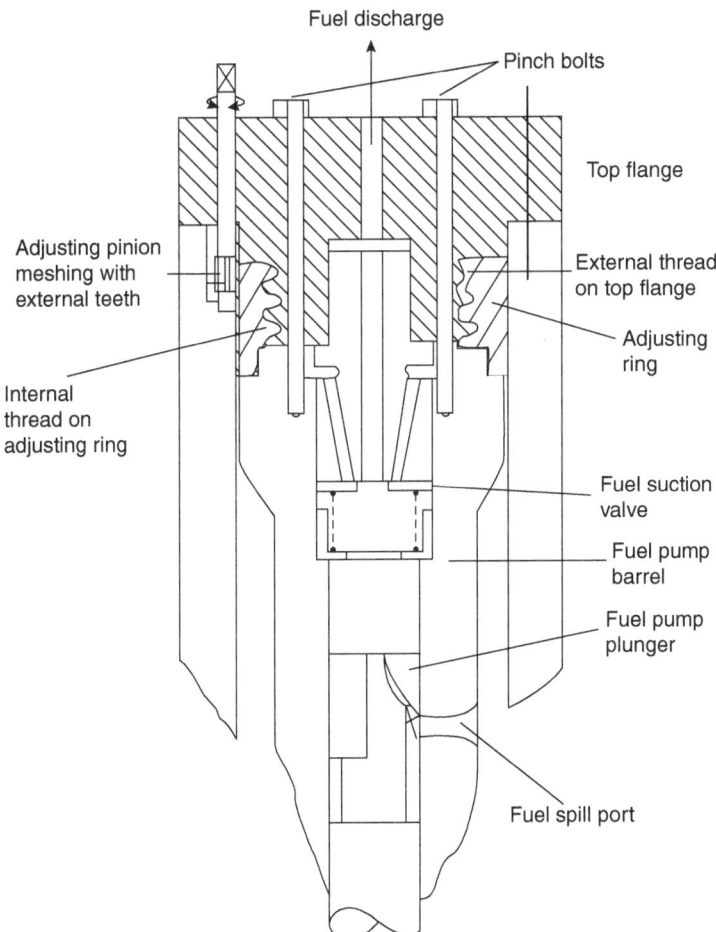

▲ **Figure 3.12** *Adjustment of injection timing by raising or lowering fuel pump barrel*

Fuel pump (valve-type fuel pump) detail

Figure 3.13 shows designs based on an earlier Sulzer design. The fuel-pump delivery is controlled by suction and spill valves. Sketch 1 shows the plunger moving down, the suction valve open and fuel oil being drawn into the barrel. Sketch 2 shows the plunger moving upwards and fuel being displaced through the still open suction valve. Sketch 3 shows the delivery commencing as the suction valve closes. Sketch 4 shows that as the plunger continues to move upwards, injection ceases when the spill valve opens.

From figure 3.13 it can be seen that the valve-type fuel pump design lends itself quite readily to control suction valve and spill valve. However, scroll-type or Bosch fuel

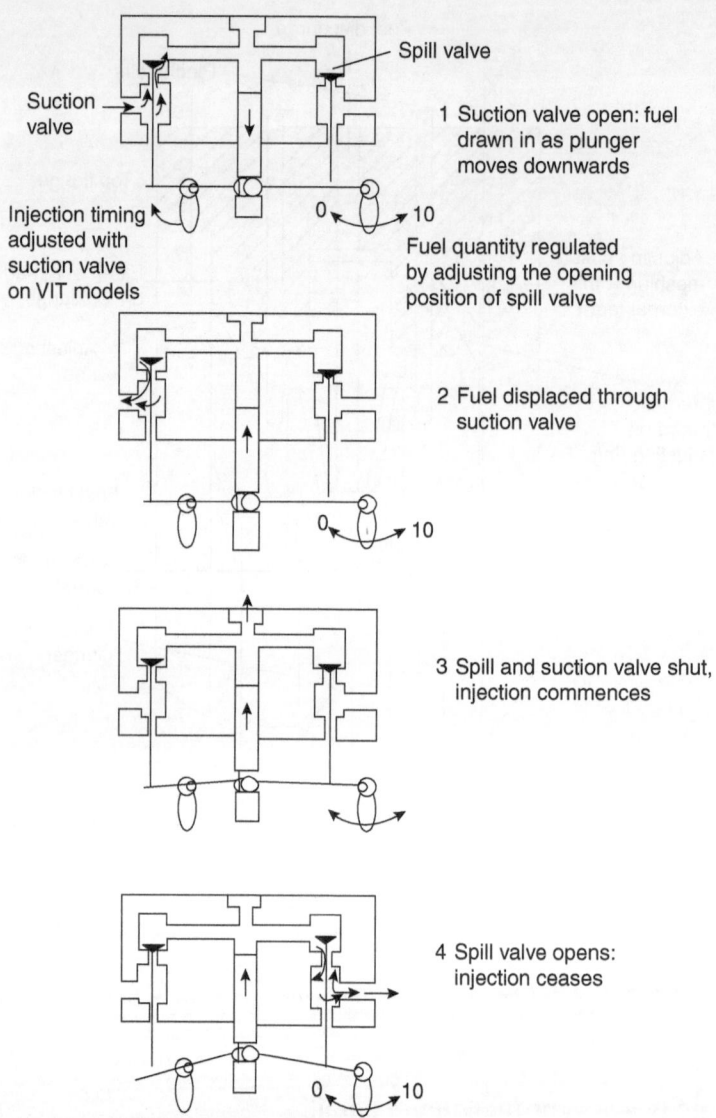

Spill valve

Suction valve

1 Suction valve open: fuel drawn in as plunger moves downwards

Injection timing adjusted with suction valve on VIT models

0 10

Fuel quantity regulated by adjusting the opening position of spill valve

2 Fuel displaced through suction valve

0 10

3 Spill and suction valve shut, injection commences

4 Spill valve opens: injection ceases

0 10

▲ **Figure 3.13** *Wärtsilä–Sulzer valve-type fuel pump*

pumps require a different method to control the beginning and the end of injection. The beginning of injection is carried out by adjusting the height of the fuel pump barrel. Referring to figure 3.14, the pump barrel has an outside threaded lower portion, which engages into the timing guide operated by a toothed rack. Movement of the rack causes the pump barrel to move vertically up or down relative to the pump plunger. The moment the plunger covers the spill port, injection commences. The duration of this process can be adjusted while the engine is in operation. The pump barrel is prevented

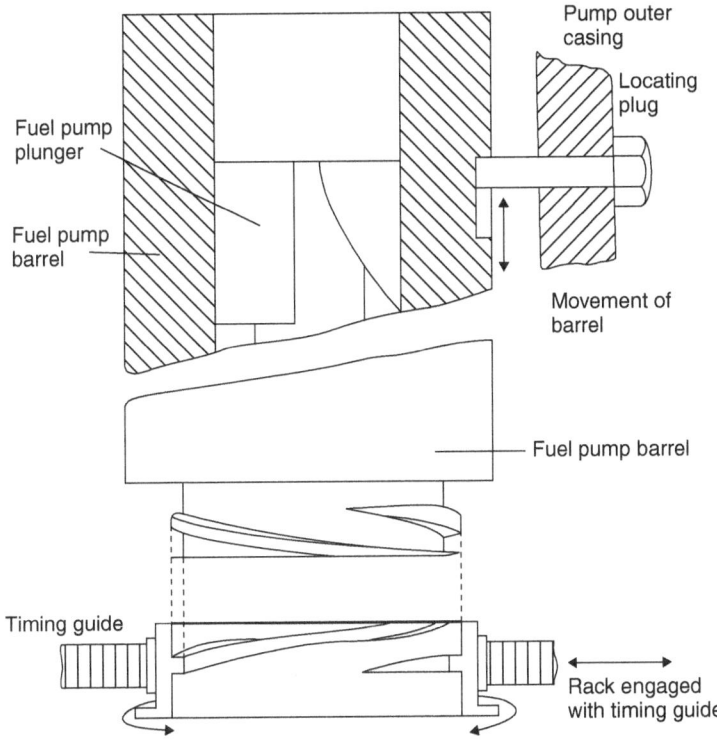

Fuel pump plunger

Fuel pump barrel

Pump outer casing

Locating plug

Movement of barrel

Fuel pump barrel

Timing guide

Rack engaged with timing guide

▲ **Figure 3.14** *Variable injection timing Bosch-type fuel pump*

from turning by a locating plug. The end of injection, and therefore the quantity of fuel delivered, is regulated by rotating the plunger, which varies the position of the helix edge relative to the spill port.

Currently the standard MAN Diesel MC/MC-C family of engines come with a chain-driven camshaft, camshaft-controlled fuel injection and exhaust valve opening systems as well as conventional fuel oil pumps, all tried and tested technology. The engine is fitted with a pneumatic/electric/hydraulic control system for engine speed control and manoeuvring. Using this system, MAN quote a specific fuel consumption of 167 g/kWh ± 5%.

Two-stage fuel injection

Efforts are continually being made to improve the reliability and economy of medium-speed engines that operate on heavier grades of fuel. One way of achieving this is by having a carefully controlled, reliable combustion process. This requires good atomisation with short injection periods but this results in high injection rates at high engine loads with commensurate high rates at intermediate and low loads where increased ignition delays may be experienced. Indeed, research has shown that, at low loads, the injection process may be completed before ignition commences. As the fuel is well mixed with the air, the combustion is very intense and almost instantaneous when ignition does eventually occur. This uncontrolled energy release will cause 'diesel knock' and possibly destructive thermal and mechanical stresses.

As a solution to this problem, Wärtsilä developed a two-stage injection process in which the fuel injected during the pilot stage is constant and independent of engine load (see figure 3.15). The quantity of fuel injected during the pilot phase is set at about 2–6% of the MCR, which is marginally less than the amount required to compensate for friction losses when the engine is idling. The pilot fuel is injected in advance of the main injection phase but the quantity involved is too small to damage the combustion chamber components. The injection and ignition of the pilot fuel minimises the ignition delay because it raises the temperature of the combustion air. The fuel injected during the main stage enters a favourable environment with combustion commencing as soon as the first fuel droplets enter the combustion chamber. This eliminates the possibility of unburned fuel being stored in the combustion chamber and hence the destructive uncontrolled release of energy.

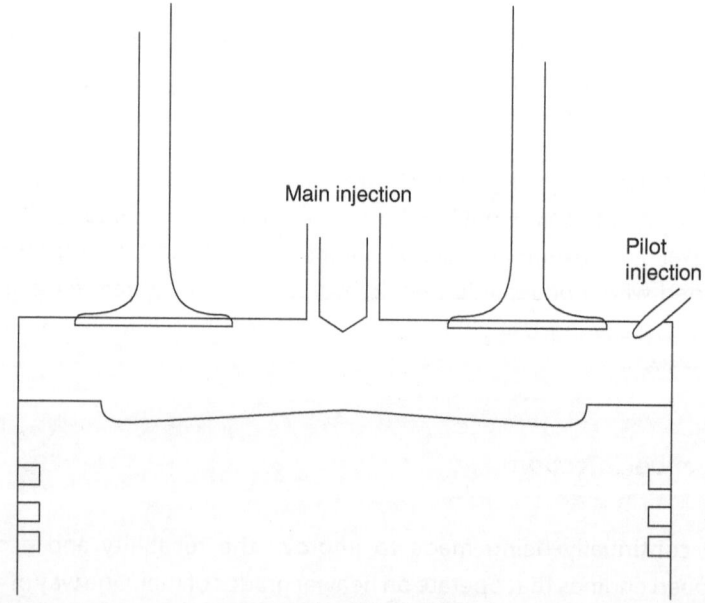

▲ **Figure 3.15** *Position of main and pilot injection in cylinder*

Both fuel valves are supplied by the same fuel pump. The fuel pump, however, has two plungers to supply pilot and main injection. The main plunger of this fuel pump is of the conventional scroll- or Bosch-type. The pilot plunger is positioned above the main plunger, but since the quantity of fuel injected in the pilot stage is constant, this plunger has no helix. As the pilot plunger covers the suction/spill port, injection commences and ceases as the lower edge of the plunger uncovers the suction/spill port (figure 3.16).

This arrangement has the advantage that the emission of NOx is reduced and allows the use of low-cetane fuels.

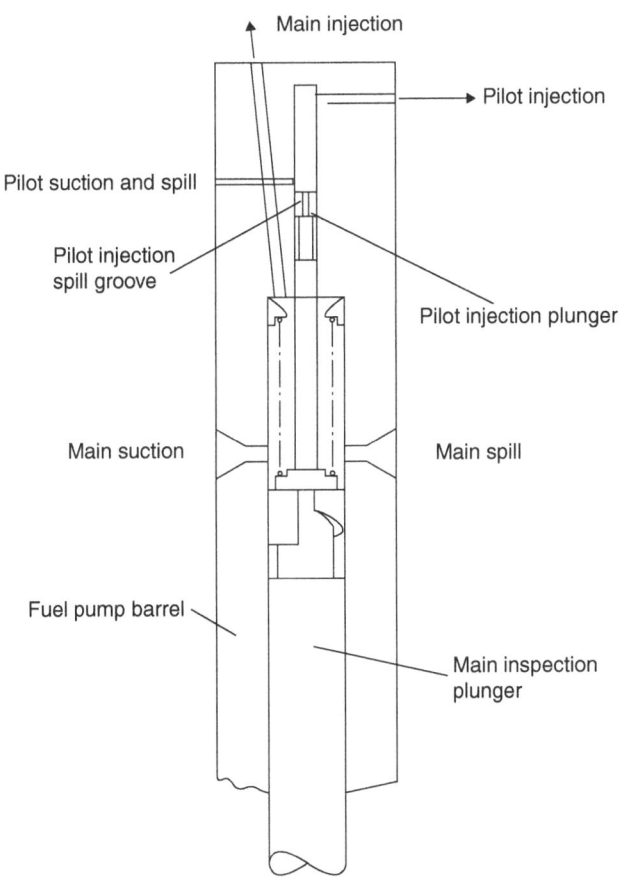

▲ **Figure 3.16** *Two-stage fuel pump*

EFI (some with common rail)

Now that the student has studied the basics of fuel injection/combustion and seen the limitations and constraints that the older mechanical systems imposed upon engine designers, let's have a look at the next steps in diesel engine design.

Recent advances in material science, linked to the continual quest to reduce costs, have led to the development of reliable and accurate measuring and sensing technology. It is now possible to measure the fuel rail pressure and the combustion pressure in real time and feed this information back to a central processing unit that will be able to continually adjust the engine's settings to give the best combustion conditions possible at all times.

Riding on the crest of this wave of development is the four-stroke medium- and high-speed diesel engine. Not only does it have an infinitely variable, common rail fuel injection system but it also has variable opening and closing of the inlet and exhaust valves, giving a very sophisticated and powerful system. However, such a system does not come easy and there has been considerable investment in development time and resources by the engine manufacturers to design reliable systems.

The changes in marine engines have been lagging behind advancement in the engines used by other industries such as road transport. Legislation is now driving the need for change and lessons learned in other industries means that the pace of change in the marine engines is considerable.

The problem is that engines that have a camshaft-controlled combustion process have a system that is linked to the speed of the engine and therefore there is little room to design a system of control that is load dependent and not speed dependent. Common rail systems permit a continuous, load-independent control of injection timing, pressure, volume and phasing. This means that common rail technology achieves the highest levels of flexibility for all engine loads and gives significantly better results than a conventional fuel injection system. Reliable and efficient CR systems have been developed for an extensive range of marine fuels, including residual fuels such as heavy fuel oil (HFO). This gives the added advantage of using a single fuel for both two-stroke and four-stroke engines.

Basic system design

The basic idea of the common rail system is quite simple. High-pressure fuel – as much as 2,000 bar – is circulated around the engine close to the fuel injectors. There will be a short length of pipe from the 'common rail' to the fuel injector. When the fuel injector is opened, the high-pressure fuel flows into the combustion space through the small holes in the injector.

The fuel will atomise well due to the very high pressure of the fuel in the fuel rail and due to the very small angled holes in the injector outlet. The major advance with this system comes from the ability to open and close the fuel injector so quickly. This enables very close control over the timing, duration and sequencing of the fuel injection process.

Therefore, so much more can be done to influence the combustion process. A two-stage phased injection is shown in figure 3.17 and has the effect of lowering the peak temperature of the main injection. The peak temperature is partly responsible for the production of NOx emissions, which are harmful to the environment. Therefore, ability to modify the injection has an immediate payback in an engine with a cleaner exhaust.

Figure 3.18 shows a further development where a third phase is introduced. This follows the main injection phase, assists the mixing of fuel and air and raises the temperature towards the end of combustion. The temperature increase promotes soot oxidation and reduces the amount of particulate matter being emitted with the flue gases.

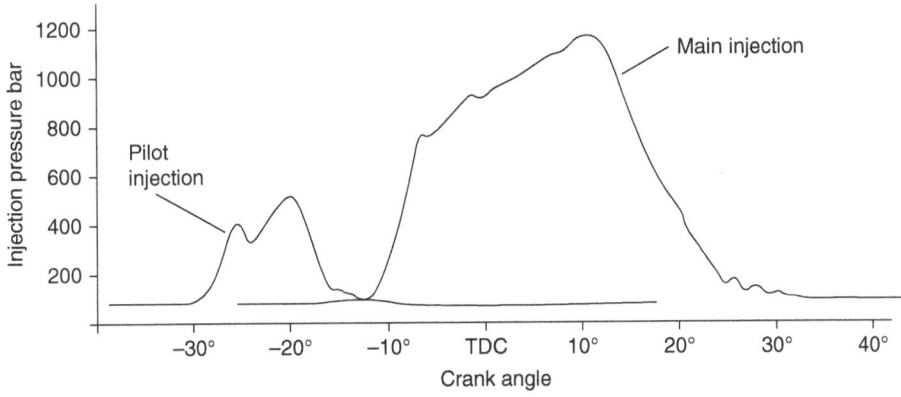

▲ **Figure 3.17** *Pilot and main injection against crank angle*

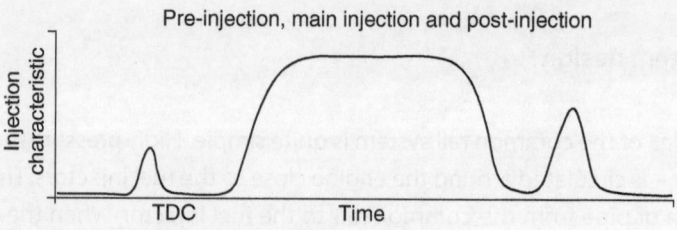

▲ **Figure 3.18** *Phased injection timing*

The fuel injection systems on modern engines is so fast that 'phased injection' can be achieved. This is also known as Stratified Fuel Injection (SFI) and was initially developed to improve the performance of older engines.

However, this technique is also being used for engines running alternative fuels such as Ammonia. The first and third phases of the injection process are taken with injecting diesel fuel and the longer middle phase is where the ammonia is introduced into the combustion space.

This is a good way to use ammonia as a marine fuel as the diesel (pilot) fuel combusts easier than ammonia due to its higher certain value and higher operating temperature.

Safety considerations

The Flag State examiner will need to be sure that any engineer gaining a certificate of competency from his/her administration has a comprehensive understanding of the safety required with engines operating the common rail fuel injection system.

The first and very obvious statement to make is that we have high-pressure fuel inside pipework that stretches over longer distances than has been the case in the past. Due to the high pressure, if any fuel does escape then it also has the potential to spray over a larger area. Given the fact that over the years fuel spraying onto hot surfaces has started a good number of serious engine-room fires, the examiner will quite rightly concentrate his/her questioning of the candidate to ensure an understanding of this point.

Yes, all the high-pressure piping will be shielded in a double casing and yes, care will be taken at the design stage to ensure that the fuel lines run in as safe a place as possible. However, it will be the candidate's responsibility to point out that this is a potential source of danger and that particular attention should be paid to the integrity of the pipework and associated fixings. Good maintenance practices must also be followed. It will be vitally important that the correct 'high-pressure' fittings, couplings and seals are used – unapproved alternatives are not acceptable. The pipework must be supported correctly after any maintenance work and it is important that any supports that were removed to ease the dismantling of the pipework are replaced.

Substandard workmanship could lead to failure in the pipework or could allow leaks of fuel out of the system or air into the system. Either of these faults will affect the operation of the engine and if the engine stopped it could endanger the vessel as well.

Different manufacturers will of course design their own modification upon the basic system. For example, MAN Diesel have completed extensive studies on the consequences of any failure of their system, before fitting it to the ME engine. Verification of their studies has been completed by a research and development (R&D) programme carried out in their test laboratories and on their research engines.

MAN Diesel, for example, have followed the following design concepts:

- The fuel injectors are only pressurised during injection.
  - o Meaning that there is no danger of uncontrolled injection, even if a control valve or injection valve leaks.
- All high-pressure components are double-walled.
  - o Any fuel from leaking or broken pipes is contained within the double-walled pipework. The leaking fuel can then be led to an accumulator that is fitted with an alarm.
- Flow-limiting valves are fitted to the fuel pipe for each cylinder.
  - o The valves will limit the quantity of fuel to be injected, even in the case of leaking or broken components.
- Non-return valves are fitted in the line for each cylinder.
  - o These valves prevent backflow from the low-pressure system into the cylinder, for example if there is a fuel valve seizure or breakage.
- There are between two and four high-pressure pumps fitted to the systems.
  - o Therefore, there is redundancy in the system and should one pump fail, emergency operation is possible.
- Pressure-limiting valves are fitted. These have the additional pressure-control function and act as a safety valve.
  - o This design characteristic means that emergency operation is possible, even in case of any failure in fuel rail pressure.
- An emergency stop valve/flushing valve is included in the design.
  - o The valve, which is actuated by compressed air, stops the engine in case of emergency.
- Surplus rail-pressure sensors and TDC speed pick-ups are fitted.
  - o No interruption of engine operation will occur due to pick-up or sensor error.

All this hard work has gone into the careful design of the new system. However, it must be understood by the engineering staff on board ship that it is now their responsibility to fully understand the system, its function and especially its manual operating mode if there is a fault in the automatic system.

The Wärtsilä approach with their RT Flex engine is to control the fuel injection process with a 'volumetric fuel injection control unit', which controls the timing and volume of the fuel to be injected. Wärtsilä do have a common rail, which is fed with heated fuel oil at high pressure (nominally 1,000 bar) ready for injection. Fuel supply is via a number of high-pressure pumps driven by multi-lobe cams.

Fuel is delivered from the common rail through the injection control unit, which is placed next to each cylinder, to the standard fuel injection valves. The fuel injectors are hydraulically operated by the high-pressure fuel oil. The control unit uses quick-acting valves to regulate the timing of fuel injection, control the volume of fuel injected and set the shape and phase of the injection. The three fuel injection valves in each cylinder cover are separately controlled so that, although they normally act in unison, they can also be programmed to operate separately as necessary.

Fuel Systems

In recent years the quality of the fuel available to the marine industry has deteriorated. This has led not only to problems with the combustion but also to problems, with the storage of the fuel. To minimise the effects from some of these problems careful consideration should be given to the design of the overall fuel system and this should include the bunkering system.

Modern residual fuels tend to have a high viscosity and may also have a high pour point so it is important that, upon the completion of the bunkering process, it is checked that the fuel has drained freely into the bunker tanks.

If the vessel is loading bunkers in cold climates, it may be necessary to include insulation on the exposed bunker lines. An indication of a fuel with a high pour point may be a high loading temperature. Here the supplier is trying to ensure that the fuel is easily pumped on board. If a waxy fuel is suspected then a pour point test should be carried out.

Due to the problems associated with incompatibility, fuels from different sources should not be mixed. The importance of segregation of fuels from different sources cannot be overstated and should be practised, wherever possible, by transferring remaining fuel into smaller tanks prior to bunkering in order that the total quantity of fuel loaded can be received into empty tanks.

Even if a vessel is equipped with adequate storage to ensure segregation, mixing may occur in the settling and service tanks when fuels are changed over. If compatibility problems are suspected then fuel changeovers should be accomplished by running down the settling tank before pumping in the next, possibly incompatible, fuel.

With the introduction of the new emission regulations and the requirements for vessels to operate on 'LSF', fuel suppliers are using more blending techniques to comply with the regulations and for this reason the risk of incompatibility will only increase. Fuels supplied to a ship must be treated before use. In fact, comment has been made that this is one of very few products that are purchased but are not 'fit for purpose' and must have additional treatment by the purchaser before they can be used.

Technically, bunker fuel is any grade of fuel that is used by the ship but the term 'main bunker fuel' has come to mean the fuel used for powering the main engine. The term comes from the days when the ships were powered by coal and this was loaded into a 'bunker'. Main engine fuel must be supplied to a specification that is set out in the ISO standard 8217. The two latest revisions from 2017 and 2024 are the most important.

The standard sets out the specification for the fuel characteristics including viscosity, density, flash point, pour point, sulphur, carbon residue, water and ash. The new (ISO8217:2024) specification addresses some of the residual fuel quality problems that have been experienced by the industry, with the inclusion of acid number limits as well as a limit on hydrogen sulphide. The distillate grades have had the inclusion of oxidation stability and a lubricity requirement introduced and the residual marine fuels have a calculated carbon aromaticity index added as an indicator of ignition delay. There is also a limit on sodium content as well as stricter limits for ash and vanadium and there has been a significant reduction in limits for aluminium and silicon, which are also known as cat fines (see Chapter 2 of Volume 8 of the Reeds series).

However, the new specification fuel does come at a price and DNV Petroleum Services (DNVPS) reported in 2011 that there was a resistance to using the ISO8217:2010 fuel specification. Some of this was due to charter party agreements but another reason was problems with availability of products meeting the new specification; however, according to the DNVPS survey, only 10% of the total respondents said they would not eventually switch to ISO8217:2010. Most are now using the latest standard.

The knock-on effect from taking on 'off spec' fuel was that filters started clogging due to sludge, sticking or seized fuel pumps and even piston ring breakages have been attributed to the quality of the fuel. When this does happen there is always a cost involved. It is the job of the engineering officers to ensure they have evidence to support any claim on insurance, or against a third party, that the shipowner might wish to make.

There are some basic precautions that the ship's staff must take during the bunkering stage. These include:

- *Communication*
  - The engineering officers must make sure that the rest of the ship's company know at which port bunkering is likely to be taking place.
  - Engineering officers must work with other relevant officers and crew during the actual operation so that they can all keep a watch on the process while carrying out their own tasks such as loading cargo or taking on stores and spare gear.
  - Communication between the ship and the bunkering vehicle (barge or road tanker) is vital to completing a safe operation.
  - Efficient lines of internal ship communication are also extremely important.
- *Resources*
  - The bunker station – which is the point where the vessel's fixed pipework is connected to the flexible bunkering hose usually provided by the bunker suppliers – should be 'manned' at all times during the operation.
  - The engine-room valve operating station or valve chest should be manned at all times and the officer stationed here should be in constant communication with the officer at the bunker station.
  - Adequate 'drip trays' should be placed under the final flange where the ship's pipework meets the flexible bunkering pipe.
  - Appropriate 'oil spill' dispersant and absorbent material should be placed close to the bunkering station.
  - Any water-freeing holes in the ship's bulwark around the site of the bunker station should be temporally blocked so that if there was a spill the oil would be retained on board where it can be cleaned up without incurring a financial penalty.
- *Other actions*
  - The Chief Engineer should check and agree the order quantity and quality with the manager in charge of the bunker barge or tanker.
  - Samples should be taken, ideally at the start and at the end of the operation but random samples could also be taken during the bunkering process.
  - The quantity being delivered needs to be checked – the traditional way has been to check the bunker fuel in the ship's tanks before and after delivery of the fuel. Alternatively, an engineer could go to the bunker barge or road tanker and check the quantities there before and after delivery (see below for the more modern approach).

○ The Chief Engineer needs to record where the bunkers from that load are stored and that the records are understood by all the ship's senior management team.

○ The Chief Engineer needs to update his/her standing orders so that the engineers know the sequence for drawing the fuel during the next voyage.

MARPOL Annex VI gives minimum values for the emissions from the flue of ships that come under the jurisdiction of countries that are signatories to IMO. Regulation 18 states that fuel of the correct standard should be available. However, it also recognises that there will be bunker ports in countries that are not party to the MARPOL agreement. When purchasing fuel from such ports, IMO recommend that ship managers have a clause inserted in their agreement detailing the specification of the fuel, ideally to meet the IMO requirements. A Bunker Receipt Note with specific contents must be issued for each delivery, together with a sample that is fully representative of the fuel delivered. These must be retained, not necessarily on board, for three years in the case of the documentation and at least 12 months in the case of the fuel sample, in case they are required as proof of compliance. Furthermore, the regulation gives steps that must be taken in the event of non-compliance.

Flag States issue guidance for ship operators about the requirements for meeting the MARPOL Annex VI regulations. Classification societies also issue assistance to their members about the necessary steps to comply with good practice.

Fuel management – on-board systems

Ideally, if ships were designed with two service tanks and two settling tanks then fuel changeovers could be accomplished with the minimum amount of mixing. This complexity of design would, however, have to be considered at the design stage and shipowners may wish to consider this when drawing up a new-build specification.

The temperature of the stored fuel must be monitored to ensure that it does not fall to near its pour point. This is important, especially when fuel is stored in double bottoms, since it is not uncommon for fuels to have a pour point of 25°C and approaching that point the fuel becomes unpumpable, which can happen when the vessel is in climates that can be considered temperate. The heating capacity of the fuel system, including the tank heating, trace heating and main system heating, should be able to deal with the viscosity of any fuel the vessel is likely to encounter. Tank heating must be able to maintain temperatures above the maximum likely pour point.

Keeping the flash point of a fuel within specification is a legal requirement. The flash point is the temperature at which any vapour that is given off will ignite when an external flame is applied. This is usually quoted as the temperature measured under standardised conditions. The fuel's flash point is defined, and kept within tolerance, to minimise fire risk during normal storage and handling. The minimum flash point for fuel in the machinery space of a merchant ship is governed by international legislation and set at the value of 60°C. For fuels used for emergency purposes, external to the machinery space the flash point must still be greater than 43°C. However, even when residual fuels are at a temperature below their measured flash point they are still capable of producing light hydrocarbons, and could still be flammable. The normal maximum storage temperature of a fuel is 10°C below the flash point, unless special arrangements are made.

Storage-tank heating as well as settling- and service-tank heating should maximise the separation of water and solid matter from the fuel and still be able to maintain the correct post-purification temperature. This is important given the requirement for fuel to be stored at 10°C below the flash point, which could be 60°C. However, the purification and clarification temperatures of high-viscosity fuels may be substantially higher than this; 100°C, for example. To comply with this regulation a post-purifier fuel cooler may be required to return the fuel to below its flash-point-related value, as it is returned back to the storage tank.

If a fuel storage temperature is allowed to drop close to its pour point at any stage during the storage, then wax can start to form, which may not readily be absorbed into the fuel again when the temperature is raised. The wax forms a sludge, which can block filters and the small passages in the fuel injection equipment.

When operating with high-viscosity fuels it may be necessary to employ high rates of heat transfer during fuel heating. This could lead to thermal cracking of the fuel, resulting in carbon deposits on the heating surfaces causing reduced heating capacity. To maintain optimum heat transfer and heating steam consumption there should be a facility to enable the oil side of the heater to be cleaned periodically by circulating with a proprietary carbon remover. A typical fuel system is shown in figure 3.19.

Many fuel-related problems will not arise if an effective 'on-board fuel management policy' is adopted and followed through closely by each of the crew who serve on board. To recap, such a policy would include the following:

1. Representative samples of fuel, in addition to the supplier's sample, taken at loading. These should be sealed, clearly labelled and retained on board for three months after the fuel has been consumed.

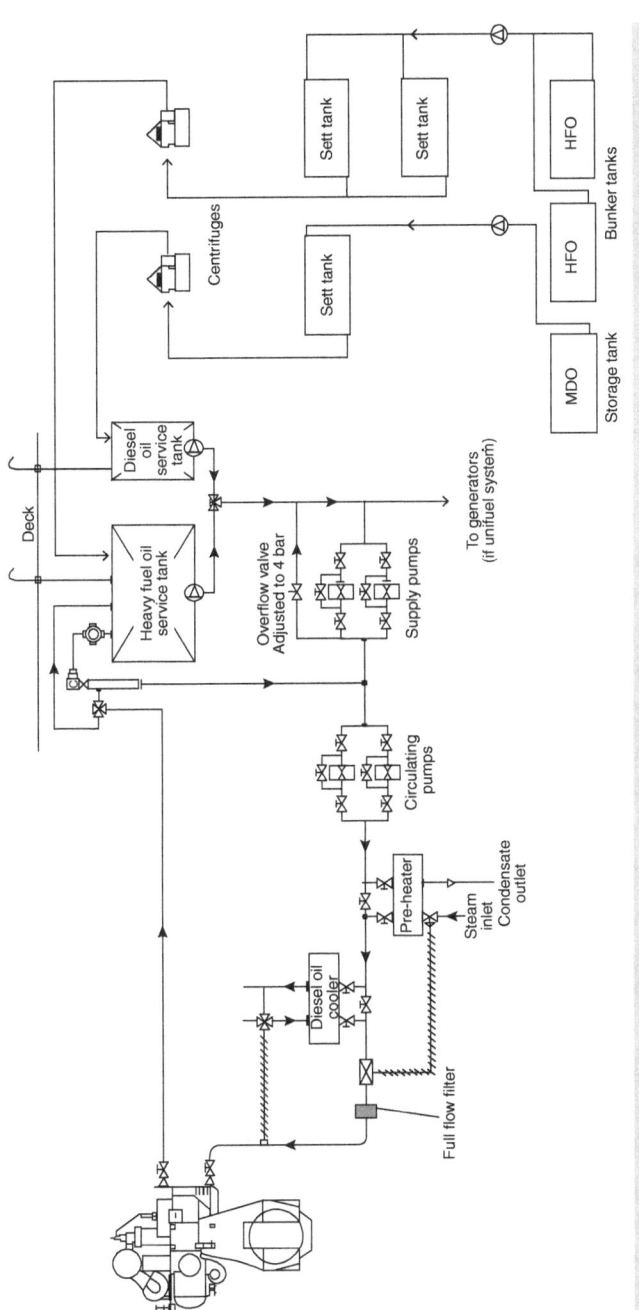

▲ Figure 3.19 General arrangement of a modern fuel system

2. Segregation of fuels from different sources by loading into empty tanks. This may involve the transfer of remaining fuels into smaller tanks prior to loading.

3. Draining bunker lines at the completion of loading. Closing all bunker valves when this is accomplished.

4. Maintaining storage temperatures at least 5°C above the pour point and 10°C below the flash point.

5. Sending a representative fuel sample for analysis and taking the appropriate action upon receipt of the results.

6. Fitting a certified mass flow meter to accurately measure the quantity of bunkers delivered free of contamination.

7. Draining water from tanks at regular intervals.

8. Monitoring fuel consumption against fuel remaining on board. This should be achieved by daily measurement of all fuel tanks; on older ships, dipping the fuel tanks will probably provide the most accurate results. However, on a vessel fitted with accurate depth gauges the remote readings will provide the data required.

9. Regular checks of the fuel purification plant to see if excess water or solid impurities are being removed by the purifiers.

10. Temperature and viscosity control of the fuel from the storage tanks to the point of use by the machinery is very important. If the fuel temperature drops close to its pour point, the fuel filters could become clogged.

11. If the bunker tanks are filled to the very top and the temperature is slightly low, then as the temperature rises so will the volume. It will not take much for the fuel to rise up the vent pipe and cover the vessel with fuel, which could spill into the water, causing the vessel to be fined or the Master and Chief Engineer arrested.

Precautions to be taken before and during the bunkering operation

Fuel and lubrication oil can be delivered to the ship by toad tanker or bunker barge. With this operation being so important due to the heightened risk of spillage, the Chief Engineering Officer may always wish to lead the team personally. However, all the supporting technical staff should also have a good working knowledge of the process. The first activity is to hold a team briefing to ensure that everyone involved is clear about their roles and responsibilities.

With reference to Fig 3.20 please note the following. Before the bunkering process can proceed, the following must be checked and reported to be correct:

1. Drip trays under the ship/shore connection should be checked for structural integrity and that they are dry (not full of water or oil).

2. Blanks on the ship/shore connection, which is not being used, should be securely fitted to prevent any leakage should the internal pipe become full of oil due, for example, to a wrong valve being opened.

3. Absorbent material and oil spill dispersant should be set close to the positions of any potential spillage.

4. Radio communications and ship's internal telephone systems should be tested for correct functioning and emergency shut off procedures discussed. This will be especially important where the language between the staff on the ship and the staff on the barge/road tanker is not common. International English should be used throughout the bunkering operation but this point needs to be checked.

5. All the ship's staff should be aware of the measurement system in use, ehether this is soundings or ullages. In addition, the contents and sequencing of the tanks to be filled should be known to everyone involved. If this is achieved, then all members of the bunkering team can be on the lookout for any potential errors during the operation.

6. When connecting the hose from the barge/road tanker then a new joint/gasket should be used and all the bolts tightened evenly so that there is no leaking from the connection during the bunkering process.

7. Once the activity has been started then constant monitoring of tank levels must be carried out and the pumping rate slowed down and stopped before tanks are completely full. If this is not done, then oil can spill onto the outside deck via the air vent.

8. An effort should be made to fill the tanks evenly (port and starboard) to minimise any listing that could occur and affect the tank readings.

Additional requirements for Alternative Fuels

Where alternative fuels are concerned, the risk involved with their use is reduced by both the design of the equipment used and the ship's structure as well as robust operating procedures. The hoses and couplings will not be the same as those used for bunkering conventional fuels. The location of the bunkering station is essential in reducing the risk from fire and exposure to the toxic elements within the fuel e.g. open and well ventilated. Procedures and plans should be in place and enforced to reduce the risks while bunkering. Space around the bunkering station may needs to be designated a 'safe space' and movement within it restricted during the bunkering operation.

Methanol (Chemical composition CH₃OH)

The first consideration with methanol is that it is a 'Low Flashpoint' fuel and is therefore subjected to the IMO's IGF code including the requirement for the technical staff to have completed the relevant IGF training. One of the attractions of using methanol as a marine fuel is that it is a liquid under atmospheric temperatures and pressures. However, in addition to having a low flashpoint, methanol burns with a flame that is only sometimes visible and makes very little smoke, making a fire difficult to detect visually. Ships that are using methanol as a fuel will have documented procedures within their safety management system relating to the burn characteristics of the fuel and relating to its toxicity and risk assessment.

The additional procedures for the bunkering of methanol will include the need to check the:

- operation of the fire detection and fighting equipment
- condition of all personal protective equipment (PPE)
- hoses need to be in-date from the last time they were formally tested
- special hoses and connections, which will need to be inspected for damage
- relevant staff checked for knowledge about any emergency action that may be necessary
- procedures relating to the venting and inerting of the tank space above the fuel
- electrical bonding equipment for security and damage
- safety zones have been marked correctly

It is important that fuel samples are taken as the specification of the fuel is important as is the quality of the delivery. Impurities that can reduce the quality are water and dirt from storage tanks that have not been cleaned thoroughly.

Ammonia (Chemical composition NH3)

Ammonia is toxic and flammable with a flash point of 38°C although the flammable range is narrower than other fuels. Ammonia is normally lighter than air but it is also hydroscopic and after absorbing water the subsequent vapour can become heavier than air. The chemical is also reactive and will affect a range of materials and therefore the whole fuel system must be built to the correct standards, for the specific ammonia handling method designed for the specific circumstances of the ship.

The problem for ship owners choosing ammonia as a fuel is that it is not a liquid at normal atmospheric temperatures and pressures. For ammonia to be a liquid at atmospheric pressure it needs to be at about -33.5°C and when it is at about one bar ammonia still needs to be cooled or partly refrigerated. Therefore, students can see that if ammonia is at an atmospheric background temperature then for it to remain a liquid the pressure would need to be between 9 and 16 bar.

The engineering officers all need to be very familiar with the actual properties and state of the ammonia that is being used onboard their particular vessel. In addition, an ammonia bunkering system will need a 'boil off' management system so that if there are any fluctuations in temperature/pressure, an expanding gas will not be able to escape into the atmosphere.

Liquid Natural Gas (LNG)

The most striking thing about LNG is the working temperature, of around -160°C, that is required when it is used as a fuel for a ship's propulsion plant. The 'cryogenic' nature of the operation brings a step change in procedures, equipment and knowledge required for the safe handling of the Liquid Gas.

Equipment

Despite good insulation, with the fuel being at such a low temperature some heat will soak through and start to raise that temperature. This can only be tolerated for a short period of time as the pressure will start to rise, presenting additional stress on equipment such as storage tanks and pipelines.

Classification societies will of course require ship builders to use the correct grades of steel, suitable for use with cryogenic fluids. The materials required must retain minimum levels of strength and toughness at the temperature range required. In addition, a low co-efficient to expansion will also be required to ensure that the temperature change does not introduce excess stress into the pipes, pumps and valves etc due to expansion/contraction. In addition, flanged joints could be a source of potential leakage if the correct materials are not used for both the flanges and the bolts. Care must also be taken to ensure that the correct tightening procedures are followed so that the flange is tight at all temperatures. It is important that the correctly specified jointing material is used to ensure a gas tight connection.

Where practicable welded joints should be kept to a minimum. However, where they are required all welds should be properly tested.

Flexible hoses – these must be specifically designed to withstand the low temperatures involved. Thay could be metal based or made of a 'composite' material specially designed to handle cryogenic fluids and designed to the relative ISO standards and approved by the classification society.

As the pressure rises, at some point it must activate a system for reducing the pressure again. The excess gas is known as 'boil off gas' (BoG) must therefore be handled in some way. Systems for this are either vent to atmosphere or process the vapour through a re-liquefaction plant. Either of these are the least desirable options and therefore the BoG must be kept to a minimum through the correct handling of the LNG.

Bunkering – during the bunkering process the lines are kept full of LNG and therefore the additional fuel delivered will start to pressurise the bunker lines and storage tanks. This means that vapour return lines must be provided as part of the bunker pipework.

Leak detection is not as straightforward as it is with the more traditional fuel oils. LNG is odourless and the gas itself is colourless, making it difficult to detect. The low temperature does, however, cause moisture in the air to freeze, giving some indication to the presence of a leak. Initially, while still liquid LNG will be heavier than air and fall but as it increases in temperature the gas will start to rise. It is very important for staff to understand these characteristics and know how to use leak detection equipment. In addition, any area that is suspected to contain any gas could, of course, also be deficient in oxygen.

Procedures

Safe areas around the bunker station need to be setup and only competent and authorised persons allowed into those areas. Full and relevant risk assessments must be made, taking into account the characteristics of the low temperature fluid as well as being a low flashpoint fuel.

Personal Protective Equipment (PPE) in all instances is very important. Staff involved in handling LNG should use PPE at all times and it is important to know that some of the essential equipment to be used and precautions to be followed are not immediately obvious. This includes equipment to protect against the results of exposure to extreme cold, oxygen depleted spaces, gas leaks and over pressurization of tanks.

Knowledge

In addition to having a general level of education and training to the IMO's minimum standard set out in the STCW regulations all staff involved with handling LNG must also have completed the specialist training about the International Code of Safety for Ships Using Gases or Other Low-flashpoint Fuels (IGF Code) regulations.

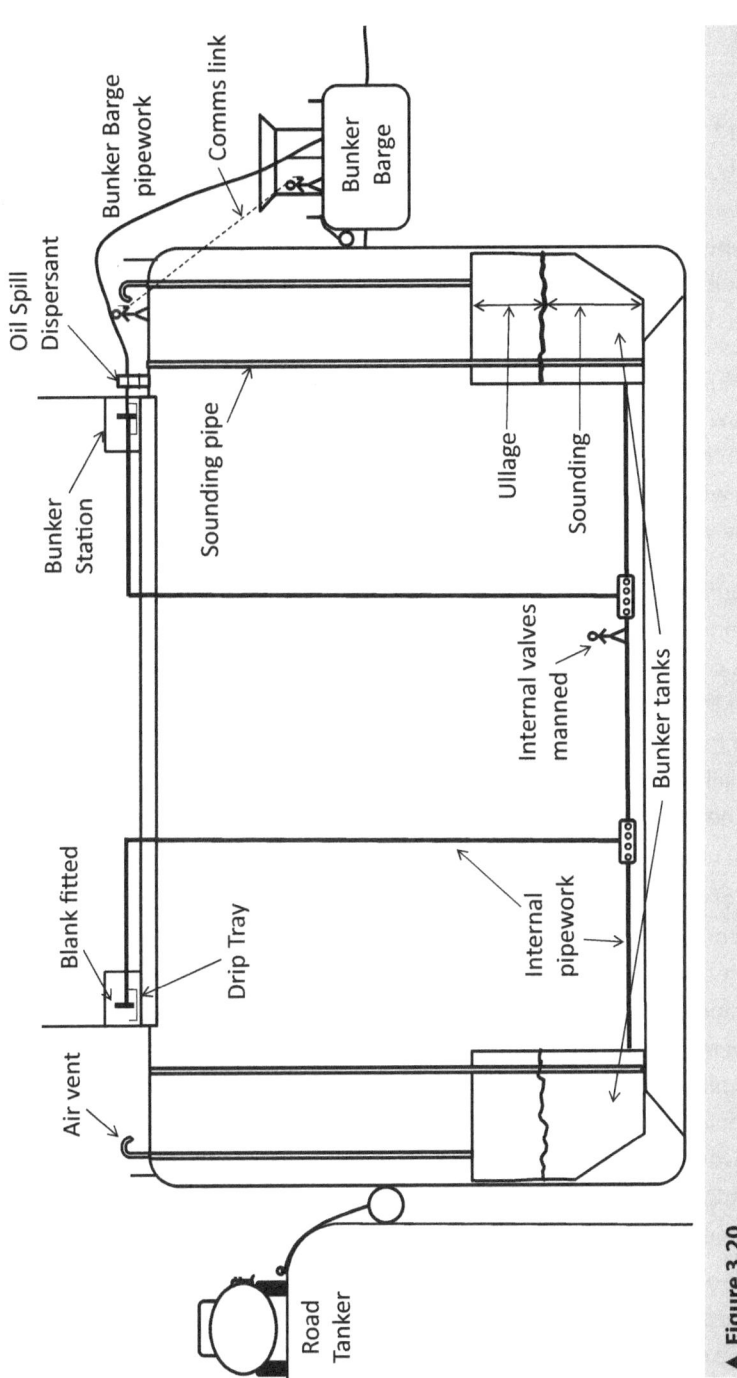

▲ **Figure 3.20**

Analysis of the fuel

As already stated, fuel should now be supplied according to the specification set out in ISO 8217 and preferably the 2024 version of the standard. However, the delivery of 'off spec' bunkers is a growing problem and is causing concern with shipowners and operators as well as engine manufacturers. Analysis of fuel oil is now recommended. Diligent operators are encouraged to seek the help of one of the oil majors and have their fuel oil samples collected and analysed at short notice. Some problems identified could be as follows:

1. *Low flash point:* Regulations require a flash point above 60°C. If the flash point is found to be below this level then the owners and classification society should be informed. A lower flash point fuel will render the vessel 'out of class'. The addition of a higher flash point fuel will not raise the flash point of the original stock. To avoid the generation of a flammable vapour, heating temperatures should be regulated carefully.

2. *High sulphur:* Sulphur is present in crude oil and the specific amount depends on the original source of crude oil used and the type of refining process. During combustion, sulphur is converted into sulphur oxides, which become corrosive upon contact with water and if left unchecked will damage engine pistons and cylinder liners. The acids produced must be neutralised by the cylinder lubricant and marine engine lubricants are especially developed with a high BN to cope with this acidity. If the correct lubricant is used, the sulphur content of a marine fuel is technically not important but the increasing environmental implications are now of great concern to the legislators.

Annex VI of MARPOL 73/78 sets out the sulphur content of any fuel oil used on board ships that originally were not to exceed 4.50% m/m max. After 2010, both Annex VI and the EU directive 2005/33/EC restricted the SOx emissions of ships sailing in the Baltic Sea Emission Control Area (ECA) to 6 g/kWh, which corresponds to a fuel oil sulphur content of maximum 1.5% m/m. In addition, the EU directive extended the 1.5% m/m sulphur limit to ferries operating to and from any EU port. The North Sea and English Channel have now become a ECA area where the 1.5% m/m sulphur limit applied. The EU directive further set a limit of 0.1% m/m max on the sulphur content of marine fuels used by ships at berth (and by inland waterways), effective from 1 January 2010, which also became a ECA area requirement from 2015.

The knock-on effect is that from 2015 the current generation of marine engines will not achieve these low levels of emission without additional 'after engine' technology such as selective catalytic reduction (SCR) (see page 347).

3. *High water content:* This may separate when heated; however, water could also form a stable emulsion, which is difficult to separate without the addition of emulsion-breaking chemicals. If the water contamination is salt water, not uncommon in the marine environment, serious problems associated with sodium-vanadium corrosion and turbocharger fouling may be experienced. Water contamination also introduces the risk of bacteria into the fuel. Bacterial growth can occur at the oil/water interface, which, if allowed to proliferate, can cause blockage of filters and the fuel system. The problem of bacterial or microbial attack is greater in fuel that is unheated, especially diesel oil, since the temperatures involved when heating high-viscosity fuels will pasteurise the fuel and thus kill off bacteria. Since prevention is better than cure, draining the water from the oil is by far the best course of action.

4. *High vanadium:* This may cause high-temperature corrosion. The use of an ash-modifying chemical additive to maintain the vanadium oxides in a molten state will prevent adhesion to high-temperature components. However, vanadium is bound chemically within the fuel and as a consequence cannot be removed. The vanadium deposits are very hard and can cause extensive damage to turbochargers.

5. *Instability and incompatibility:* Instability refers to tendency of the fuel to produce a sludge by itself. Incompatibility is the tendency of the fuel to produce a sludge when blended with other fuels. These sludges form when the asphaltene content of the fuel can no longer stay in solution and so precipitates out, sometimes at a prodigious rate. The deposited sludge blocks tank suctions, filters and pipes and quickly chokes purifiers. In engines, the blockage of injector nozzles, late burning and coking can result in damage to pistons, rings and liners. Therefore fuels from different sources should not be mixed on board.

6. *High aluminium content:* This contamination is a result of carry-over of 'catalytic fines' from the refining process of the initial oil. These 'fines' are an aluminium compound ranging in size from 5 mm to 50 mm and are extremely abrasive. Very low levels of aluminium indicate the presence of catalytic fines in the fuel, which, if used, will lead to high levels of abrasive wear in the fuel system, piston, rings and liner in an extremely short period of time; 30 ppm of aluminium is generally considered as the maximum allowable level in fuel oil bunkers before purification. As a result of the small size of these compounds they are difficult to remove completely by centrifuge. The purification plant, in correct operation, will reduce the aluminium content to about 10 ppm before it is used in the engine. It has been found that if the aluminium content is above 30 ppm, difficulties will be experienced in attaining a safe level of 10 ppm after purification. Due to the problem of 'cat fines', the 2010 version of the ISO 8217 specification for fuel oil was introduced.

4

SCAVENGING AND PRESSURE CHARGING

The Gas Exchange Process

It used to be very simple. For maximum performance and economy to be maintained, it was essential that during the gas exchange process the cylinder was completely purged of residual gases at the completion of the exhaust phase of the cycle and a fresh charge of air introduced into the cylinder ready for the following compression stroke, and this is still the ultimate aim for the current generation of engines. However, even the most efficient systems still leave behind unburned hydrocarbons from the previous cycle and as we shall see later the most fuel-efficient engines are using techniques such as exhaust gas recirculation (EGR), which is a technique to reduce harmful emissions from the engine.

The gases in each cylinder need to be replaced ready for the next individual combustion to take place. Even the large 2 stroke engines running at 100 rpm will only have just over half a second to expel the exhaust gases and replace them with a new charge. It is remarkable to note that in these large engines the stroke is up to two and a half metres.

This exchange takes energy and is referred to as 'pumping losses'. Using the Miller cycle and exhaust gas recirculation can reduce these losses. The EGR process recirculates some of the exhaust gas so that less fresh air is introduced into the cylinder and therefore less nitrogen thus reducing the nitrogen oxides in the exhaust gas. However, the oxygen level needs to be maintained for stoichiometric combustion to take place. The EGR action does raise the overall temperature inside the combustion chamber and therefore also makes combustion easier.

The modern two-stroke marine engines use the uniflow scavenging process (see figure 4.2). A significant reason for this is that the gas only has the one direction to move, from the bottom of the cylinder to the top.

In the case of four-stroke engines, purging the cylinder of the gases from the previous cycle is relatively easy and carried out by careful timing of inlet and exhaust valves where, because of the time required to fully open the valves from the closed position and conversely to return to the closed position from fully open, it becomes necessary for opening and closing to begin before and after dead centre positions if maximum gas flow is to be ensured during exhaust and induction periods. Typical timing diagrams are shown in figure 4.1 for both normally aspirated and pressure-charged four-stroke engine types. Crank angle available for exhaust and induction with normally aspirated engines is seen to be of the order of 420°–450° with a valve overlap of 40°–60° depending upon precise timing – with more modern pressure-charged engines, this increases to around 140° of valve overlap. The basic object of overlap, that is, exhaust and inlet valves opening together, is to assist in final removal of any exhaust gases from that cylinder so that contamination of charge air is minimal. The extension of overlap in the case of pressure-charged engines serves to: (a) further increase this scavenge effect and (b) provide a pronounced cooling effect, which either reduces or maintains mean cycle temperature to within acceptable limits even though loading may be considerably increased. Consequent upon (b) it becomes clear that thermal stressing of engine parts is relieved and, with exhaust gas turbocharger operation, prolonged running at excessively high temperatures is avoided. This latter process would have an adverse effect on materials used in turbocharger construction and could also contribute towards increased contamination.

In some cases an apparent anomaly exists between the temperature of exhaust gas leaving the cylinder and the temperature at the inlet to the turbocharger, being as much as 90° higher. This is partially explained by the fact that over the latter part of the gas exchange process the relatively cold scavenge air will have a depressing effect on

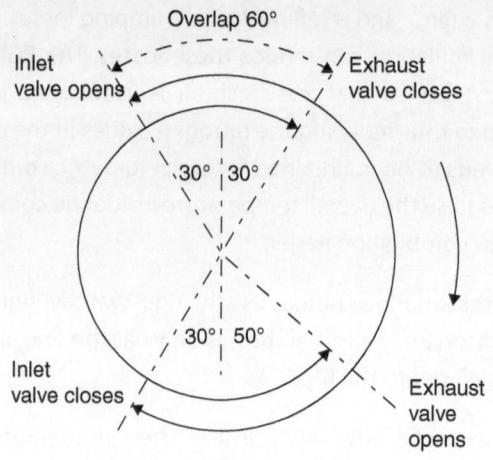

Naturally-aspirated four-stroke

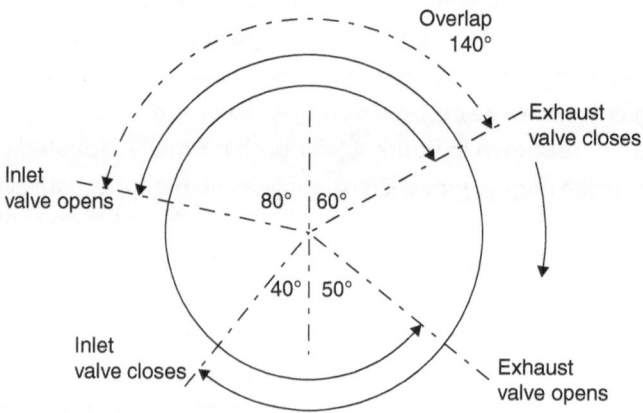

Pressure-charged four-stroke

▲ **Figure 4.1** *Typical timing diagrams*

the temperature indicated at that cylinder outlet, which will tend to indicate a mean value over the cyclic exchange.

More probably the increase may be largely attributed to the change of kinetic energy into heat energy and an approximately adiabatic compression of the gas column between the cylinder and turbine inlet. Two-stage turbocharging is now being introduced to increase the charge pressure following a reduction when the 'Miller' cycle is used (see Chapter 1).

Two-Stroke Cycle Engines

The two-stroke engine has only one revolution in which it has to complete the combustion cycle and therefore the time available for clearing the cylinder of residual exhaust gases and recharging with a fresh air supply is very much reduced compared with the four-stroke engine. Of necessity, the gas exchange process is carried out around the BDC where the positive displacement effects of the piston cannot be exploited as much as they can be in the four-stroke cycle. The total angular movement of the two-stroke piston seldom exceeds 140° compared to well in excess of 400° with the four-stroke piston's operation. This comparison gives some indication of the need for a high-efficiency scavenging process, in the two-stroke engine, if cylinder charge is not to suffer considerable and progressive contamination and subsequent loss of performance as well as increased temperature and thermal loading. Prior to the introduction of turbocharging to two-stroke machinery there was a need for a low degree of pressure charging of 1.1–1.2 bar to ensure adequacy of the gas exchange process. This was carried out by the use of a scavenge pump or, in smaller engines, by the use of under-piston pressurisation of the inlet air that was fed through a transfer port to the mail cylinder.

Modern two-stroke engines are all turbocharged, making the scavenging process much easier. The scavenging arrangement of two-stroke engines is generally described as: (a) uniflow or longitudinal scavenge and (b) loop and cross scavenge (figure 4.2).

All the latest large two-stroke marine engines currently under construction have been designed to work with the uniflow system, which employs a poppet-style exhaust valve situated at the cylinder head. Very early versions of the uniflow system were opposed piston engines where the exhaust (upper piston) uncovered the exhaust port while at the top of its travel. The uniflow system is the most efficient system, which is why it has been adopted for new engines.

In the uniflow system the charge air is admitted through ports at the lower end of the cylinder and as it sweeps upwards towards the exhaust discharge area, almost complete evacuation of residual gases is obtained. By suitable design of the scavenge ports or the provision of special air deflectors, the incoming charge air can be given a swirling motion; this intensifies the purging effect and also promotes a degree of turbulence within the charge, which is required for good combustion when fuel injection takes place.

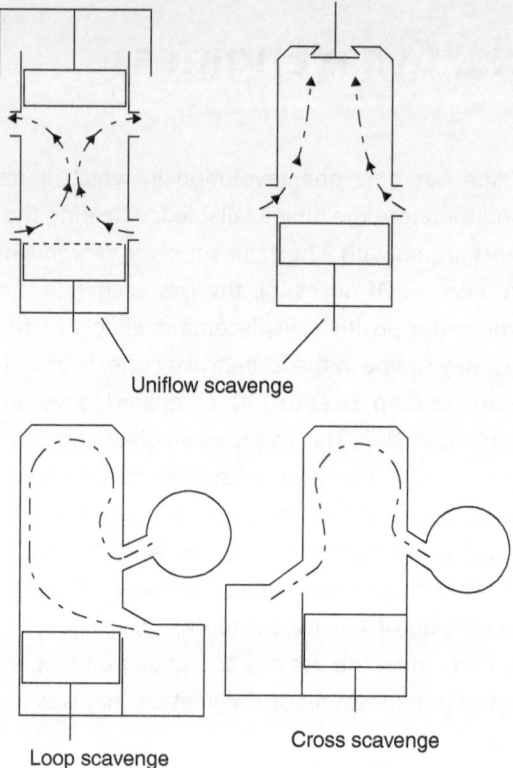

Uniflow scavenge

Loop scavenge

Cross scavenge

▲ **Figure 4.2** *Scavenging of two-stroke engines*

Both cross and loop scavenge systems have exhaust and scavenge ports arranged around the periphery of the lower end of the liner and in so doing eliminate the need for cylinder head exhaust valves or upper exhaust ports and the associated operating gear. This simplifies the engine construction considerably and, in the past, it might also have led to a reduction of maintenance. Due to a simplified cylinder head construction, the cylinder combustion space can be designed for optimum combustion conditions. However, the scavenging efficiency is now so much lower than with the uniflow system due to the more complex gas–air interchange and the possibility of charge air passing straight to exhaust with little or no scavenging effect. Careful attention to port design did reduce this problem but not enough to stop its fall from favour. There will of course still be engines working to these old designs so the marine engineering student should be familiar with their operation.

The gas exchange process itself may be divided into three separate phases:

- blowdown
- scavenge
- post-scavenge.

During 'blowdown' the exhaust gases are expelled rapidly – the process being assisted by generously dimensioned exhaust ports or valves arranged to open rapidly. At the end of this 'blowdown' period when the scavenge ports begin to uncover, the cylinder pressure should be at or below the charge air pressure so that the scavenge process that follows effectively sweeps out the residual gases without any resistance from a pressurised charge in the cylinder. With scavenge ports closed again, the post-scavenge period allows completion of the gas exchange process. The engine design should ensure that the exhaust discharge mechanisms close as quickly as possible to prevent undue loss of charge air and maximise the trapped air ready for the beginning of compression, giving the highest possible density of charge ready for combustion.

Although some loss of charge air is unavoidable, it should be borne in mind that the air supply is in excess of that required for combustion and the cooling effect of the air passing through the system has the result of keeping mean cycle temperatures down so that service conditions are less exacting. The production of NOx during combustion happens at the peak temperatures during the process. Therefore, if these peak temperatures are reduced then so is the volume of environmentally harmful NOx gases. In the latest engines this is accomplished by using the 'Miller' cycle (see Chapter 1), which modifies the timing of the inlet and exhaust valves to ensure that there are no peak temperatures produced during the combustion process. This can only be done with an engine that has full control over the inlet, exhaust and the start and stop of the fuel injection. Also with this system, the pressure charging is increased with the use of two-stage turbocharging.

The increased cylinder pressures encountered with modern turbocharged machinery may result in exhaust opening being advanced so that sufficient time is given for cylinder pressure to fall to or below charge air pressure when the scavenge ports uncover. A complementary aspect of earlier opening to exhaust is the increased pulse energy obtainable from the exhaust gas, which can be utilised to improve turbocharger performance. In many cases this is the main criterion that influences exhaust opening, since the loss of expansive working is more than offset by the gain in turbocharger output.

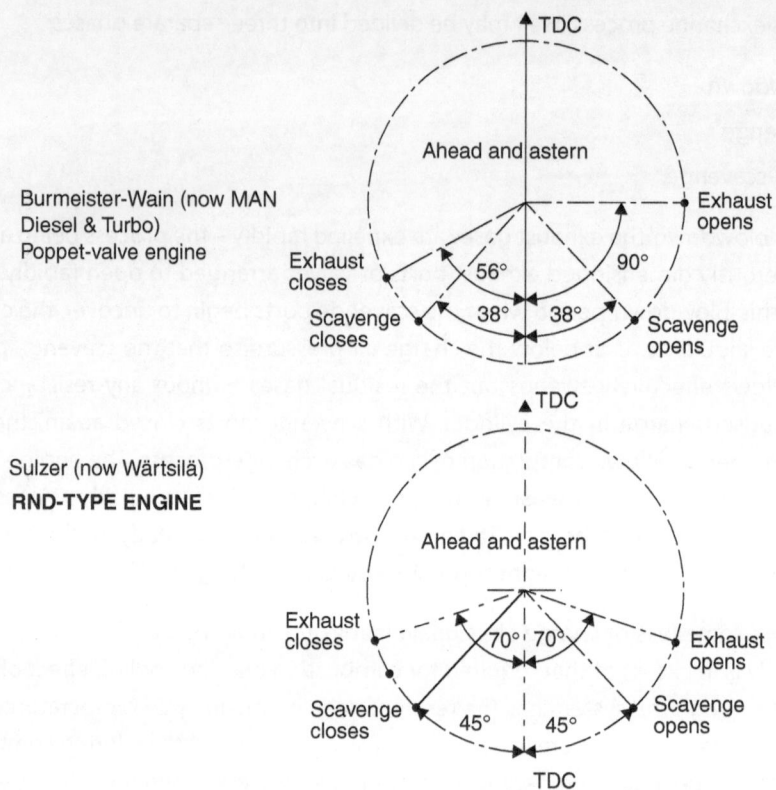

▲ **Figure 4.3** *Timing for some direct drive slow-speed diesels*

Obviously in the case of reversing engines there may be some slight penalty incurred if prolonged operation in the astern direction is considered. Figure 4.3 shows the timing for some of the present generation of direct drive slow-speed diesels.

In addition, with fully electronic 'camshaftless' engines the timing of the injection and exhaust valve can be optimised for ahead and astern operation without the need for complicated mechanical equipment.

Pressure Charging

By increasing the density of the air, and therefore the mass of oxygen present in the cylinder at the beginning of compression, a corresponding greater mass of fuel can be burned, giving a substantial increase in power developed. The degree of pressure charging required, which determines the increase in air density, is achieved by the use of free-running turbochargers, driven by the energy left in the exhaust gases expelled from the main engine. About 20% of the energy available in the exhaust gas is utilised in this way. In the past it was usual practice to employ some form of scavenge assistance either in series or in parallel with the turbochargers. This was accomplished by engine-driven reciprocating scavenge pumps, under-piston effect or independently driven auxiliary blowers. Only the under-piston effect and auxiliary blowers would be used on the engines of ships still in service today.

The turbocharger provides charge air at 70–95% of required pressure, with under-piston effect or series pump making up the balance (figure 4.4b,c). There is an increase in temperature of air delivered to the engine since air cooling is carried out after the turbocharger only. With parallel operation, air supply to the engine is increased by air delivery from pumps with proportionate increase in output, resulting in greater exhaust gas supply to the turbocharger and improved turbo-charger performance (figure 4.4d). Figure 4.4e shows two-stage turbocharging used on the latest engines.

The advantages of pressure charging may be summed up as:

- substantial increase in power for a given speed and size
- better mass:power ratio, that is, reduced engine mass for given output
- improved mechanical efficiency with reduction in specific fuel consumption
- reduction in cost per unit of power developed
- an increase in air supply has a considerable cooling effect, leading to less exacting working conditions and improved reliability.

Due to increasing power output and fuel economy, the diesel plant is now almost universally chosen for applications that were once dominated by steam turbine plant.

Therefore, research and development in the maritime industry has now become firmly focused on improving main propulsion efficiency. Owners are asking for better fuel consumption and greater flexibility from their engines due to the diverse nature of their business and changing circumstances of their operation. Furthermore, legislation is driving the quest for a reduction in emissions from marine engines and the importance of the turbocharger in this quest is significant.

The turbocharger manufacturers have invested in new research using the latest computation fluid dynamics (CFD) techniques and as the knowledge base and advances in material science move forward, so does the efficiency of the new generation of

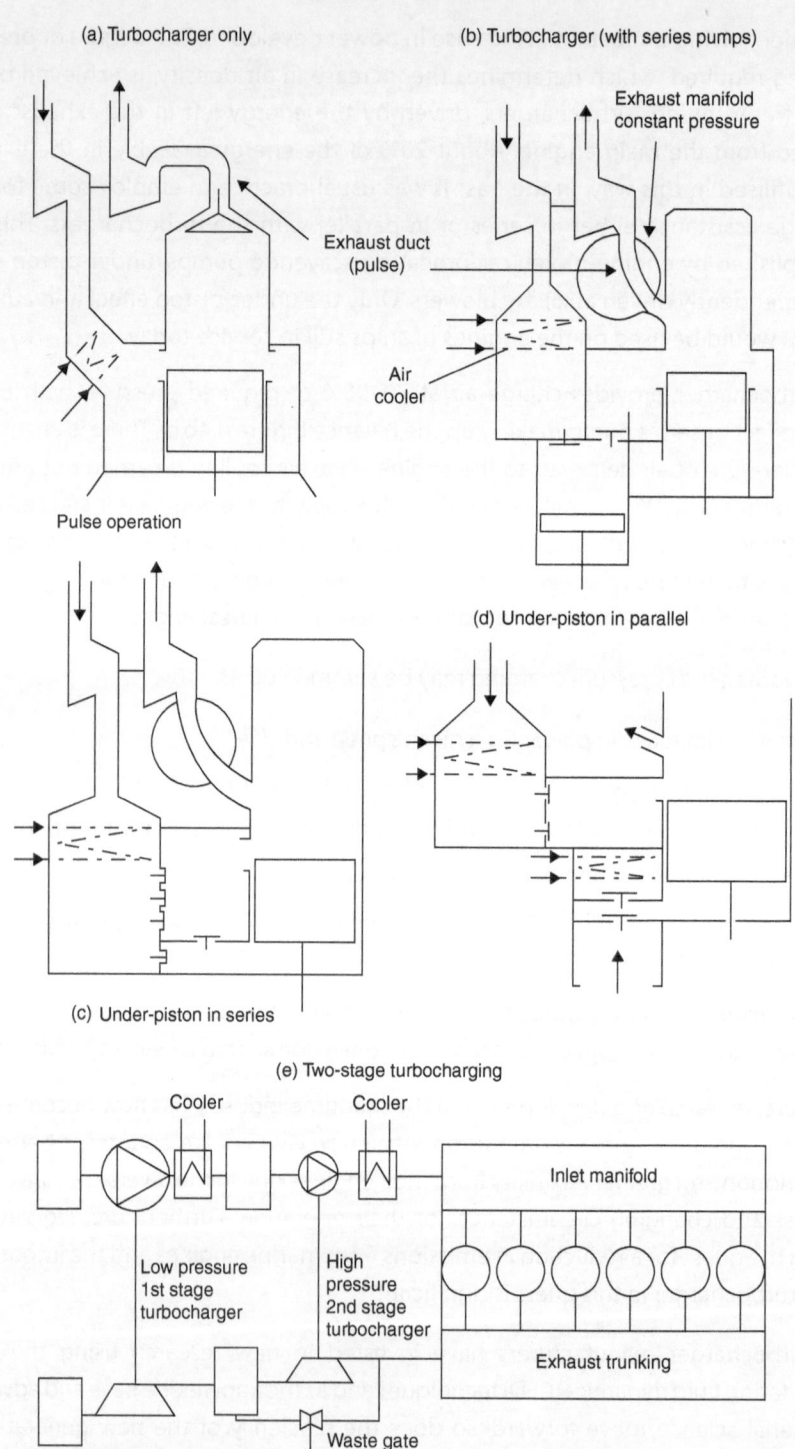

(a) Turbocharger only

(b) Turbocharger (with series pumps)

Exhaust manifold constant pressure

Exhaust duct (pulse)

Air cooler

Pulse operation

(d) Under-piston in parallel

(c) Under-piston in series

(e) Two-stage turbocharging

Cooler

Cooler

Inlet manifold

Low pressure 1st stage turbocharger

High pressure 2nd stage turbocharger

Exhaust trunking

Waste gate

▲ **Figure 4.4** *Pressure charging*

turbocharger. While further work still has to be completed in this area to meet the full requirements of IMO regulations, the industry now has other challenges as the total energy efficiency of ships starts to focus the minds of naval architects and ship and engineering designers.

However, energy efficiency does not just depend upon the efficiency of the main engine; increasing the thermal efficiency of the power plant takes into account waste heat recovery systems (WHR), see Chapter 10, which also contributes to the vessel's overall efficiency and reduction in life costs. The increasing efficiency of the turbocharger is the key in allowing the development of sophisticated WHR systems. The more efficient turbochargers use less heat energy to drive the turbocharger for the operation of the engine, therefore more heat energy is left over for use by additional WHR.

The improvements in design include the ability to control the turbine's output to match the operation of the engine and this is set to provide owners with engines that are flexible and can therefore be optimised for vessel requirements. These improvements are resulting in a high overall efficiency of up to 70%, impacting fuel consumption and firing temperatures. The latest turbocharger design features include: air-cooled operation, which will reduce the cost, complexity and installation requirements of the turbocharger; a cartridge-style construction aimed at improving on-site maintenance procedures and reducing operational downtime; advances in bearing technology, contributing to a reduction in life costs and extending service life.

Alternative compressor options are also enabling better turbocharger optimisation for specific applications. Compressors are now made from aluminium, which is lighter and therefore takes less energy to drive.

Variable vane technology is set to increase the turbo-charger's flexibility, improving the operational range and giving the engine that ability to perform efficiently over a wider operational envelope. High-pressure ratio capability of up to 5:1 can be achieved in a single stage using an aluminium compressor, without compromising design life. This enables higher power densities to be achieved and the possibility of reducing emissions.

Turbocharger speeds have increased due to the improvements in material science but 'maintenance-induced failure' has started to work its way into the system and carrying out work on these advanced machines, by non-service engineers, is no longer recommended.

Constant pressure and pulse operation

In general, the manner in which the energy contained within the exhaust gases is utilised to drive the turbocharger may be described in two ways:

1. The pulse system of operation
2. Constant pressure operation.

Pulse operation

This makes full use of the higher pressures and temperatures of the exhaust gas during the blowdown period and with rapidly opening exhaust valves or ports the gases leave the cylinder at high velocity as pressure energy is converted into kinetic energy to create a pressure wave or pulse in the exhaust leading to the turbocharger. For pulse operation it is essential that the exhaust leading from the cylinder to the turbine entry are short and direct, without unnecessary bends, so that volume is kept to a minimum. This ensures optimum use of available pulse energy and avoids the substantial losses that could otherwise occur with a corresponding reduction in turbocharger performance. Of necessity, exhaust ducting must be arranged so that the gas exchange processes of cylinders serving the same turbocharger do not interfere with each other to cause pressure disturbances that would affect purging and recharging with an adverse effect upon engine performance. With two-stroke engines the optimum arrangement is three-cylinder grouping with 120° phasing, which gives up to 10% better utilisation of available energy than cylinder groupings other than multiples of three. Due to the small volume of the exhaust ducting and direct leading of exhaust to the turbine inlet, the pulse system is highly responsive to changing engine conditions, giving good performance at all speeds. Theoretically, turbocharging on the pulse system does not require any form of scavenge assistance at low speeds or when starting. In practice, however, the use of an auxiliary blower or some other means of assistance is employed to ensure optimum conditions and good acceleration from rest.

Constant pressure operation

In this system the exhaust gases are discharged from the engine into a common manifold or receiver where the pulse energy is largely dissipated. Although the pulse energy is lost, the gas supply to the turbine is at almost constant pressure so that optimum design conditions prevail since, under normal conditions, gas flow will be steady rather than intermittent. Furthermore, as the engine ratings increase, the constant pressure energy contained in the exhaust gas becomes increasingly dominant so that sacrifice of pulse

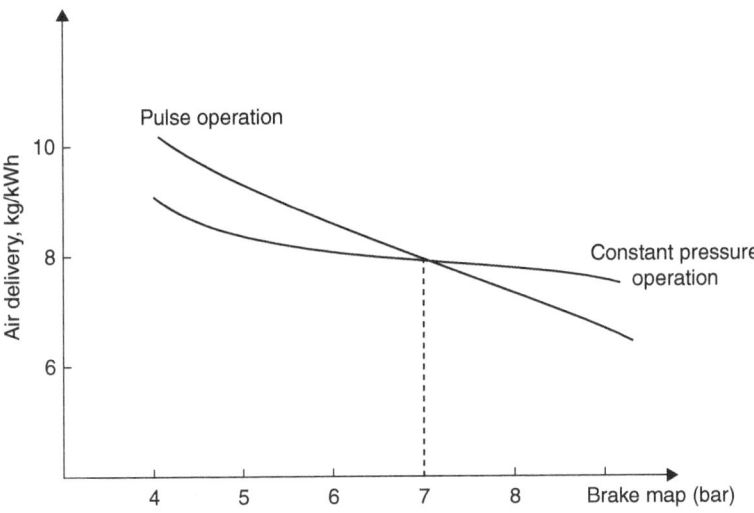

▲ **Figure 4.5** *Air delivery*

energy in a large volume receiver is of less consequence. Figure 4.5 shows the results of tests carried out on a Wärtsilä two-stroke engine, which indicate that up to a BMEP of around 7 bar, the advantage lies with the pulse system but as the BMEP increases beyond this figure the constant pressure system becomes more efficient, giving greater air throughout and some slight reduction in the fuel rate.

Due to the much larger volume of the exhaust system associated with constant pressure operation, the release of exhaust gas is rapid and earlier opening to exhaust is generally only necessary to ensure that cylinder pressure has fallen to or below the charge air pressure when the scavenge ports begin to uncover. With a possible reduction in exhaust lead expansive working can be increased, which is a further contributory factor in reducing the fuel rate. A major drawback to constant pressure operation is that the large capacity of the exhaust system gives poor response at the turbocharger to changing engine conditions, with the energy supply at slow speeds being insufficient to maintain turbocharger performance at a level consistent with efficient engine operation. Some form of scavenge assistance such as under-piston scavenging is often utilised. To offset this, however, the number of turbochargers required as compared to pulse operation can be reduced, a greater flexibility exists in the case of turbocharger location and exhaust arrangement and no derating of engine need be considered for cylinder groupings other than multiples of three. For this reason, most large slow-speed two-stroke engines tend to be of the constant pressure configuration.

Figure 4.6 shows the diagrammatic arrangement of the Wärtsilä RTA scavenge engine, which operates with constant pressure supercharge. In normal operation air is drawn into under-piston space B from common receiver A and compressed on the downstroke of the piston to be delivered into space C so that when scavenge ports uncover, purging is initiated with a strong pressure pulse. As soon as pressure in spaces B/C falls to common receiver pressure in space A, scavenge continues at normal charge air pressure. For part-load operation the auxiliary fan is arranged to cut in when charging pressure falls below a preset value. Air is drawn from space A and delivered into space F and this, together with under-piston effect, ensures good combustion and trouble-free operation under transient conditions. See figure 4.7 for the MAN arrangement.

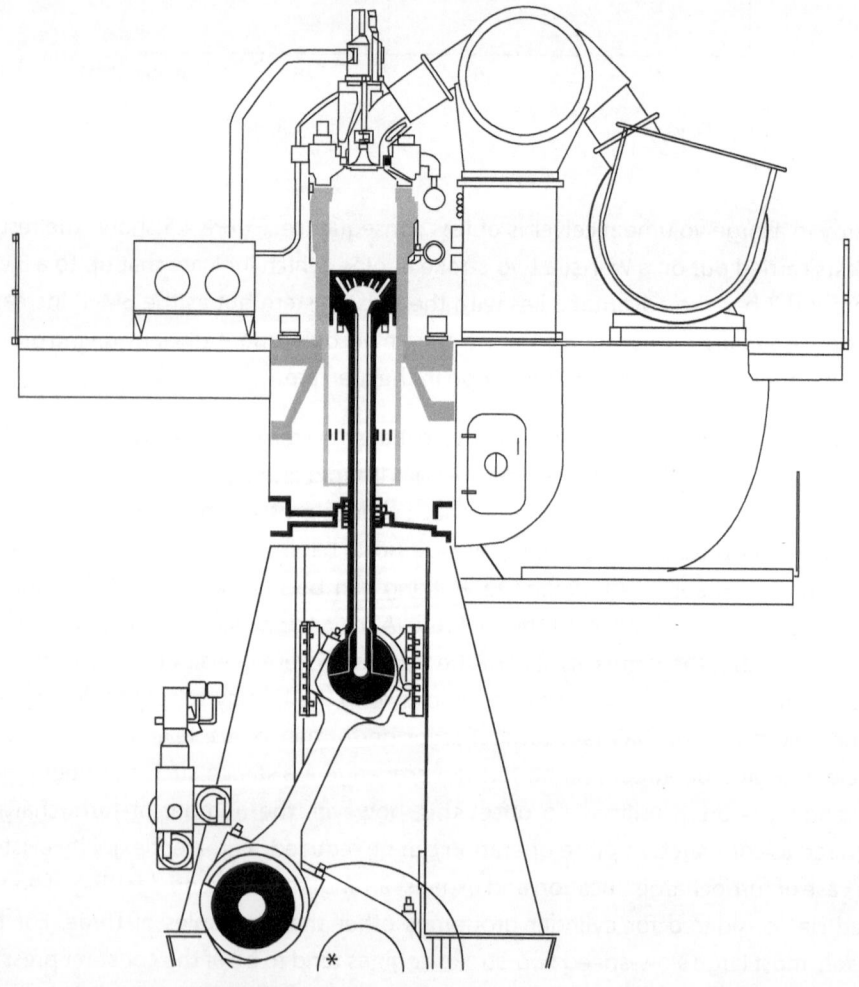

▲ **Figure 4.6** *Wärtsilä RTA scavenge arrangement*

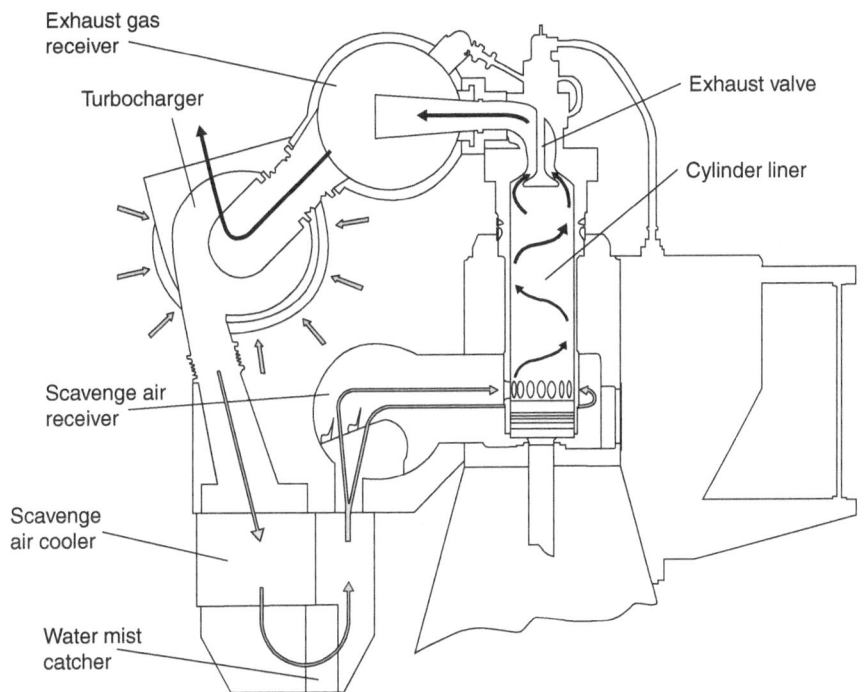

Exhaust gas
receiver

Turbocharger

Exhaust valve

Cylinder liner

Scavenge air
receiver

Scavenge
air cooler

Water mist
catcher

▲ **Figure 4.7** *MAN Diesel 'M' series scavenge arrangement*

Air Cooling

During compression of the air at the turbo-blower, which is fundamentally adiabatic, the temperature may increase by about 60–70°C with a corresponding reduction in density. This means that the air must be passed through a cooler on its passage to the engine in order to reduce its temperature and restore the density of the charge air to optimum conditions. Correct functioning of the cooler is therefore extremely important in relation to efficient engine operation. Any fouling that occurs will reduce heat transfer from air to cooling medium and it is estimated that the 1°C rise in temperature of air delivered to the engine will increase exhaust temperature by 2°C. Reduction in air pressure at the cooler outlet due to increased resistance is also a direct result of fouling. It is therefore imperative that air coolers are kept in a clean condition. It is preferable that this is accomplished on a regular basis rather than changing a dirty cooler since progressive fouling will have an adverse effect on engine performance. Regular cleaning should be included into the ship's routines and can be carried out by spraying with a commercial air cooler cleaning solvent. Under conditions of high humidity, precipitation at the cooler may be copious. Carry-over of this water to the engine can have a number of detrimental effects. Water contamination of cylinder lubricating oil may reduce its viscosity and hence its ability to withstand the imposed loads, leading to increased cylinder and piston ring wear. Water contamination may also lead to corrosion of engine components. To prevent the carry-over of water, a water separator is fitted. Figure 4.8 shows a water separator fitted on the outlet side of an air cooler. This separator utilises the difference in the mass of water and air. As the moist air flows into the vanes, its direction is changed. Due to its lower mass the air is able to change direction easily to flow around the vanes. The water, however, because of its greater mass and therefore momentum, is not able to change direction so easily and flows into the water trap to be removed at the drain. The water separator should also be sprayed with cleaning solvent when cleaning the air cooler. It must be noted that the vapour given off by cleaning solvents is harmful and by spraying into air coolers may contaminate the atmosphere throughout the engine. The air coolers should not be cleaned when personnel are working within the engine.

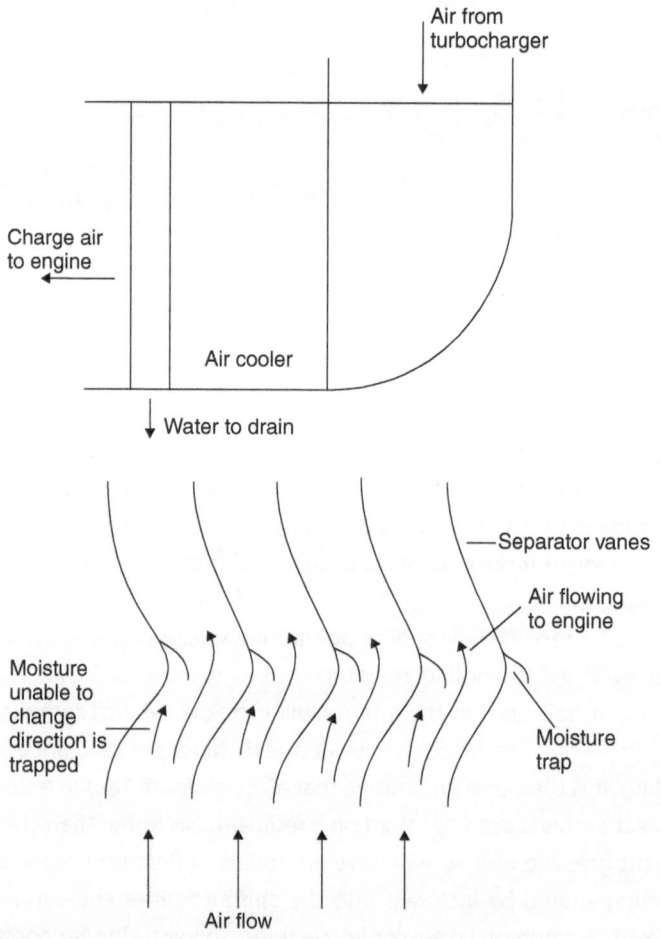

▲ **Figure 4.8** *Water separator*

Some engine manufacturers are introducing water injection into the combustion process. This is different from the water contamination described above because it is a carefully designed system that has been developed following an extensive research and development programme where the effect on all the engine components and fluids will have been considered and any adverse effect will have been removed as part of the engine's design. The reason for the water injection is to reduce the peak temperatures of combustion, thus reducing the harmful NOx.

Another very important point that needs to be understood by the students presenting themselves for examination is the reason for not overcooling the charge air. The temperatures of the inlet air, combustion and exhaust have all been calculated carefully by the engine's designers. This is not only done so that the correct density of air can be achieved but also so that the gases do not fall to their 'dew point' where water will be formed from any steam in the system. The water of course will combine with any oxide of sulphur to form sulphuric acid, which in turn will damage the engine or other components.

Turbochargers

The majority of marine turbochargers are still single-stage axial flow turbine wheel driving a single-stage centrifugal air compressor via a common rotor shaft to form a self-contained free-running unit. Expansion of the exhaust gas through the nozzles results in a high-velocity gas stream entering the moving blade assembly. Due to the high rotational speeds, perfect dynamic balance is essential if troublesome vibrations are to be avoided.

However, despite the high level of balancing, the effect of external vibrations being transmitted via the ship's structure to the turbochargers is a further problem to be resolved. This is achieved by mounting the bearings in resilient housings incorporating laminar spring assemblies to give both axial and radial damping effects. Another aspect of this arrangement is to prevent flutter or chatter at bearing surfaces when they are stopped so that incidental bearing damage is prevented. Lubrication of the bearings may be by separate or integral oil feed, but whatever arrangement is adopted it must be fully effective at a steady axial tilt of up to 15° and support a temporary tilt of 22½° as may occur in a heavy sea. The bearings themselves may be a combination of ball and roller bearings or separate sleeve (journal)-type bearings.

The various claims of superiority as to the effectiveness of the different types of bearings centre around the mechanical efficiency of the bearing configuration. The manufacturers of turbochargers equipped with rolling element bearings claim a distinct mechanical efficiency advantage across the whole operating range. On the other hand, manufacturers of turbochargers equipped with sleeve-type bearings claim comparable efficiency under full-load conditions but admit to lower efficiency at lower engine loads. With high speeds of operation, the mechanical efficiency factor does seem to favour rolling element bearings. Against this, however, is the fact that periodic replacement of ball and roller assemblies is essential if trouble-free service is to be maintained – this is due to the fact that rapid and repeated deformation with resultant stressing causes surface metal fatigue of contact surfaces with the result that failure will occur. The effects of vibration, overloading, corrosion or possible abrasive wear lead to premature failure, which emphasises the need for isolation of bearings from external vibrations together with use of the correct grade of lubricant and effective filtration. Plain bearings should, however, have a life equal to that of the blower provided that normal operating conditions are not exceeded. Ball bearings can end up with tiny indentations in the rolling surface caused by vibrations from the vessel when the turbocharger is at rest for longer lengths of time. There has also been a trend towards 'inboard' plane bearings, which enable the rotor to be supported without any bending as is the case when the bearings are at either end of the rotor.

Referring to figure 4.9, it can be seen that the blower end of the turbocharger consists of a volute casing of light aluminium alloy construction; this houses the inducer,

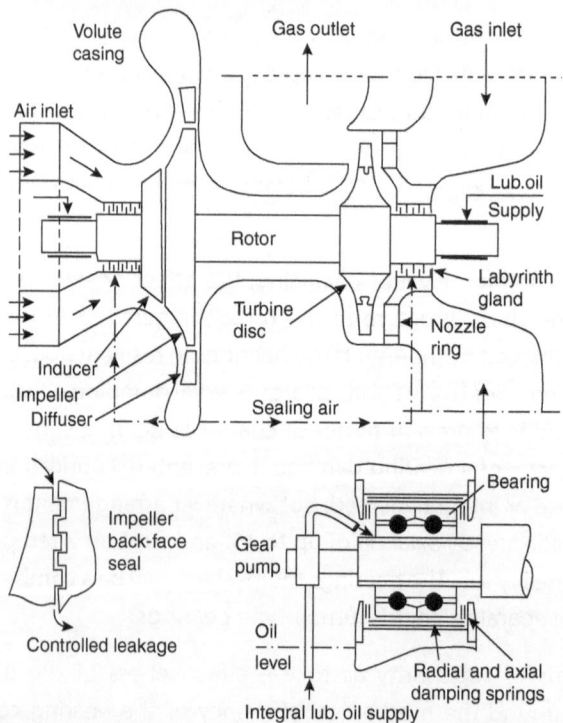

▲ **Figure 4.9** *Turbocharger*

impeller and diffuser, which are also of light alloy construction. The function of the inducer is to guide the air smoothly into the eye of the impeller where it is collected and flung radially outwards at ever-increasing velocity due to the centrifugal effect at high rotational speed. At discharge from the impeller it passes to the diffuser where its velocity is reduced in the divergent passages, thus converting its kinetic energy into pressure energy. The diffuser also functions to direct air smoothly into the volute casing, which continues the deceleration process with further increase in air pressure. From here the air passes to the charge air receiver via the air cooler.

The turbine end of the turbocharger consists of casings that house the nozzle-ring turbine wheel and blading, etc. In older designs casings were water cooled but in turbochargers for modern large slow-speed two-stroke engines, with relatively low exhaust gas temperatures, the casings are uncooled. Uncooled designs retain more heat energy in the exhaust gas in the waste heat boiler, thus improving the overall plant efficiency. Figure 4.10 shows the temperature advantage of uncooled designs.

The components in the high-temperature gas stream, that is, the nozzle ring, turbine wheel, blades and rotor shaft, are manufactured from heat-resisting nickel-chrome alloy steel to withstand continuous operation at temperatures in excess of 450°C.

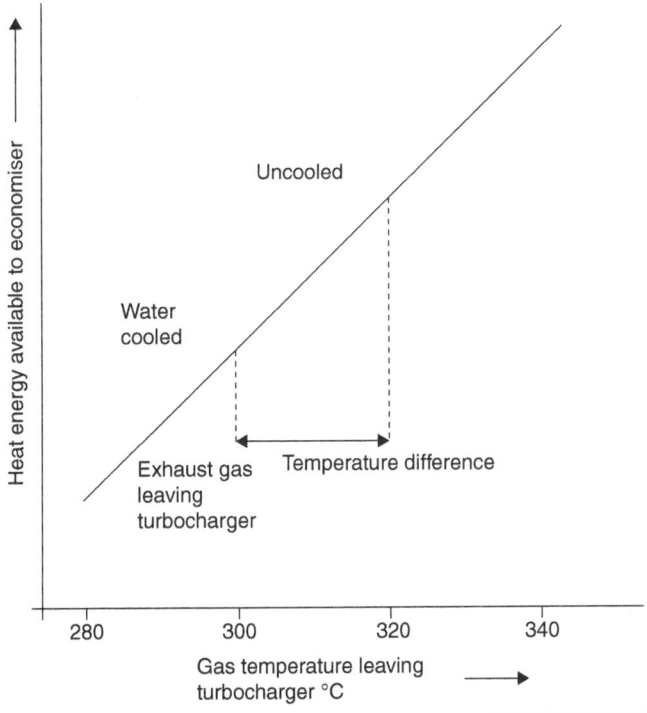

▲ **Figure 4.10** *Advantages of uncooled turbocharger*

Some degree of cooling may be given by controlled air leak-off past the labyrinth seal, between the back of the impeller and volute casing, which flows along the shaft towards the turbine end.

The cooling medium for cooled exhaust gas casings is generally from the engine jacket water-cooling system, although in some cases sea water has been employed. In both cases anti-corrosion plugs are fitted to prevent or inhibit corrosion on the water side. With water-cooled casings, experience has shown that under light load conditions when low exhaust temperatures are encountered it is possible that precipitation of corrosion-forming products – mainly sulphuric – will occur on the gas side of the casing. This results in serious corrosive attack, which is more marked at the outlet casing because of lower temperatures. Methods of prevention such as enamelling and plastic coatings, etc have been tried to alleviate this problem with varying degrees of success. A particularly effective approach to the problem is the use of air as the cooling medium with the result that this particular instance of corrosive attack is virtually eliminated.

Some manufacturers utilising sleeve-type bearings mount them inboard of the compressor and turbine. This has several advantages:

1. A short, rigid shaft is possible.
2. It allows large-volume turbine and compressor inlet casings, free of bearing housings.
3. The main casing, bearing housings and turbo-machinery form one module, allowing the rotor to be withdrawn from the turbine casing without disconnecting engine ductwork.

The oil for the bearings is supplied from the main engine lubricating oil system or a separate oil feed, as shown in figure 4.11. The oil level in the high-level tank should be maintained about 6 m above the turbochargers. This will ensure that the oil pressure reaching the bearings should never fall below a pressure of around 1.6 bar. If the level of oil falls below the mouth of the inner drain pipe, an alarm condition is initiated. After an alarm it takes about 10 min to empty the high-level tank, which is sufficient to ensure adequate lubrication of the turbochargers as they run down after the engine is stopped.

As discussed earlier, sleeve-type bearings suffer the disadvantage of having a lower mechanical efficiency at part-load conditions. The effects of this can be minimised by careful design. To reduce friction, the bearing length is reduced. A thrust bearing is incorporated into the main bearing but axial thrust is taken by this only at start-up, shutdown and very low loads, the main thrust being taken by sealing air acting on the turbine disc. Figure 4.12 shows sealing air from the compressor outlet being

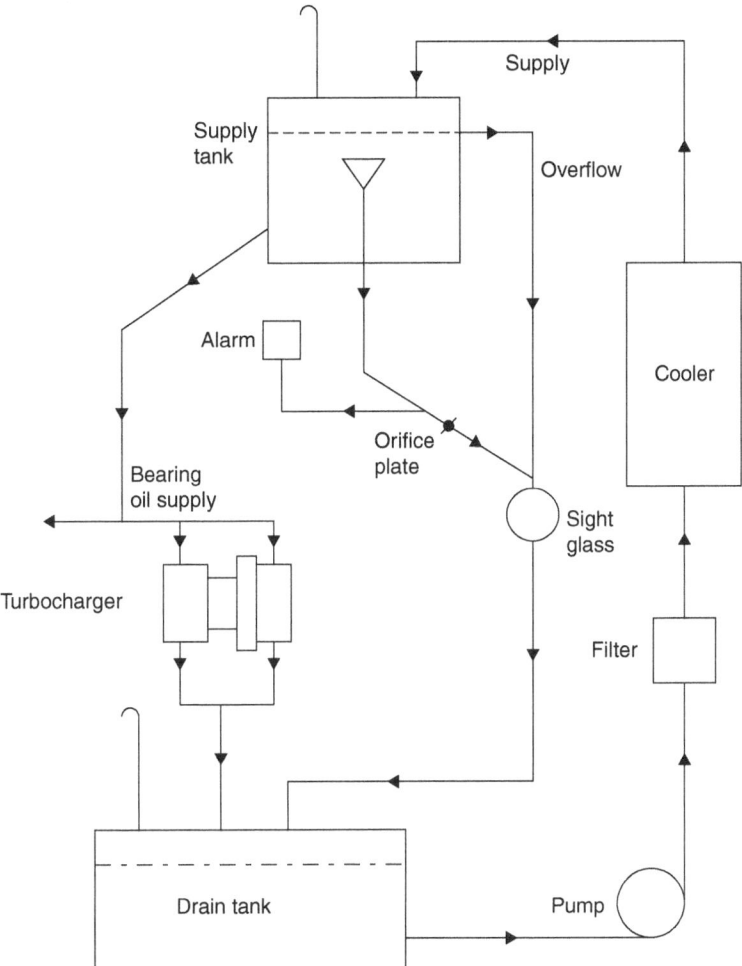

▲ **Figure 4.11** *Turbocharger lubrication system*

fed to the chamber behind the turbine disc. This air flows past the leak-off labyrinth at a rate dependent upon the clearance. As the turbocharger load increases, so does the axial thrust. This has the effect of moving the rotating element towards the compressor end, which causes the clearance at the leak-off labyrinth to decrease, reducing the flow of air. The air pressure acting on the turbine disc increases and imposes an opposing force to the axial force. The makers of turbochargers claim that engines utilising this type of turbocharger can run down to 25% load unassisted by auxiliary fans.

Recent developments have increased the overall efficiency of turbochargers by improving the aerodynamic performance and increases in pressure ratio. One

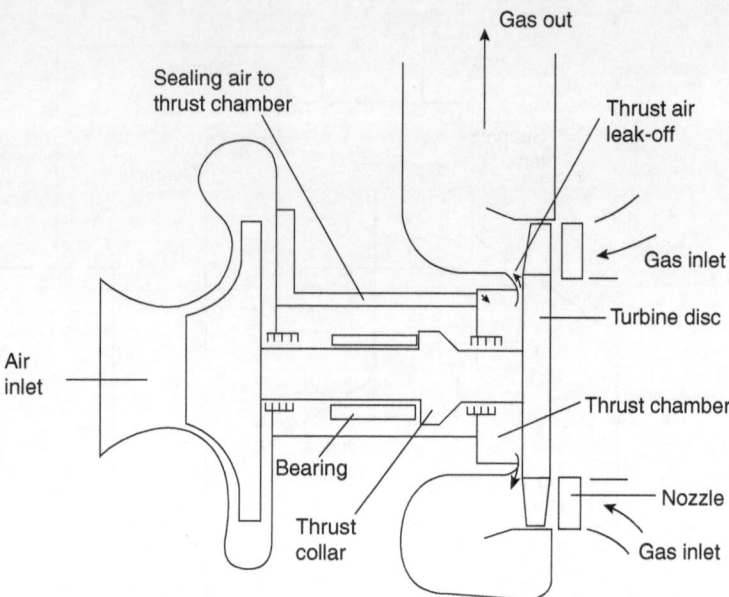

▲ **Figure 4.12** *Turbocharger with plain bearings*

improvement attained is as a result of the general adoption of constant pressure charging for large slow-speed two-stroke engines. This eliminates the excitation of blade vibration by exhaust gas pulses. Excitation of blade vibration is still possible but with careful attention to the choice of nozzle vane number and natural frequencies of vibration of blades it is possible to dispense with the need for rotor blade damping wire. Not only does this give greater turbine aerodynamic efficiency, but greater resistance to contamination by heavy fuel combustion products.

Radial flow turbines

For smaller higher-speed diesel applications (690–6,700 kW range), the use of radial flow turbochargers is common (figure 4.13). The casings are uncooled but require insulation. Bearings are of the sleeve type and are lubricated from the engine lubricating oil system. The turbine wheel is a one-piece casting of a design that gives acceptable efficiencies over the entire operation range. The compressor is also of a one-piece design of backswept vane giving stable operating characteristics. At high airflows the efficiency tends to decrease due to losses at the turbine exit. A comparison between the efficiencies of axial and radial turbines can be seen in figure 4.14.

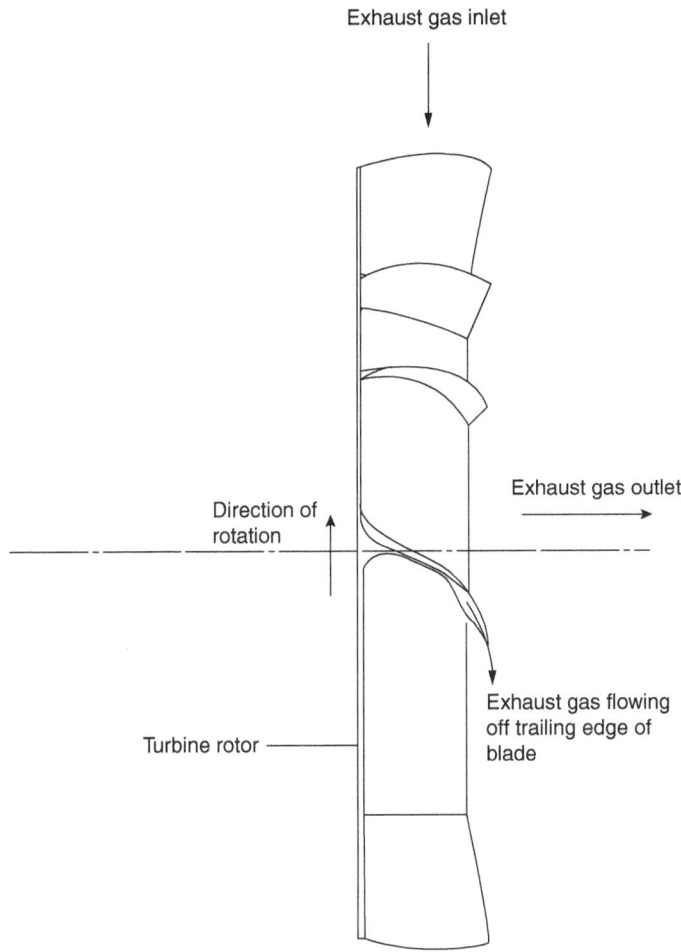

▲ **Figure 4.13** *Radial flow turbine*

MAN Diesel & Turbo state the following characteristics for their NR series of radial flow turbines:

- For engine outputs from 450 to 5,400 kW per turbocharger
- Maximum pressure ratio 4.5
- Maximum permissible temperature 650–720°C
- Suitable for heavy fuel, diesel or biofuel and gas operation
- Radial flow turbine
- Uncooled casings
- Inboard plain bearings

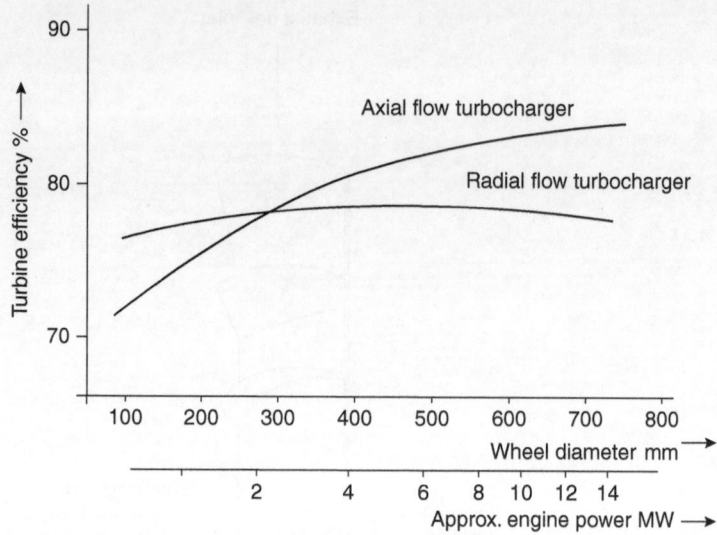

▲ **Figure 4.14** *Comparison of radial and axial flow turbocharger efficiency*

- Lubricated by the engine's lube oil system
- Easy maintenance.

The turbocharger manufacturer ABB (whose name was devised from the joining of ASEA Electric motors and Brown Boveri turbochargers) developed the Axial Turbine for the TPL series of turbochargers primarily to increase the flow capacity of the gases through the turbine.

Variable turbine blading

Turbochargers are now so efficient that more than sufficient air is produced than can be used by the engine. However, a turbocharger also needs to supply sufficient air as soon as the engine starts to operate. This means that even at moderate speeds/loads too much air is supplied to the engine. Early designs incorporated a system where the excess air pressure was bled off via a boost pressure relief valve, which was commonly known as a 'wastegate'. These systems were effective but did not make the best use of the turbocharger as energy was also wasted along with the excess pressure. The solution was to design a turbocharger with variable turbine blading, which is now a common feature in the design of some turbochargers (T/C).

Incorporating this feature into the T/C enabled the turbocharger to have a variable output maximizing the output at low loads and not producing too much boost at the higher engine loads. This maximises efficient operation with minimal energy requirements over a large operating range.

As a consequence the turbocharger does not require all the exhaust gas to be able to supply enough air to the engine and therefore unused exhaust gas can be utilised in a waste heat recovery system (see chapter 10 for more details about these systems).

The mechanism incorporates a method of altering the angle of the blading and thus changing the way that the gases move through the turbine.

Electrically powered turbochargers

The significant issue with a conventional turbocharger is that it is not operating when the engine is stationary as it needs the exhaust gases from combustion to operate. Therefore, at the starting phase of diesel engine operation, the combustion is deficient of oxygen (air) to support the increasing speed of the engine. Once the turbocharger has started working then excess air is supplied for combustion and the engine settles down to more efficient running.

This means that during the starting phase where incomplete combustion happens a darker, sootier and more polluting exhaust gas is produced. The problem is most acute for the large two-stroke engines as they are required to start from stop on a frequent basis. In addition, the manoeuvring activity takes place in or close to a port or land facility when inefficient combustion is least desirable and even Variable Vane technology in the turbocharges is not sufficient to reduce the problem enough.

To reduce this effect an electrically powered air blower is used to supply the quantity of air required for delivering the correct level of oxygen for efficient combustion during the start-up procedure.

Two-stage turbocharging (T/C)

This is a term used to describe two turbochargers, of different sizes, placed one after another in sequence. The reason for this is to raise the scavenge or inlet air pressure to a value above that achievable by a single turbocharger.

The exhaust leaving the engine first passes through the turbine driving the smaller and higher-pressure of the two turbochargers. The exhaust from this T/C is directed to the second (larger) low-pressure T/C. Following this, the exhaust is allowed to flow to the funnel via the exhaust gas boiler (economiser). The low-pressure turbine draws in air

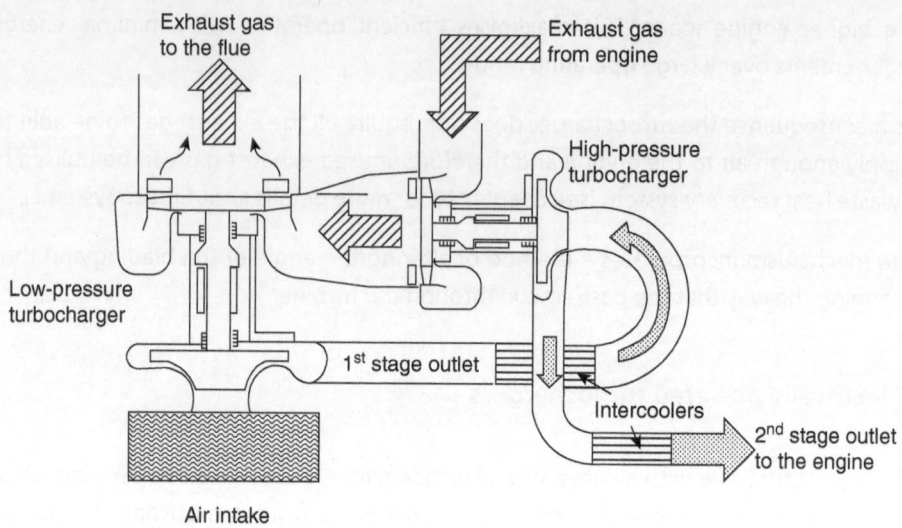

Exhaust gas
to the flue

Exhaust gas
from engine

High-pressure
turbocharger

Low-pressure
turbocharger

1st stage outlet

Intercoolers

2nd stage outlet
to the engine

Air intake

▲ **Figure 4.15a** *Two-stage turbocharging arrangement*

from the atmosphere and provides the first rise in pressure. The air then passes to the second, high-pressure T/C where the air is provided with a second boost in pressure.

This system is being utilised in engines designed to meet the IMO Tier III standard, where it is linked to an engine running the Miller inlet timing (see page 13) and fitted with exhaust gas re-circulation (EGR). This allows high efficiencies to be achieved, leading to lower specific fuel consumption. The main benefit, however, is that the increased power is achieved with lower peak temperatures and thus significantly lower NOx emissions are produced.

Turbocharger Fouling and Cleaning

Turbocharger fouling

Turbochargers that have contaminated turbines and compressors will have less efficiency and lower performance than their design specification, which results in higher exhaust temperatures. In four-stroke applications the charging pressure can increase due to the constriction of the flow area through the turbine, resulting in unacceptable high-ignition pressures. To maintain turbocharger efficiency, it is important to ensure that all the other engine operating parameters are kept within the manufacturer's recommendations. If the compressor draws its air from the machinery space, steps must be taken to maintain as clean an atmosphere as possible since leaking exhaust gas and/or oil vapour will accelerate the deterioration of efficiency. In some installations the turbochargers draw air through ducts from outside the engine room.

Water washing – blower side

On the air side of the turbocharger, dry or oily dust mixed with soot and possibly salt from a salt-laden marine atmosphere can lead to deposits, which are relatively easy to remove with a water jet, usually injected at full load with the engine warm. A fixed quantity of liquid (1–2½ l, depending upon blower size) is injected for a period of from 4 to 10 s, after which an improvement should be noted. If unsuccessful the treatment can be repeated but a minimum of 10 min should be allowed between wash procedures. Since a layer of a few tenths of a millimetre on impeller and diffuser surfaces can seriously affect blower efficiency, the importance of regular water washing becomes obvious. It is essential that the water used for wash purposes comes from a container of fixed capacity – under no circumstances should a connection be made to the fresh water system because of the possibility of uncontrolled amounts of water passing through to the engine. A cupful of water could be mixed with a general cleaning agent and carefully poured into the air filter as the turbocharger is running.

Water washing – turbine side

This must be carried out at reduced speed by rigging a portable connection to the domestic fresh water system and injecting water via a spray orifice before the protective grating at the turbine inlet for a period of 15–20 min, with drains open to discharge excessive moisture and/or deposits that do not evaporate off. Since water washing may not completely remove all deposits, and can interact with sulphur, causing a resultant corrosive attack, chemical cleaning may be used in preference. This effectively removes deposits at the turbine and moreover is still active within the exhaust gases passing to the waste heat system so that further removal of deposits occurs, which maintains heat transfer at optimum condition and keeps back-pressure of the exhaust system well within the limits required for efficient engine operation.

Dry cleaning – turbine side

Instead of water, dry solid bodies in the form of granules are used for cleaning. About 1.5–2 kg of granules are blown by compressed air into the exhaust gas lines before the gas inlet casing or protection grid. Agents particularly suited to blasting are natural kernel granules, or broken or artificially shaped activated carbon particles with a grain size of 1.2–2.0 mm.

The blasting agents have a mechanical cleaning effect, but it is not possible to remove fairly thick deposits with the comparatively small quantity used. Therefore, this method must be adopted more frequently than for cleaning with water. Dry cleaning is carried out every 24–50 h. The main advantage of this type of turbine cleaning is that it can be carried out at full or only slightly reduced load. The cleaning equipment configuration is shown in figure 4.15b.

Turbocharger manufacturers recommend that heavily contaminated machines that have not been cleaned regularly from the very beginning or after overhaul should not be cleaned by water washing or granulate injection. This is because the dangers of incomplete removal of deposits may cause rotor imbalance. These turbochargers need careful dismantling, and cleaning with the machine apart is recommended.

Surging

Surging is a phenomenon that affects centrifugal compressors when the mass flow rate of air falls below a sustainable level for a given pressure ratio. Consider the system in figure 4.16 where a constant speed compressor supplies air through a duct. The outlet of the duct is regulated by a damper. With the damper fully open the pressure ratio across the compressor will be at its lowest value with the largest mass flow rate of air. As the damper is closed the resistance increases, as does the pressure ratio, but the mass flow of air decreases. If the damper is closed further, a point will be reached where, because of the resistance, there will be such a low mass flow rate and high pressure ratio across the compressor that flow breaks down altogether. When this occurs, the pressure downstream of the compressor is simply relieved to atmosphere, backwards, through the compressor. This is known as surging and is accompanied by loud sounds of 'howling and banging'. The events leading to the surging can be followed on a graph of pressure ratio against mass flow. This graph is known as a compressor map (figure 4.17b).

Surging may occur in heavy weather when the propeller comes out of the water and the governor shuts the fuel off almost instantaneously. To obtain efficient and stable operation of the charging system, it is essential that the combined characteristic of the engine and blower are carefully matched. The engine operating line, as indicated in figure 4.17a, is mainly a function of these characteristics and taking into account the fact that blower efficiency decreases as the distance between surge and operating lines increases, the matching of blower to engine becomes a compromise between acceptable blower efficiency and a reasonable safety margin against surge. An accepted practice is to provide a safety margin of around 15–20% to allow for deterioration of service conditions such as fouling and contamination of turbochargers and increasing

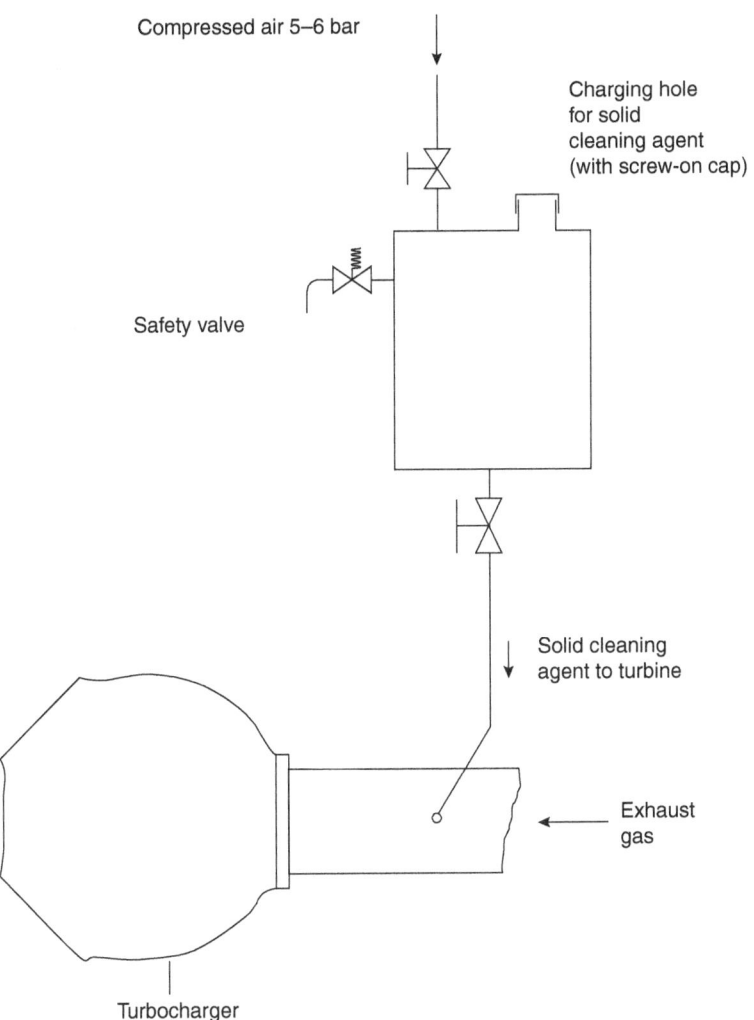

Compressed air 5–6 bar

Charging hole
for solid
cleaning agent
(with screw-on cap)

Safety valve

Solid cleaning
agent to turbine

Exhaust
gas

Turbocharger

▲ **Figure 4.15b** *Dry turbocharger cleaning equipment*

resistance of ship's hull, etc. Apart from fouling of the turbocharger, other contributory factors to surging are contamination of exhaust and scavenge ducting, ports and filters. Since faulty fuel injection leads to poor combustion and greater release of contaminants, the need to maintain fuel injection equipment at optimum conditions is essential. Other related causes are variation in gas supply to turbochargers due to unbalanced output from cylinder units and mechanical damage to turbine blading, nozzles or bearings, etc.

During normal service the build-up of contaminants at the turbocharger can be attributed to deposition of airborne contaminants at the compressor which, in general are easily removed by water washing on a regular basis. At the turbine, however, more active contaminants resulting from vanadium and sodium in the fuel, together with the products of incomplete combustion, deposit at a higher rate, which increases with rising temperature. A further problem arises with the use of alkaline cylinder lubricants, with the formation of calcium sulphate deposits originating from the alkaline additives in the lubricant. Again, water washing on a regular basis is beneficial in removing and controlling deposits, but particular care needs to be taken to ensure complete drying out after the washing sequence since any remaining moisture will interact with sulphurous compounds in the exhaust gas stream, with damaging corrosive effect; however, this effect should only be slight as the turbine dries out quite quickly after the water is removed.

Turbocharger breakdown

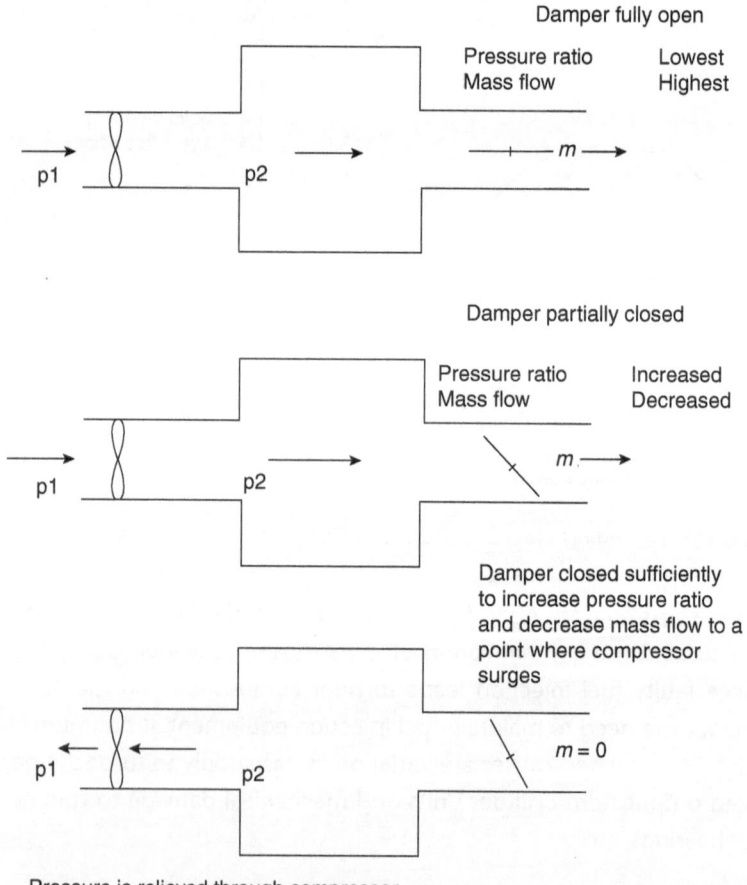

▲ **Figure 4.16** *Surging of turbocharger compressor*

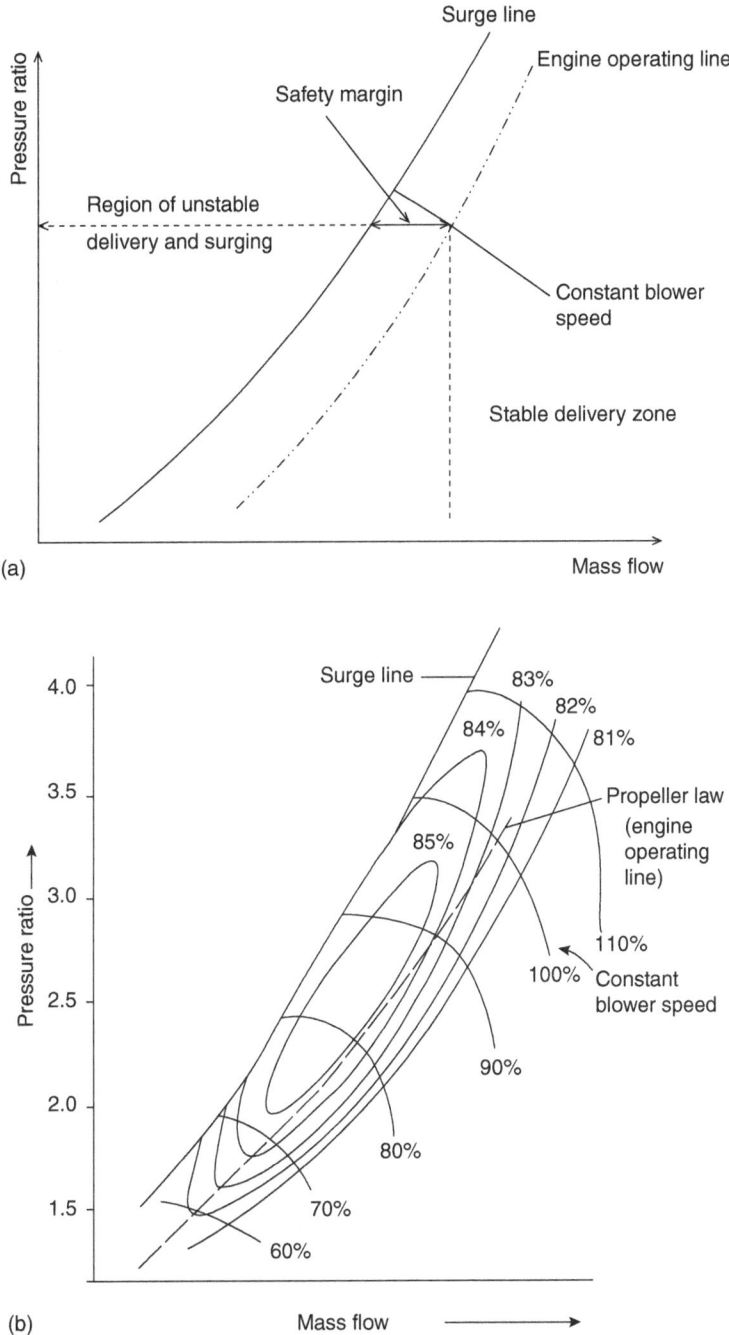

▲ **Figure 4.17** *(a) Providing a safety margin; (b) Compressor map*

Correct operating and shutdown procedures, depending upon engine type, will be found in the engine builder's and/or turbocharger manufacturer's recommended practice. As a general rule, however, in the event of damage to the turbochargers, the engine should be stopped immediately so that the damage is limited and broken fragments do not cause further damage elsewhere within the engine. Under conditions where the engine cannot be stopped without endangering the ship, engine speed must be reduced to a point where the turbocharger speed has dropped to a level where any vibration that is usually associated with a malfunction is no longer perceptible.

If the engine can be stopped but lack of time does not permit in situ repair or possible replacement of the defective charger, it is essential that the rotor of the damaged unit is locked and completely immobilised. If exhaust gas still flows through the affected unit once the engine is restarted, the coolant flow through the turbine casing needs to be maintained but due to the lack of sealing air at the shaft labyrinth glands, the lubricating oil supply to the bearings will need to be cut off – with integral pumps mounted on the rotor shaft, the act of locking the shaft ensures this – otherwise contamination of the lubricant together with increase in fouling will occur. For rotor and blade cooling, a restricted air supply is required and can be achieved by closing a damper or flap valve in the air delivery line from the charger, to a position that gives limited flow from the scavenge receiver back to the damaged blower. Alternatively, a blank flange incorporating an orifice of fixed diameter can be fitted at the outlet flange of the blower.

Where only a single turbocharger has been affected, out of a number associated with the engine, the power developed by the engine will obviously depend upon charge air pressure attainable. At the same time a careful watch must be kept upon exhaust condition and temperature to ensure efficient engine operation with good fuel combustion. In the event of all turbochargers becoming defective, it is possible to remove blank covers from the scavenge air receiver so that natural aspiration supplemented by any under-piston effect or parallel auxiliary blower operation is possible – if this method of emergency operation is carried out, protective gratings must be fitted in place of blind covers at the scavenge air receiver. In all cases when running at reduced power, special care must be taken to ensure any out-of-balance forces, due to variation in output from affected units, do not bring about any undue engine vibration.

5

STARTING AND REVERSING

Introduction

The large two-stroke marine engines are 'direct drive' propulsion systems. This means that if the engine rotates then so does the drive shaft and the propeller. This also means that if the engine needs to be turned for maintenance reasons, then the propeller must be checked to be clear *before* the operation.

The propeller of choice for this arrangement is a 'fixed pitch' propeller and therefore to drive the ship astern, the propeller must be rotated in the opposite direction from when it is driving the ship ahead. As the engine is bolted directly to the transmission shaft and the propeller, then to turn the propeller in the astern direction, the engine must also be reversed.

As students will appreciate, from the timing diagrams on pages 24–25, the injection of fuel and the opening/closing of the exhaust valve will be different when the engine is operating in the two different directions. In addition, to ensure the operation of the engine in the correct direction the starting air system will also need to be adjusted.

A further possible danger for the marine engineering officer to consider is 'hydraulic lock'. This is where an incompressible medium, such as water, excess fuel or lubricating oil, enters one or more of the cylinder spaces while the engine is stationary. If the engine is started under these conditions, then considerable damage can be imparted to the internal parts of the engine.

The technique for checking that the cylinders are clear is to first turn the engine slowly, with the indicator cocks open, using the turning gear. The engineering officer should check each indicator cock for any kind of liquids. When satisfied that these are clear the next step is to 'blow' the engine over, using a brief burst of compressed air via the air start system. Again, the indicator cocks will be open and checked to be clear as the engine is turned over on air. The indicator cocks can then be closed, and the engine can then be started.

As this is such an important subject, manufacturers have developed additional systems designed to give an additional layer of protection.

Starting Air Overlap

There must be some overlap between the operation of starting air valves to the different cylinders of an engine, so that as one cylinder valve is closing another one is opening just at the correct moment to ensure a continued rotation of the engine before the fuel is introduced. This is essential to ensure a positive angular motion of the engine crankshaft with sufficient momentum to give a positive start. The usual minimum amount of overlap provided in practice is 15°. Starting air is admitted on the working stroke and the period of opening is governed by practical considerations, with three main factors to consider:

1. *The firing interval of the engine.*

$$\text{Firing interval} = \frac{\text{Number of degrees in engine cycle}}{\text{Number of cylinders}}$$

For example, with a four-cylinder two-stroke engine the firing interval is 90°, that is, 360/4, and if each cylinder valve covered 90° of the cycle then the engine would not start if it had come to rest in the critical position with one valve fractionally off closure and another valve just about to start opening.

2. The valve must close before the exhaust commences. It is rather pointless blowing high-pressure air straight to exhaust and it could be dangerous.

3. The cylinder starting air valve should allow the air to enter the cylinder after its associated piston has passed TDC to give a positive turning moment in the correct direction. In fact, some valves are arranged to start to open as much as 10° before TDC because the engine is past this position before the valve is effectively open and the compressed air is having an effect. Any reverse turning effect is negligible as the turning moment exerted on a crank very near dead centre is small indeed.

Consider figure 5.1a for a four-stroke engine. With the timings as shown, the air starting valve opens 15° after dead centre and closes 10° before exhaust begins. The air start period is then 125°. The firing interval for a six-cylinder four-stroke engine is 720/6 = 120°. The period of overlap is 5°, which is insufficient. Although this example could easily be modified so as to give sufficient (say 15°) overlap by reducing the 15° after dead centre and the 10° before exhaust opening, it can become very difficult to arrange with very early exhaust opening on turbocharged engines. A seven-cylinder four-stroke engine is much easier to arrange.

Consider figure 5.1b, which represents a two-stroke engine. This has an air start period of 115°. Firing interval for a three-cylinder two-stroke engine = 360/3 = 120°. This means no overlap. Modification can arrange to give satisfactory starting with this example but for modern turbocharged two-stroke engines having exhaust opening as early as 75° before BDC (outer) it becomes virtually impossible. A four-cylinder two-stroke engine is much easier to arrange and would be adopted. Consider figure 5.1c, which is a cam diagram for a two-stroke engine with four cylinders. The air open period is 15° after dead centre to 130° after dead centre, that is, a period of 115°. This gives 25° of overlap (115 – 360/4), which is most satisfactory. Take care to note the direction of rotation and this is a cam diagram so that, for example, No. 1 crank is 15° after dead centre when the cam would arrange to directly or indirectly open the air start valve. The firing sequence for this engine is 1 4 3 2. This is very much related to engine balancing and no hard and fast rules can be laid down about crank firing sequences as each case must be treated on its merits.

It may be useful to note that for six-cylinder two-stroke engines a very common firing sequence is 1 5 3 6 2 4 and similarly for seven and eight cylinders 1 7 2 5 4 3 6 and 1 6 4 2 8 3 5 7 respectively are often used. The cam on No. 1 cylinder is shown for illustration as it would probably be for operating, say, cam-operated valves; obviously the other profiles could be shown for the remaining three cylinders in a similar way. The air period for cylinder numbers 1, 4, 3 and 2 are shown respectively in full, chain dotted, short dotted and long dotted lines and the overlap is shown shaded.

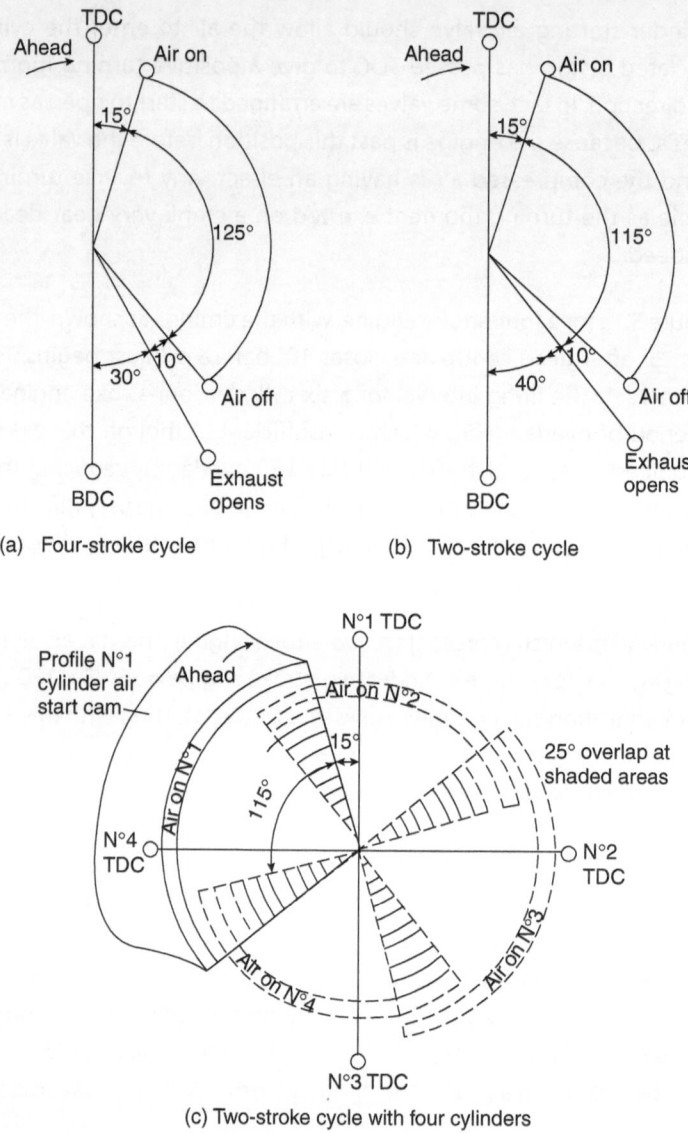

(a) Four-stroke cycle

(b) Two-stroke cycle

(c) Two-stroke cycle with four cylinders

▲ **Figure 5.1** *Air start cam and crank timing diagrams*

Starting air valves

Each engine cylinder is fitted with a starting air valve (figure 5.2), which is operated pneumatically by the air released from the operation of one of the starting air distributor control valves (figure 5.2). These are arranged radially around the starting air distributor cam as shown in figure 5.3a. At the correct engine position, compressed air from the

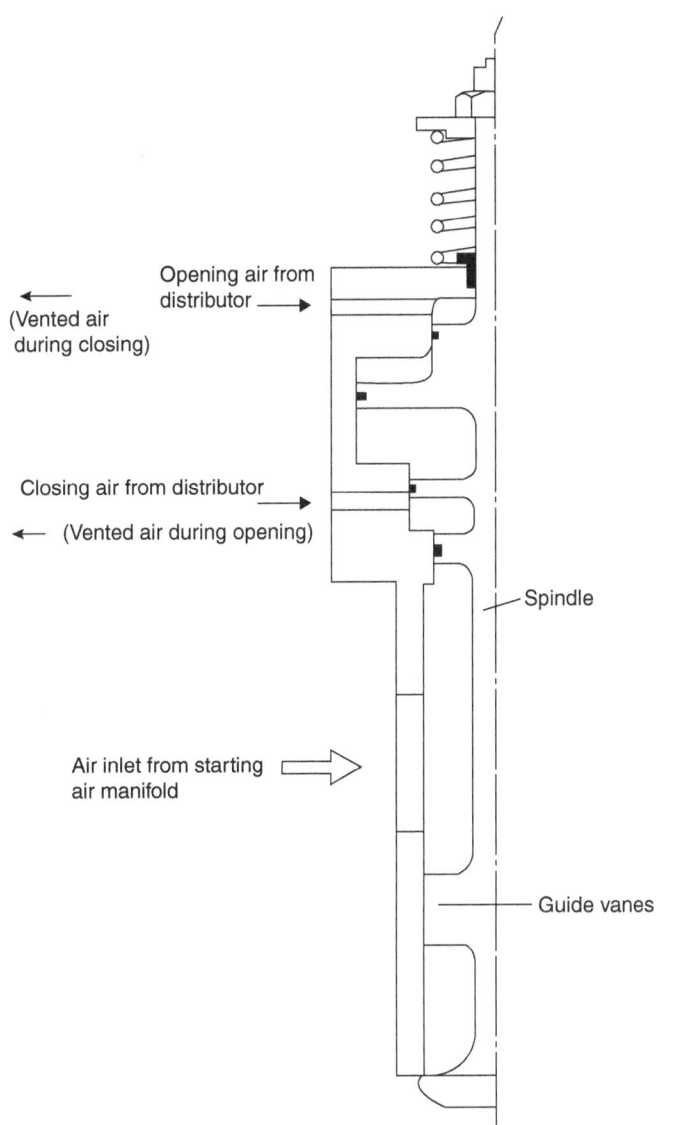

Opening air from distributor →

(Vented air during closing) ←

Closing air from distributor →

(Vented air during opening) ←

Spindle

Air inlet from starting air manifold

Guide vanes

▲ **Figure 5.2** *Starting air valve*

control air system is released by the control valve and directed to the cylinder starting air valve's upper chamber. Here it acts on the piston to open the main starting air valve. As this is happening, the air from the lower chamber is vented to atmosphere through the control valve. At the end of the starting air admission period, control air is redirected to the lower chamber to close the valve while the upper chamber is vented to atmosphere through the control valve. The valve opens and closes quickly with air

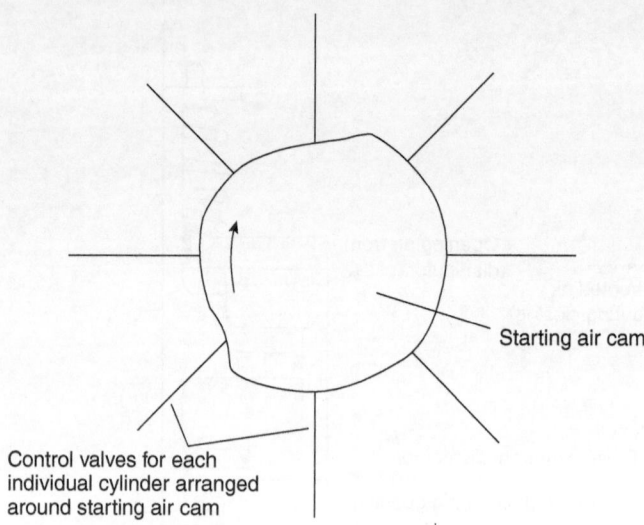

(a) Radial arrangement of starting air valve around starting air cam

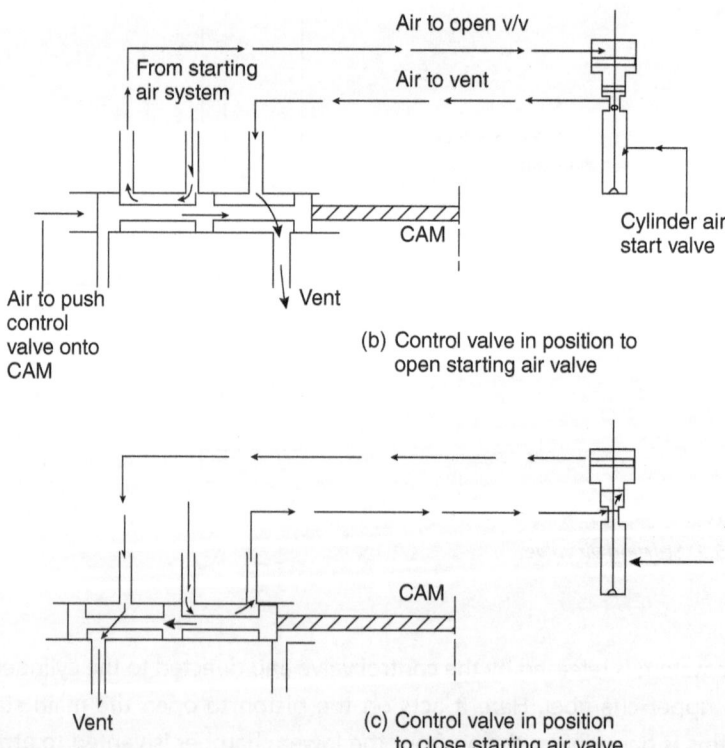

▲ **Figure 5.3** *Starting control valve arrangement for Wärtsilä (Sulzer) engine*

cushioning at the end of the closing motion to reduce shock on the valve seat. If the pressure in the cylinder is substantially higher than the starting air pressure, the valve will not open. This prevents hot gases entering the starting air manifold.

During engine operation the air inlet to the starting valve should be regularly checked. A hot inlet would indicate a leaking starting air valve allowing hot combustion gases to enter the air manifold, which may lead to an explosion if starting air is admitted.

Starting air distributor

There are many designs of air distributor all with the same basic principles, that is, to admit air to the pistons of cylinder relay valves in the correct sequence for engine starting. Valves not being supplied with air would be vented to the atmosphere via the distributor. Some overlap of timing would obviously be required.

One type of starting air distributor is shown in figure 5.3a. This is based on a design where each cylinder has its own starting control valve. The starting control valves are arranged radially around the starting air distributor cam, which is driven via a vertical shaft from the camshaft. When the engine starting lever is operated air is admitted to the distributor, forcing all control valves, against the return spring, onto the cam. The control valve of a cylinder that is in the correct position for starting will be pushed into the depression in the cam and assume the position shown in figure 5.3b. In this position, air from the starting system will be directed to the upper part of the cylinder starting air valve, causing the valve to open. At the same time air from the lower chamber of the cylinder starting air valve will be vented to atmosphere. At the end of the cylinder starting air period, the distributor cam moves the control valve to the position shown in figure 5.3c. In this position, air from the starting air system is directed to the lower chamber of the starting air valve, causing the valve to close. Air from the upper chamber is vented through the control valve to atmosphere. The starting control valves are held off the distributor cam by springs when starting air is shut off the engine.

General reversing details

Most reversible engines are direct drive two-stroke engines. The general trend in four-stroke practice is to utilise a unidirectional engine, coupled, via a reduction gearbox, to a controllable pitch propeller. The need for reversing mechanisms on the four-stroke engines is, therefore, no longer required. For this reason the two-stroke reversing mechanism principle will be considered in greatest detail.

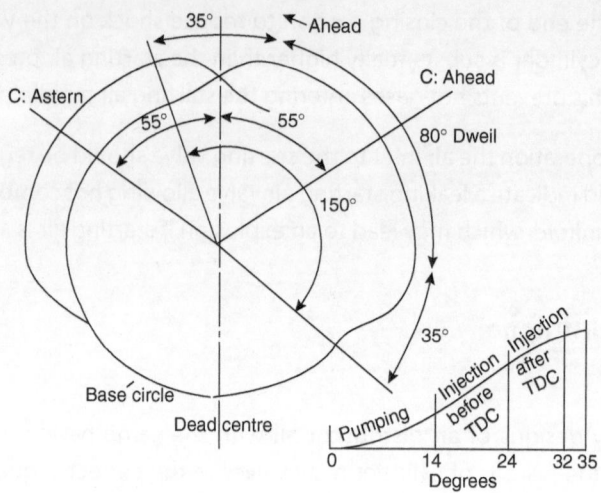

▲ **Figure 5.4** *Lost motion cam diagram*

Two-stroke reversing gear

One of the easiest ways of changing the settings on a two-stroke engine is to reposition the fuel, exhaust valve and starting air cams on the camshaft, with their associated equipment, so that the engine operating in reverse can utilise one cam. This avoids the complication of moving the camshaft axially but it also means that it is necessary to provide a 'lost motion clutch' on the camshaft and the arrangement of such a clutch is described in the next few paragraphs.

Referring to figure 5.4, consider the engine piston to be positioned at TDC in the ahead mode with the fuel cam peak centreline at 55° after this position for correct injection timing ahead and assuming anticlockwise ahead rotation. If now the engine is to run astern (clockwise), the cam is 55° + 55° = 110° out of phase. Either the cam itself must be moved by 110° or while the engine rotates 360° the cam must only rotate 250° (110° of lost motion). Note that the symmetrical cam 75° of each side of the cam peak centreline is made up of 35° rising flank and 40° of dwell.

The flank of the cam is shown on an enlarged scale in figure 5.4. It will be noted that the 35° of cam flank is utilised for building up pressure by the pumping action of the rising fuel pump plunger (14°) for delivery at injection 10° before firing dead centre to 8° after firing dead centre, and 3° surplus rise of flank for later surplus spill variation. It can be seen that the lost motion is required with cam-driven fuel pumps, in which a period of pumping is necessary before injection starts. The following points are worth specific mention:

1. Older MAN Diesel & Turbo two-stroke engines have ahead and astern cams on the same shaft. To change from one set of cams to the other, the camshaft is moved axially and so no lost motion is required. However, the latest engines are designed to move the position of the follower so that when the direction is reversed the closing cam profile becomes the opening profile and vice versa.

2. The dwell period is not normally necessary from the fuel injection aspect alone, that is, about 30° lost motion would be adequate and is provided as such on British Polar and older Sulzer engines.

3. Dwell, in which the fuel plunger is held before return, is often provided to give a delay interval. For example, with older B&W engines about 80° dwell gives a rotation (total) of the camshaft of about one-third of a revolution, which allows an axial travel with a screw nut arrangement of reasonable size and pitch to change over for reverse running.

4. Older loop scavenged Wärtsilä Sulzer engines have about 98° lost motion as the distributor repositioning for astern is from the same drive shaft as the fuel pumps, but via a vertical direct drive shaft.

Refer now to figure 5.5. The design in this figure, which is based on older Wärtsilä Sulzer engine practice, has a lost motion on the fuel pump camshaft of about 30°. When reversal is required, oil pressure and drain connections are reversed. Oil flowing laterally along the housing moves the centre section to the new position, that is, anticlockwise as shown in figure 5.5. The oil pressure is maintained on the clutch during running so that the mating clutch faces are kept firmly in contact with no chatter.

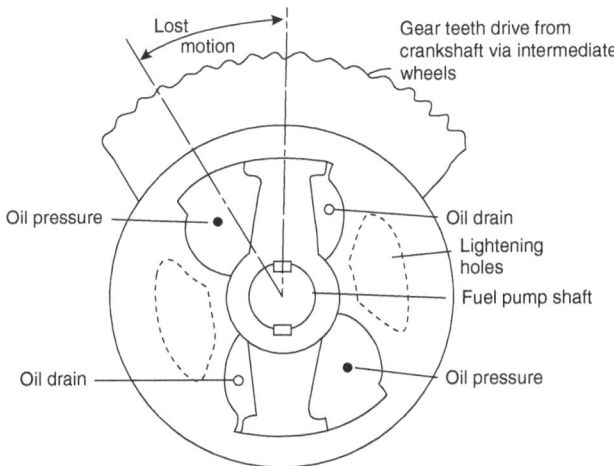

▲ **Figure 5.5** *Lost motion clutch*

There are a number of variations on this design but the principle of operation is similar, although not all types rotate the clutch to its new position before starting and merely allow the camshaft to 'catch on' with the crankshaft rotation when lost motion is completed. It is worth pausing for a moment and reflecting upon the limitations of the mechanical designs. For example, think how difficult it would be to arrange a fuel cam with a profile that gave a pre-ignition and post-ignition phase to the injection process.

Practical Systems

Having described the basic principles of starting and reversing, the actions are now combined to give a selection of systems that have been used on the various engine types.

Starting air system

Consider first the air off position. Air from the storage bottle passes to the automatic valve, which, however, remains shut as air passes through the pilot valve (1) to the top of the automatic valve piston. All cylinder valves and distributor valves are venting to atmosphere via the automatic valve. If now the lever is moved to the position shown in figure 5.6, the air pressure on top of the automatic valve is vented through the pilot valve (1) by the linkage shown. This causes the automatic valve to open as the up force on the larger piston is greater than the down force on the smaller valve with the spring force. The lower vent connection is closed and air flows to all cylinder and distributor valves.

The cylinder valves are of the air piston relay type described earlier and, despite main air pressure on them, will be closed except for one valve (or possibly two). This distributor has the piston pilot valves mounted around the circumference of a negative cam. Only one distributor pilot valve can be pushed into the negative cam slot, that is, No. 6, and hence air flows through the No. 6 distributor pilot valve only to the upper part of the piston for the No. 6 cylinder air starting valve, which will open. All other starting valves are shut and venting to atmosphere. The position shown for illustration is air onto No. 6 cylinder of a six-cylinder engine running ahead. When the lever is moved forward onto fuel, the whole system is again vented to atmosphere through the automatic valve.

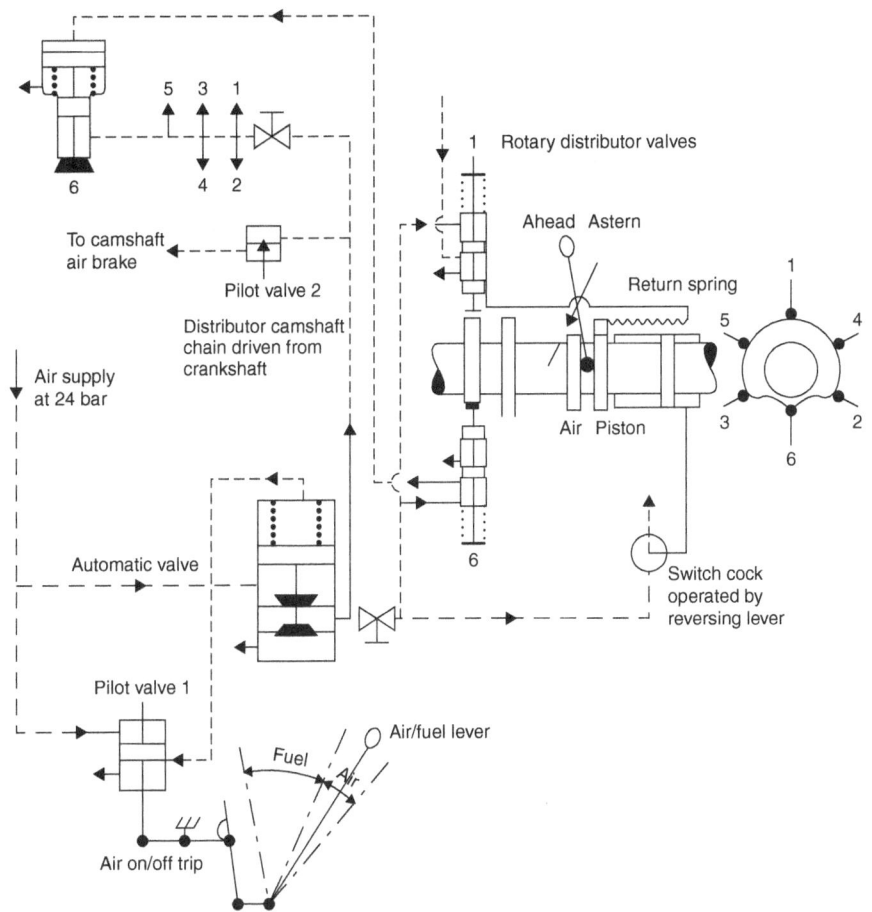

▲ Figure 5.6 *Typical starting air system*

For astern running the reversing lever is moved over, which allows air to pass, via a switch cock, to push the light distributor shaft along by means of an air piston (alternatives are scroll, direct linkage, etc), thus putting the astern distributor cam into line with the distributor pilot valves. Distributor pilot valves are kept out by springs during this operation. The air-fuel lever is then operated as previously described for the engine to run astern. Air start timing for a two-stroke engine, upon which the above system is typical, is 5° before firing dead centre to 108° after firing dead centre (122° after for astern). B&W engines also employ a revolving plug type of distributor on some engine designs. Again, some types of these engines utilise an air brake on the main camshaft so that air pistons pressure against the pilot valve (2), operated from the reversing lever, while the lost motion is being travelled by the engine. The main camshaft is therefore kept stationary and just before the lost motion is complete the air pressure is released to atmosphere, thus releasing the brake.

Consider figure 5.7. Air from the starting receiver at 30 bar maximum flows to the pre-starting valve (via the open turning gear blocking valve shown), and directly to the automatic valve. At the automatic valve, air passes through the small drilled passage to the back of the piston and this, together with the spring, keeps this valve shut as the pilot valve is shut with air pressure on top and atmospheric vent below.

If the air starting lever is operated with control interlocks free, the opening of the pre-starting valve allows air to lift the pilot valve, vent the bottom of the automatic valve and cause it to open as shown. This allows air to pass to the cylinder valve via non-return and relief valves and also to the distributor. The distributor will allow air to pass to the appropriate cylinder valve, causing it to open due to air pressure on the piston top. In this design, when the piston top of the cylinder valve is connected

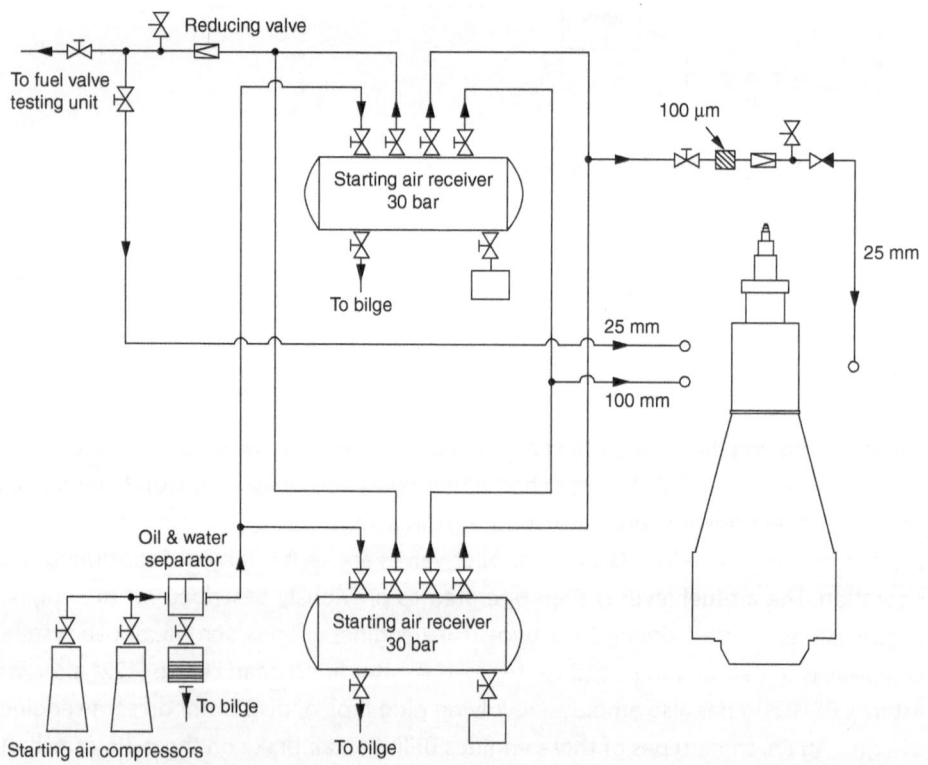

▲ **Figure 5.7** *Starting air supply system for a two-stroke marine engine*

to the atmosphere for venting, the bottom of the valve is connected to air pressure and this ensures a rapid closing action. The distributor of this engine is very similar in principle to that shown for the B&W engine previously except that a positive cam is used by Sulzer.

A mechanical interlock is provided as a blocking device from the telegraph as shown. There is also a connection to the reversing oil servo and an interlock connection from the reversing system to the air start lever via a blocking valve. These are described in figure 5.8.

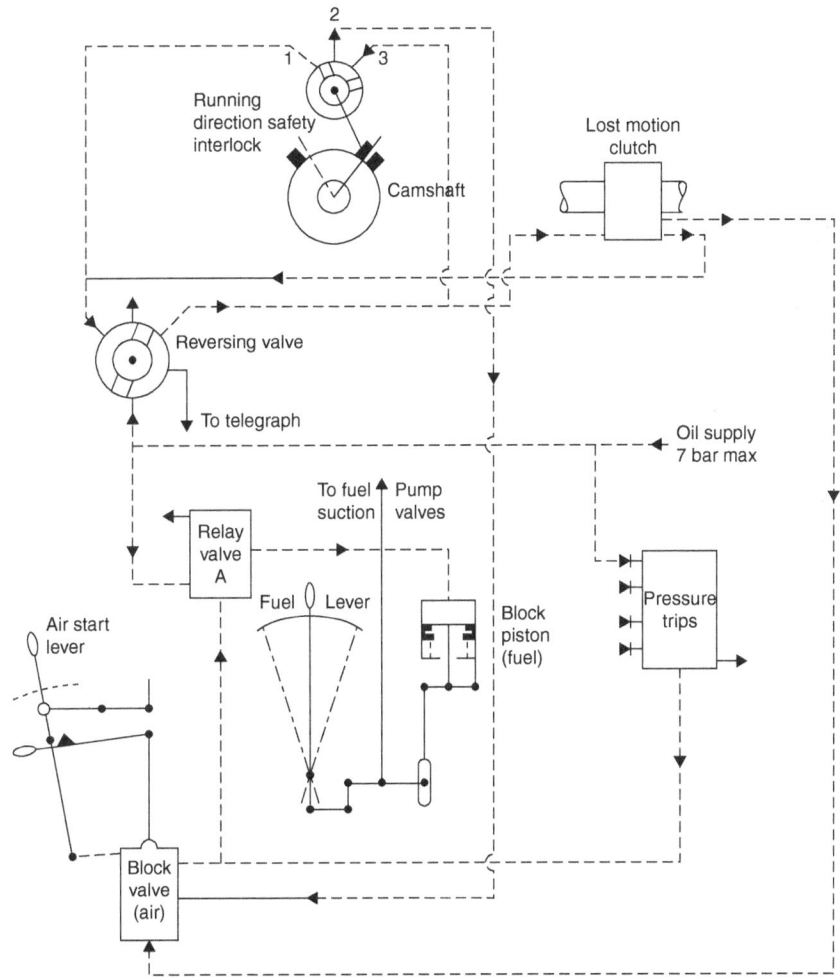

▲ **Figure 5.8** *Hydraulic control system*

Consider a reversing action from ahead to astern: oil pressure from left of reversing valves to right of the clutch and under relay valve A and the air block valve. The telegraph reply lever on the engine telegraph is first moved to stop and the fuel lever moved back to about notch 31/2; the starting lever is mechanically blocked by the linkage shown in figure 5.7.

The telegraph linkage to the reversing valve moves this valve and releases oil pressure from the lost motion clutch. This drop in pressure causes relay valve A to move down by spring action, which relieves pressure on the block piston (fuel) thus cutting off fuel injection. The pressure on the block valve (cam) is also relieved, which serves also to lock the starting lever.

Consider now the situation as shown in figure 5.8. When engine speed reduces, the telegraph lever can be moved to astern. This allows pressure oil to flow from the right through the reversing valve, as shown on the sketch, to the left of the lost motion clutch to reposition them for astern.

When the servo has almost reached the end of its travel, pressure oil admitted to the block valve (air) releases the lock on the air start lever. (The mechanical lock on the air lever with the telegraph had been released when the telegraph lever was moved to the astern running position.) Pressure oil also acts on relay valve A admitting oil to the block piston (fuel), thus allowing the fuel control linkages to the fuel pumps to assume a position corresponding to the load setting of the fuel lever.

If the pressure trips act in the event of low oil pressure (supply and bearings) or low water pressure (jacket or piston) then a trip piston moves up under preset spring pressure, thus connecting the oil pressure to drain. This pressure drop causes the block piston (fuel) to rise up under its spring force and shut off fuel injection.

Connections 1 and 3 from the running direction safety interlock to the reversing valve only allow fuel to the engine if the rotation agrees with the telegraph position. If not, the block piston (fuel) is relieved of pressure via the block valve (air) and relay valve A.

Movement of the air starting lever can now be carried out as both locks have been cleared and, subject to no trip action and satisfactory correspondence between rotation direction and the telegraph reply lever indication, fuel can be admitted following the full sequence of air starting as described previously, and as illustrated in figure 5.7. It is obvious that this system has a large amount of auto control and is easily adjusted for bridge control.

Control gear interlocks

There are many types of safety interlocks on modern IC engine manoeuvring systems. The previous few pages have picked out a number relating to the Sulzer RND engine and these will be adequate to cover most engine-type designs as the principles are all very similar.

Consider the interlock systems illustrated in figures 5.8 and 5.9. The telegraph and turning gear interlocks are straight mechanical linkages. In the former case, rotation of the telegraph lever from stop position causes the pin to travel in the scroll and unlock the air start lever as well as reposition the reversing valve. The turning gear blocking valve can be seen to close when the pinion is placed in line with the toothed turning gearwheel of the engine. The interlock exerted on the block piston (fuel) is also a fairly simple principle working on the relay valve A from the pressure trips and is as described previously.

Similarly, the block valve (air) operates mechanically via the lever lock on the air start lever and horizontal operating lever, which rises to unlock under the oil pressure acting

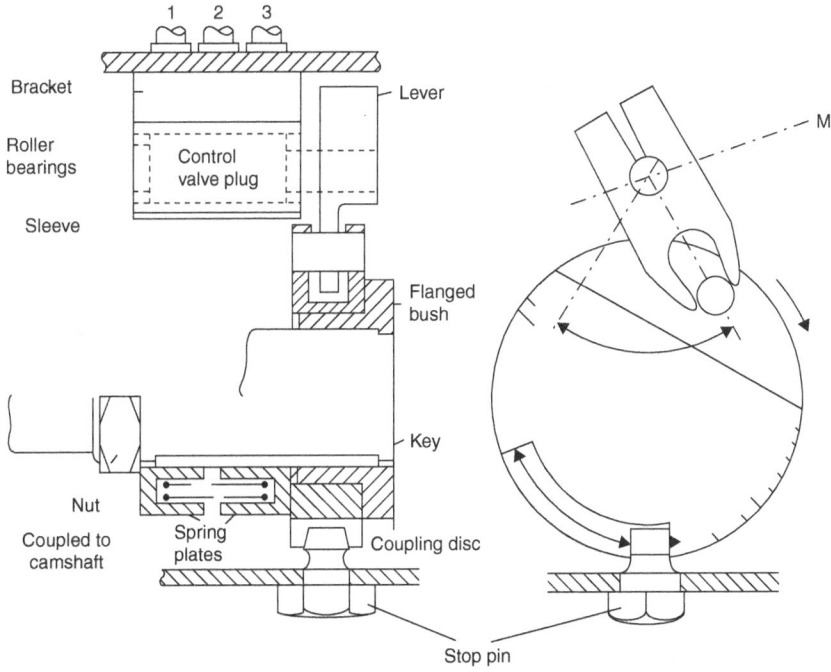

▲ **Figure 5.9** *Safety lock for correct rotation*

through the servo on the block valve (air) after the clutch reversals have taken place. (The pressure trips are merely spring-loaded pistons moving against low oil or water pressure to relieve control oil pressure, just like conventional relief valves.) It is perhaps appropriate here to describe one trip in detail and the direction safety lock will now be considered briefly. The function is to withhold fuel supply during manoeuvring if the running direction of the engine is not coincident with the setting of the engine telegraph lever. Refer now to figure 5.9.

At the camshaft forward end, the shaft is coupled to the camshaft and carries round with it, due to the key, a flanged bush and spring plates, which cause an adjustable friction pressure axially due to the springs and nut. This pressure acts on the coupling disc, which rotates through an angular travel T until the stop pin prevents further rotation. This causes angular rotation of a fork lever and the repositioning of a control valve plug in a new position within the sleeve. Oil pressure from the reversing valve can only pass to the block valve (air) and unlock the air start lever and the fuel control if the rotation of the direction interlock is correct. If the stop pin were to break, the fork lever would swing to position M and the fuel supply would be blocked.

Modern reversing systems

In the previous sections, reversal was carried out by utilising lost motion or by moving the camshaft axially to utilise a different set of cams. It can be seen that these methods involve added complication in the running gear and control systems of engines. In order to eliminate undue complications and improve engine response, B&W have designed a much simpler reversing system, eliminating the need to change the relative positions of exhaust cam and crankshaft.

In the latest L-MC designs, although the exhaust valve opening is not symmetrical about bottom centre (see figure 1.13d), the engine is able to operate in both directions without the necessity of changing exhaust valve position. However, astern operation is somewhat impaired because late closing of the exhaust valve allows a loss of combustion air to exhaust.

In order to change the fuel pump timing for astern running, MAN Diesel & Turbo engines utilise a movable fuel pump guide roller. Figure 5.10a shows the principle of the action for the guide roller in the ahead position. To change to the astern position a pneumatic cylinder, controlled from the engine starting and reversing system, moves the guide roller to the position shown in figure 5.10b. This has the effect of correctly positioning the fuel pump for astern running.

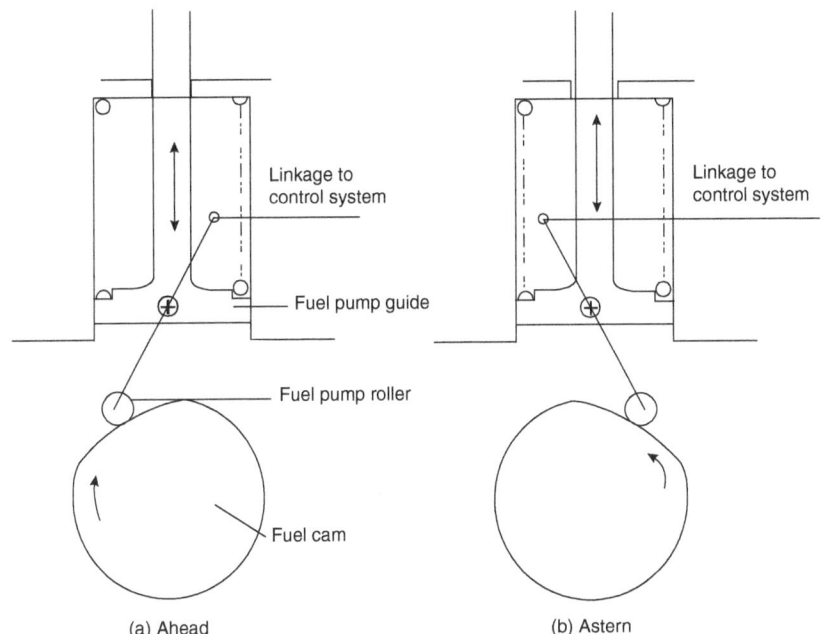

Linkage to
control system

Linkage to
control system

Fuel pump guide

Fuel pump roller

Fuel cam

(a) Ahead

(b) Astern

▲ **Figure 5.10** *Reversing mechanism of modern MAN Turbo Diesel B&W engines*

Modern engines can use an air distributor, which activates each air start valve in the correct sequence. The distributor allows the control-air to travel along the control air pipes to each air start valve situated within each cylinder head. Alternatively each air start valve can be operated by a solenoid on/off valve. This system allows the use of the air distributor to be discontinued.

Some manufacturers have also developed a system for turning the engine slowly on air before introducing the main bulk of the starting air. This is designed to reduce the risk from 'hydraulic lock'.

All engineers should be aware of the dangers from fire and explosions in air start systems. One of the very important checks to be made by the engineering watchkeeping officer is to feel the air start lines close to each air start valve on each cylinder head of the engine. Any pipe with a raised temperature could indicate a leaking air start valve. The potential risk here is that oily products could enter the air start line and then if there is sufficient heat present a fire could occur.

The advantage of the 'electronically controlled' engines such as the RT Flex is that the mechanical devices used to control these actions is completed by the software and not by the mechanisms required for these non-electronic solutions.

6

SENSORS, INSTRUMENTS AND MACHINERY CONTROL

Introduction

Instrumentation and control mechanisms have always been part of engineering machinery and systems. However, these mechanisms have been restricted by the limitations of mechanical, electrical or pneumatic devices. Over the past 20 to 30 years an almost silent revolution has taken place with the development of electronic devices that can be operated in different ways by altering a software program that is controlling the function of the device.

There will always be a place for mechanical safety equipment such as steam boiler safety valves or cylinder relief valves and overspeed trips. Programmable Logic Controllers (PLC) and microprocessors introduce another level of control ability that was not possible in the past.

The electronically controlled fuel injection system (page 128), for example, needs to be operated in a fraction of a second, which was not possible using the older mechanical fuel injectors.

It is intended in this chapter to give a brief overview of the control equipment and systems associated with the main power plant and support systems. Instrumentation and control systems are covered in chapter 11 of volume 8 (*General Engineering Knowledge for Marine Engineers*) and in greater depth in volume 10 (*Instrumentation and Control Systems*) of the Reeds Marine Engineering Series.

It should also be noted that this move to a new order means that new practices become much more important. These include completing regular software updates and thinking much more about data and cyber security.

Governing of Marine Diesel Engines

The term 'governor', when applied to a diesel engine, is used to describe the mechanism for controlling the engine's speed under differing conditions of load.

There must exist a clear distinction between the function of a governor and an overspeed trip. Therefore, for an engine's protection, governors should not be the only line of defence. While governors control the engine speed between close limits and under different loads, separate independent overspeed protection is necessary to shut down the engine in the event of the instantaneous shedding of load, and resultant rapid increase in speed, or governor malfunction.

In the past, diesel engines, especially when driving electrical generators, have invariably utilised a mechanical governor that works on the principle of weights (flyweights) being thrown out by centrifugal force (figure 6.1). A change in speed resulted in variation of the position of the flyweights and alteration of fuel supply. Old variations of the larger, slow running direct drive diesel engines were not generally fitted with such a governor but they invariably were fitted with an overspeed trip. This trip was arranged to allow full fuel supply under normal operating conditions but in the event of revolutions rising about 5% above normal the fuel was totally shut off until revolutions dropped to normal again. At about 15% above normal revolutions, the trip would stay locked, with the fuel shut off, and this would continue until reset by hand.

The centrifugal mechanical governor must perform two separate functions. These are to:

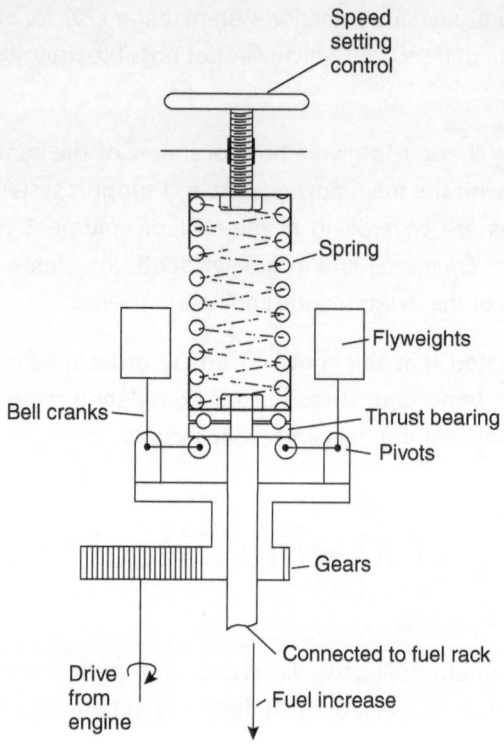

▲ **Figure 6.1** *Mechanical governor*

1. act as a speed measuring device and
2. supply the necessary power to move the fuel-controlled system.

The mechanical governor system is arranged to be self-regulating, where the output shaft controlling the fuel rack is designed to reduce the fuel if the speed rises above the desired value and increase the fuel if the speed reduces. Students will be able to study this early concept as it shows the heart of the more sophisticated modern mechanical/hydraulic, centrifugal governors (see figure 6.4) made by companies such as Woodward and found on many ships in service today.

The latest versions employ microprocessor control for the mechanical/hydraulic amplified controlling mechanism. This gives the power required to physically move the fuel control 'racks' on large or small diesel engines.

Figure 6.2 shows, in block diagram form, the arrangement of such a centrifugal or flyweight mechanical governor system. It is regarded as an open loop system as there is no signal, representing the value of the output condition, being fed back to act as a comparison with the system's input action.

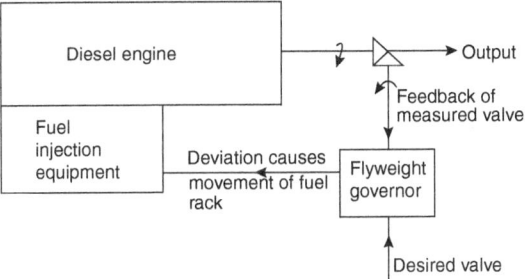

▲ **Figure 6.2** *Closed loop control*

A closed loop control system is one in which the control action is dependent on the output. The measured value of the output, in this case the engine speed, is fed back to the controller, which compares this value with the desired value of speed. If there is any deviation between the values, measured and desired, the controller produces an output that is a function of the deviation. In this case the controller output would be proportional to the deviation, that is, proportional control.

In control terminology, deviation is sometimes called error, since it is the difference between measured and desired values, and desired value is sometimes called set value. Proportional control suffers from offset. In the example, if a speed change occurs the flyweights take up a new equilibrium position and the fuel supply will be altered to suit the new conditions. However, the diesel is now running at a slightly different speed than before. If the original speed was the desired value then the new speed is offset from the desired value because the new controlled condition is in proportion to the change.

When considering governors on diesel engines we have to think about the use of the term 'speed droop' (figure 6.3) or just droop. It is used to define the change in speed between no load and full load conditions. If speed droop did not happen and the governor was continually trying to keep the engine at one speed then, due to the delay and mechanical inertia, the governor would overcompensate for a given change in load. In this case the diesel would hunt while the governor constantly made changes to keep the steady running condition.

A set of circumstances without droop is known as an isochronous condition and an engine fitted with an isochronous governor will hunt, usually overcompensating more and more until the over-speed trip activates. However, the term isochronous has taken a new meaning, as we will find out later.

The forces involved in determining the mechanical governor's movement are inertia, friction and spring. Considerable effort may therefore be required to cause movement

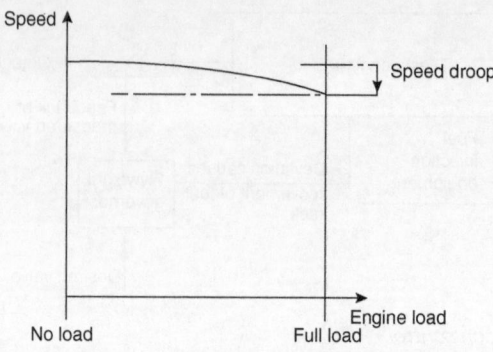

▲ **Figure 6.3** *Speed droop*

one way or the other and this would result in a change of speed without any alteration of the governor's control position.

The control is at best very slow and insensitive as various equilibrium speeds are possible. For simple systems, these various equilibrium speeds are not an embarrassment, but if we require a system to be more finely controlled then the two functions that the mechanical governor has to perform would be better separated into:

1. speed measurement and
2. servo-power amplification.

Figure 6.4 shows a flow diagram of the basic arrangement. A load increase would cause a momentary speed droop. The speed measuring device would obtain a measured value signal from the diesel and compare this with a desired value from the speed-setting control. The deviation would be converted into an output that would bring into action the servo-power amplifier, which would position the fuel rack, increasing the supply of fuel to meet the increase in load.

Since the speed measuring device does not have to position the fuel rack – in fact it could be near zero loaded – it can be very responsive, minimising the time delay between load alteration and fuel alteration utilising a closed loop process. The servo-power amplifier is usually a hydraulic device that simply, quickly and effectively provides the necessary muscle to move the fuel rack.

A proportional action governor is diagrammatically shown in figure 6.5. The centrifugal speed measuring unit is fitted with a conically shaped spring, unlike that shown in figure 6.1; this gives a spring rate that varies as the square of the speed (figure 6.6) and gives a linear movement to the speed measuring system, that is, the

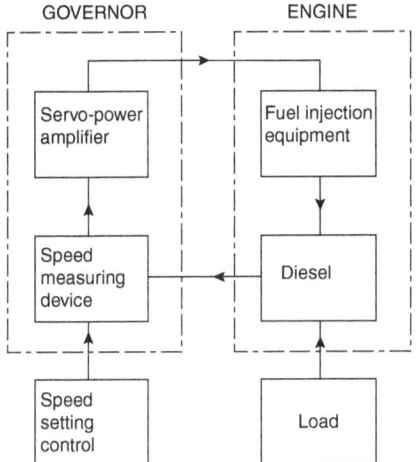

▲ **Figure 6.4** *Basic arrangement*

response is directly proportional to the change in speed. If we consider an increase in load on the engine, the pilot valve will move down due to the speed drop. The piston in the servo-amplifier will move up and increase the fuel supply to the engine. The feedback link reduces the force in the speeder spring so that the flyweights can move outwards to a new position, thus raising the pilot valve and closing off the oil supply. If for some reason the oil supply system should fail then the spring-loaded piston in the servo-cylinder would be moved down and fuel to the engine would be cut off. This is called fail-safe action. Any oil that leaks past the servo-piston will be drained off to the oil sump tank. If this were not so, the servo-piston would eventually lock in position.

Flywheels and their effect

Flywheels are required due to the pulsing nature of the torque that is delivered to the crankshaft from the firing of individual cylinders. The exact dimensions of each engine's flywheel will be determined by the number and configuration of the cylinders.

The flywheel soaks up the forces delivered by each energy pulse and carries over the rotating masses by giving up momentum when there is no input from the burning of fuel. The result is that the engine runs in a smooth rotation and not in a set of jerking movements.

Any fluctuating forces causing changes in speed would normally be outside the control of the governor. However, if the speed has to remain nearly constant during changes of

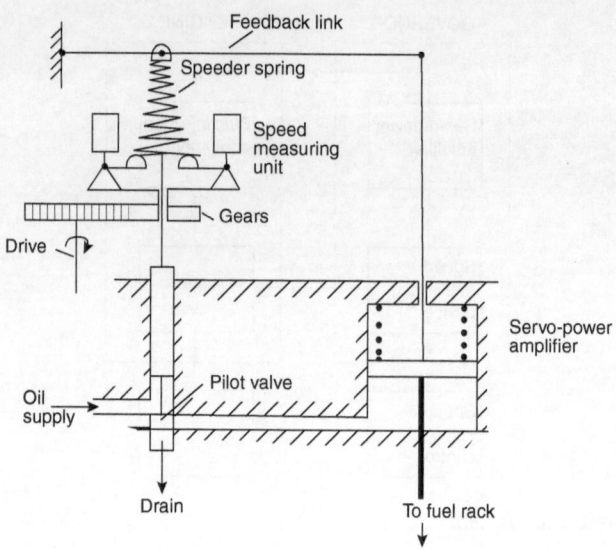

▲ **Figure 6.5** *Proportional action governor*

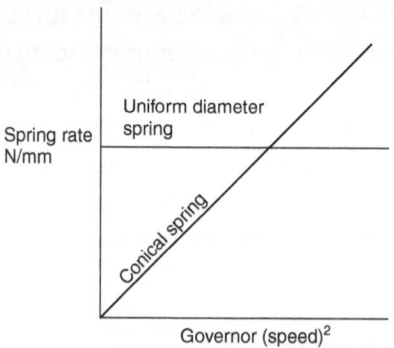

▲ **Figure 6.6** *Governor (speed)*

load it may be decided to fit a larger flywheel, which increases the moment of inertia of the system and gives an integral effect – this must not be taken to extremes or instability may occur. Flywheels, however, are not cheap and a less expensive solution to the problem may be to fit a better governor.

Integral effect or, as it is often called, 'reset action' reduces offset to zero, that is, during load alteration the speed will move from the desired value but the reset action will be applied to return the speed to the desired value, so that after the load change the speed is the same as before.

Governor with proportional and reset action

Figure 6.7 shows diagrammatically the type of governor that will, after an alteration in engine load, return the speed of the engine back to the value it was operating at before the alteration. If an increase in engine load is considered, the flyweights will move radially inwards and the pilot valve will open to admit oil to the servo-piston. The servo-piston will move up the cylinder compressing the spring and at the same time it will cause (a) the fuel rack to be repositioned to increase fuel supply to the engine, (b) rotate the feedback link 'A–B' anticlockwise about the pivot point 'A' (this point 'A' would initially be locked due to equal pressures on either side of the reset piston), (c) rotate link 'C–D' will move the reset piston control valve down and some oil will drain from the reset piston cylinder. As the reset piston moves down to a new equilibrium position, the feedback link 'A–B' will pivot about 'B' and the link 'C–D' will be rotated clockwise, closing the drain from the reset piston cylinder (and thus locking the reset piston in a new position), returning the point 'D' to its original position. This means that the engine is now running at its original speed

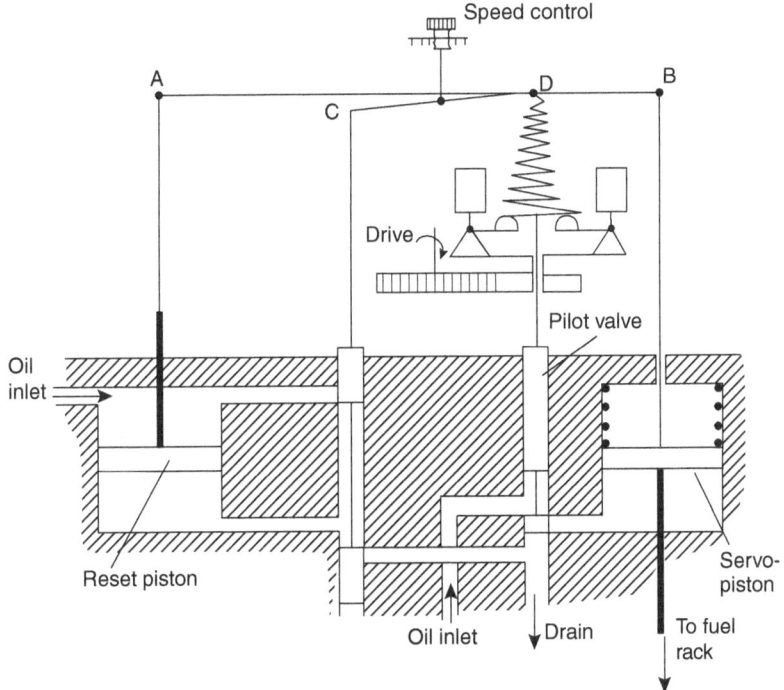

▲ **Figure 6.7** *Governor with proportional and reset action*

but with increased fuel supply. Speed droop that took place during the change of the relative positions of the two pistons was transient. This type of governor, which has proportional and reset action, is called in governor parlance an 'isochronous governor'.

Electric governor (figure 6.8)

Here, we are still describing an old-style governor, which happens to be electric; full electronic control is described later in this chapter. This type of governor has proportional and reset action with the addition of load sensing. A small permanent magnet alternator is used to obtain the speed signal; the advantage to be gained is that there will be no slip rings or brushes with their attendant wear. The speed signal taken from the frequency of the generated a.c. voltage is converted into a d.c. voltage proportional to the speed. A reference d.c. voltage of opposite polarity, which is representative of the desired operating speed, is fed into the controller from the speed setting unit. These two voltages are connected to the input of an electric amplifier. If the two voltages are equal and opposite, they cancel each other out and there will be no change in amplifier voltage output. If they are different, then the amplifier will send a signal through the controller to the electro-hydraulic

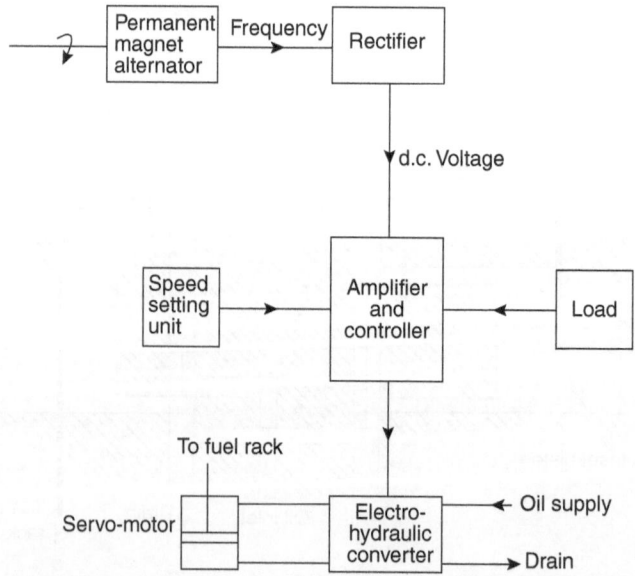

▲ **Figure 6.8** *Electric governor*

converter, which will in turn, via the servo-motor, reposition the fuel rack. In order that the system is isochronous, the amplifier controller has internal feedback.

Load sensing

The purpose of including load sensing into the governor is to correct the fuel supply to the prime mover before a speed change occurs. Load sensing governors are therefore anticipatory governors, that is, they anticipate a change in speed and take steps to prevent, as far as possible, its occurrence.

Load sensing could be achieved by mechanical means but it would be a complicated and relatively costly system. For this reason, load sensing governors tend to be of the electric type. The output of, for example, a main generator would be monitored and if a load alteration takes place a signal is fed to the governor. It must be remembered that the speed of response of the load sensing element must be better than that of the speed sensing element. The speed sensing element would be used to correct small errors of fuel rack position.

Students will now be able to see that by replacing the a.c. and d.c. systems with modern microprocessors, the older governors can be easily upgraded to give an existing engine much better performance and improved emissions.

In addition to load sensing, adding microprocessor control of mechanical-hydraulic governors gives the following advantages over the older systems:

1. Software selectable speed settings – making generator synchronisation easier.
2. Electronic governors can be mounted in positions remote from the engine, thereby eliminating the need for governor drives.
3. Controls and indicators available from electronic governors make automation easier.
4. Control functions, such as fuel limitation, acceleration and deceleration schedules and shutdown functions, ie low lubricating oil pressure, are accepted as input signals to the microprocessor within the governor.
5. Advanced algorithms give an improved response time to a change from the steady state condition. This leads to tighter control of the transient load state, which in turn helps to reduce the harmful emissions in the exhaust.

The move towards PLC and full electronic control is covered more fully under Chapter 11 of Volume 8 of the Reeds series.

Geared diesels

Two diesels geared together must run at the same speed, but if the governors of the two are not set equally then they will not carry equal shares of the load.

Figure 6.9 shows the governor droop curves for two diesels A and B. Governor A has a higher speed setting than that of B, but since they must both run at a common speed the load carried by A will be greater than that of B. Actual load carried is given by the intersection of the common speed line and the droop curves. By adjusting the speed settings, both droop curves could be made to coincide at the intended load, although this would be difficult to achieve in practice.

Shown in figure 6.10 are two sets of droop curves with the same difference in speed settings but with different amounts of speed droop. The difference in load sharing at

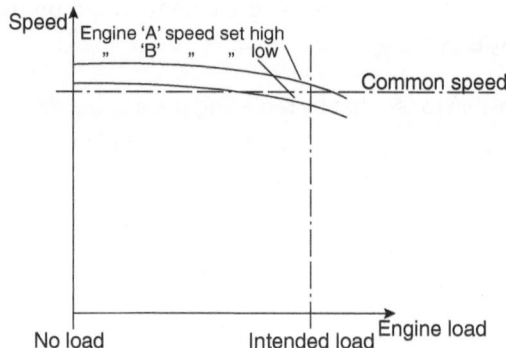

▲ **Figure 6.9** *Load sharing between two engines*

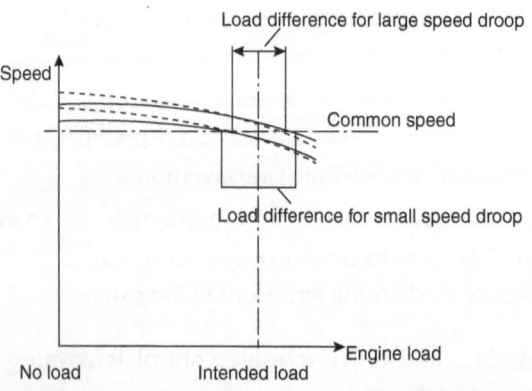

▲ **Figure 6.10** *Load difference*

the common speed is less for the larger speed droop curves than for the smaller. Hence speed droop and fine control over the desired level of speed are necessary for effective load sharing.

Bridge Control of Direct Drive Diesel Engines

A minimum and basic bridge control system first started to become common in ship designs during the 1970s and 1980s. This was the era when control systems had become reliable enough to be trustworthy. Until that time the link between the bridge and the engine room was the ship's telegraph. Once the desired speed of the engine was decided upon by the person responsible for navigating the ship, this value was indicated to the Engineering Officer in Charge via the telegraph. The Engineering Officer would then have to answer the telegraph and move to the engine controls to start or change the speed of the engine. It was therefore a significant improvement when the Navigator could alter the speed themselves. These early control systems still had to incorporate a mechanism to ensure that the engine was not accelerated too hard or run within speeds that promote resonating forces, due to balancing imperfections. Prior to the bridge control systems, these actions would have been undertaken by the engineering officer responsible for controlling the engine response to the telegraph order.

Two consoles would be provided, one on the bridge and the other in the engine room. For the bridge console the minimum possible alarms and instruments would be provided commensurate with safety and information requirements, for example low starting air pressure and temperature, sufficient fuel oil, fuel oil pressure and temperature, etc. The engine room console would give comprehensive coverage and overriding control over that of the bridge. More duplication can also be provided for the Chief Engineer and the Master. Information can also now be sent to remote locations such as the owner's office or the offices of the company supplying the technical management and/or diagnostic backup for the officers on-board.

In figure 6.11a, for simplification, all normal protective devices are assumed and subsidiary control loops are not considered. The selector would be in the engine room console and the operator can select either engine room or bridge control; with one selected the other is inoperative. Assuming bridge control, a programme would be

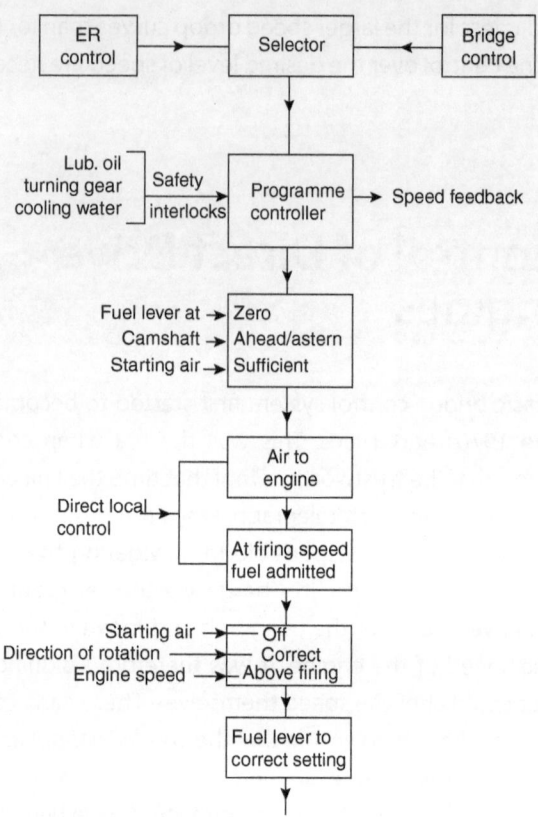

▲ **Figure 6.11a** *Engine control programme*

selected, say half ahead, then providing all safety blockages such as no action with turning gear in, etc are satisfied, the programme can be initiated and could follow a sequence of checks and operations such as:

1. Fuel control lever at zero.

2. Camshaft in ahead position.

3. Sufficient starting air.

4. Starting air admitted.

5. Adjustable time delay permits engine to reach firing speed.

6. Fuel admitted.

7. Starting air off, checks on direction of rotation and speed.

8. Fuel adjusted to set value.

Essential safety locks, such as low lubricating oil pressure or cooling water pressure, override the programme and will stop the engine at the same time as they give warning. Direct local control at the engine itself can be used if required in the event of an emergency.

Further protective considerations are as follows:

1. Governor, including overspeed trip.
2. Non-operation of air lever during direction alteration.
3. Failure to fire requires alarm indication and sequence repeat with a maximum of, say, four consecutive attempts before overall lock.
4. Movement of control lever for fuel for a speed out of a critical speed range if the bridge speed selection is within this range.
5. Emergency full ahead to full astern timing and setting.

Outline description

The following is a brief description of one type of electronic-pneumatic bridge control for a given large single-screw direct coupled IC engine to illustrate the main essentials. The IC engine lends itself to remote control more easily than turbine machinery.

Movement of the telegraph lever actuates a variable transformer, thus giving signals to the engine room electronic controller, which transmits, in the correct sequence, a signal series to operate solenoid valves at the engine. One set of solenoid valves controls starting air to the engine while a second set regulates fuel supply, the latter via the manual fuel admission lever, is coupled to a pneumatic cylinder whose speed of travel is governed by an integral hydraulic cylinder in which rate of oil displacement is governed by flow regulators. This cylinder also actuates a variable transformer giving a reset signal when fuel lever position matches telegraph setting.

With the engine on bridge control the engine control box starting air lever is ineffective and the fuel control rack is held clear of the box fuel lever. Engine override of bridge control is provided.

The function of the electronic controller is to give the following sequence for, say, start to half ahead: ensure fuel at zero, admit starting air in correct direction, check direction, time delay to allow engine to reach firing speed, admit fuel, time delay to cut off air, time delay and check revolutions, adjust revolutions. Similar functions apply for astern or movements from ahead to astern directly. Lever travel time to full can be varied from stop to full between adjustable time limits of ½ min and 6 min. Fault and alarm circuits and protection are built into the system.

Modern engine control systems use electrical and electronic components to gather information from the plant. This means that a very large amount of data can be collected and processed by the control systems. The information in a digital form can be used in a variety of ways. Initially the data will be used to control the power plant and all its associated systems. The information will relate to the temperatures, pressures and flow rates of fluids. This data will be processed and used to open and close control valves and alter the speed of the associated pumps to control the flow of fluids around the various systems. The sensors will be monitoring different values and alarm settings will alert the watchkeeping officers if any measured readings stray over these preset parameters. Much more data can be recorded and used to good effect than with any other system in the past. Data loggers can be used to record each alarm but in a time of difficulty more alarms could be triggered than the watchkeepers can keep track of. If this is the case, then 'fail safe' design features of the system should be activated to, for example, start back-up systems and machinery automatically.

In addition to controlling the performance of the power plant and its support systems the data collected can be used to inform the engineering officers about the condition of the equipment. For example, the time taken to operate a valve could be measured and compared to a standard. If the measurement changes, then the valve may need to be overhauled. This means that equipment can be maintained to its correct working condition and not to timescales or running hours, making the maintenance system more cost effective. This branch of engineering is called mechatronics and more information about this subject can be found on page 234.

Marine engineers should be aware that the harsh environment found on ships at sea means that sensors, instruments and other parts of the control systems must be robust and built to the correct standards. Therefore, any spares ordered for replacements must come from a well know source where the quality can be assured. Low quality parts will not stand the vibration and atmosphere found in the marine environment.

Data 'bus' systems, as mentioned on page 235, are the standard industrial method of handling the large amounts of information required to run these systems correctly. The network of equipment all communicating with each other need common working protocols, to ensure that all the sensors, valves and pumps etc obey the correct commands. See page 236 for more information about communication systems.

Integrated Bridge Systems

The requirement of a modern ship is, as it always has been, for the safety of navigation. However, the conditions for practising 'Safe Navigation' are changing. The volume of traffic is increasing and traffic separation schemes (TSS) are very busy. In addition the increased accuracy of navigation, due to the use of the Global Positioning System (GPS) means that ships on an ocean passage are also closer to each other. This, together with Unmanned Machinery Space (UMS) operation of the engine room, means that full control of the ship's speed and direction should be under the control of the Navigating Officer all the time that power is available.

This has led to the development of Integrated Bridge Systems, where all the information and functions required to have complete control of the ship are placed in a central position for use by the Navigating Officer.

Development of these systems has relied on handling a vast amount of data and this has only been possible due to the increased use of high-speed networks and data bus transmission systems such as CANBus or Modbus.

Students will see that the system shown in figure 6.11 would be a subset within the fully integrated systems shown in figure 6.11a.

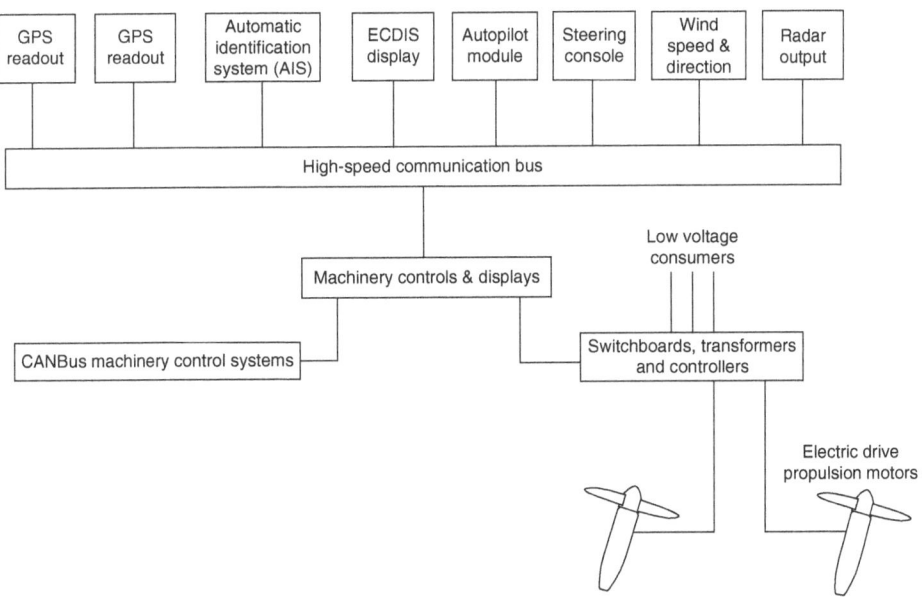

Integrated bridge system

▲ **Figure 6.11b** *Integrated bridge system*

The control systems described during the rest of this chapter would also be operated as local systems and would only interact with the main system under an alarm condition for reporting temperatures and pressures as requested by the main condition monitoring system.

Piston Cooling and Lubricating Oil Control

Simple single-element control loops can be used for most of the diesel engine auxiliary supply and cooling loops; however, during the manoeuvring of diesel engines considerable thermal changes take place with variable time lags, which the single-element control may not be able to cope with effectively. (*Note:* a single-element control system is one in which there is only one measuring element feeding information back to the controller.)

For piston cooling and lubricating oil control, the use of a cascade control system caters effectively for manoeuvring and steady-state conditions. Cascade control means that

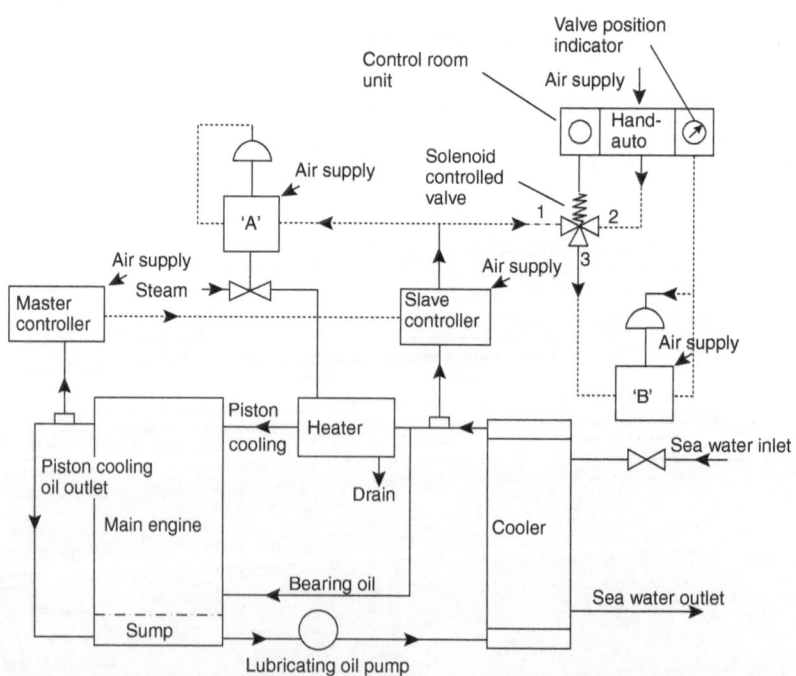

▲ **Figure 6.12** *Cooling and lubricating oil control*

one controller (the master) is being used to adjust automatically as required the set value of another controller (the slave). In figure 6.12, the two main variables to consider are sea water inlet temperature and engine thermal load. For simplicity we can consider each variable separately:

1. Assuming the engine thermal load is constant and the sea water temperature varies. The slave controller senses the change in lubricating oil outlet temperature from the cooler and compares this with its set value. It then sends a signal to the valve positioner 'B' to alter the sea water flow.

2. Assuming the sea water temperature is constant and the engine thermal load falls. The master controller senses a fall in piston cooling oil outlet temperature and compares this with its set value. It then sends a signal to the valve positioner 'B' so that the salt water flow will be reduced and the lubicating oil temperature at inlet to the piston is increased.

If the engine thermal load is low or zero then valve positioner 'A' will receive a signal from the slave controller, which will cause steam to be supplied to the lubricating oil heater. This means that the slave control is split between valve positioners 'A' and 'B' – this is called 'split range control' or 'split level control'.

Slave controller output range is 1.2–2.0 bar.

Valve positioner 'A' works on the range 1.2–1.4 bar.

Valve positioner 'B' works on the range 1.4–2.0 bar.

Hence the range is split in the ratio 1.3.

Since the piston cooling oil outlet temperature could be offset from the desired value by 8°C upwards or more, the master controller must give proportional and reset action. In order to limit the variety of spares that must be carried, the slave controller would be identical to the master controller (figure 6.13).

It may be necessary to change over from automatic to remote control. This is achieved by position control of the three-way solenoid-operated valve and regulation of the air supply to the valve positioner 'B' at the control room unit. The solenoid-operated valve would be positioned to communicate air lines 2 and 3, closing off 1.

Hand regulation of the supply air pressure to valve positioner 'B' enables the operator to control the sea water flow to the cooler. Position of the sea water inlet control valve is fed back to the control room unit. Lubricating oil temperatures would be indicated on the console in the control room.

An alternative and often preferred arrangement, using a single measuring element, is to have full flow of sea water through the cooler and operate a three-way valve (two inlets, one outlet) in the engine fresh water, or cooling circuit that bypasses the cooler.

Valve selection for such duties is most important. Maximum pressure and temperature, maximum and minimum flow rate, valve and line pressure drops, etc, must be carefully assessed so that valve selection gives the best results. With correct analysis of the plant parameters and careful valve selection, simple single-element control systems can be employed. This would avoid the extra cost of sophisticated control loops and their attendant increased maintenance and fall in reliability.

For mixing and bypass operations, a three-way automatically controlled valve with two inlets and one outlet of the type shown diagrammatically in figure 6.14 could be used. An increase in controller output pressure p causes the flapper to reduce outflow of air from the nozzle; the pressure on the underside of the diaphragm increases and the

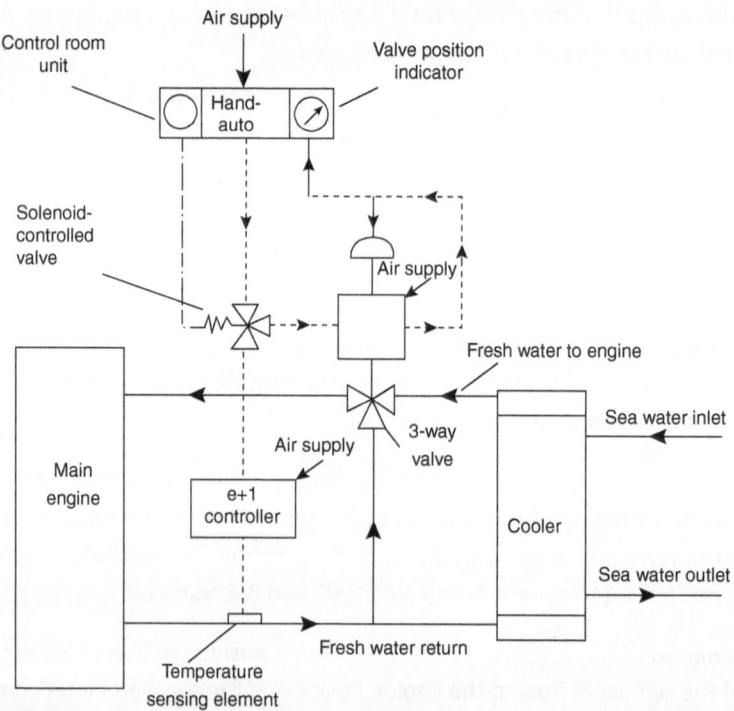

▲ **Figure 6.13** *Jacket (or piston) temperature control*

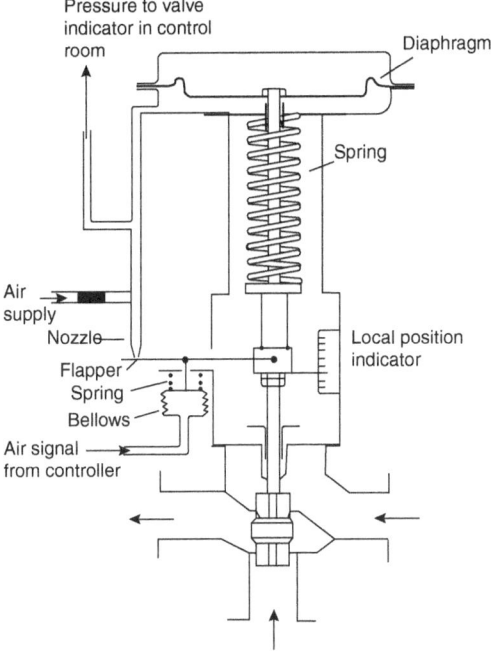

▲ **Figure 6.14** *Three-way valve and positioner*

valve moves up. As the valve moves, the flapper will be moved to increase outflow of air from the nozzle and eventually the valve will come to rest in a new equilibrium position.

Indication of valve position is given locally and remote, in the latter case by feeding back the diaphragm loading pressure to an indicator possibly situated in the control room, the valve positioner gives accurate positioning of the valve and provides the necessary muscle to operate the valve against the various forces.

Pressure alarm

The alarm diagrammatically shown in figure 6.15 can be used for either high- or low-pressure warning. It can also be used for high- or low-level alarm of fluids in tanks since pressure is a function of head in the tank.

To test the electrical circuitry and freedom of movement of the diaphragm and switches, the hand testing lever can be used. Setting is achieved, for low-pressure

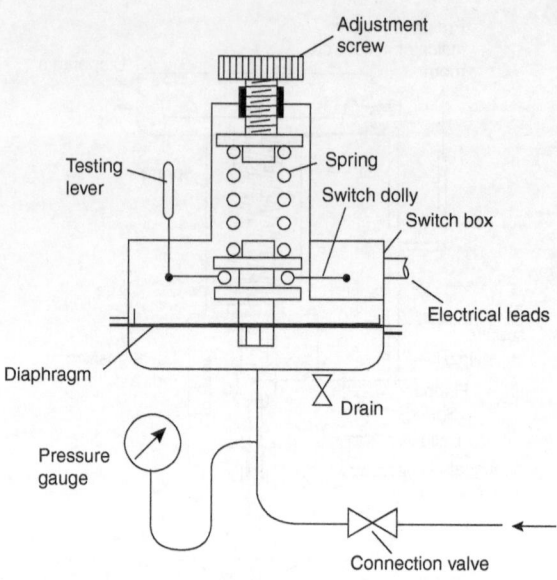

▲ **Figure 6.15** *Pressure alarm*

alarm, by closing the connection valve and opening the drain. When the desired pressure is reached, as indicated on the gauge, the alarm should sound. If high-pressure alarm is required the unit can be set by closing the connection valve and coupling a hydraulic pump to the drain connection.

Electrical and Electronic Control Systems

This section is a brief overview of the newly emerging technology that is due to take the industry by storm and students should study this is in parallel with Chapter 11 in Volume 8 the Reeds series.

Sensors

Modern control technology is based on the reliability and accuracy of sensors. These are devices that detect physical and chemical properties and transmit an electrical output signal that is proportional to the detected value.

The marine industry has been slower to change away from hydraulic and pneumatic control systems because in the harsh environment at sea these systems have proved most reliable. This, together with very little need for highly accurate measurement (due to lack of engine emission control legislation), means that the added sophistication of electronic control systems has been slow to have an impact on the industry.

However, the advantages of digital control along with the ever growing need for reducing engine emissions has meant that there is now a rush to embrace the technology. The reliability of miniaturisation processes used in the manufacture of sensors is becoming better all the time with the following concepts being used:

- Substrate and hybrid technology for pressure and temperature sensors
- Semi-conductor technology used for monitoring rotational speed
- Micromechanics for measuring pressure and acceleration
- Micro-optics for measuring light, for example, boiler alight confirmation
- Mechatronics (described in the next section).

The advancement in microelectronics and micromechanics is the driving force behind the changes and the measurement of pressure has been one of the problematic areas for the advancement of control. Strain gauges can measure small movements, which makes them useful for measuring pressure or torque. However, until recently these instruments have not been available in a micro format.

The micromechanical structure – strength and hardness, etc – of silicon can be compared with steel. However, the other properties of silicon open up a whole new technology. The silicon is lighter and its electrical and thermal conductivities are quite different from steel. Single crystal silicon wafers are used because they have almost perfect physical response characteristics. Hysteresis and creep are negligible and the single crystal material is very brittle and ruptures if the elastic limit is exceeded.

Bulk micromechanics is where the whole wafer is built up using etching techniques and can be used to produce diaphragms of between 5 and 50 μm for measuring pressure. Surface micromechanics is where the silicon wafer is used as a substrate and moving structures can be formed on the surface of the silicon wafer.

Measuring the high pressures generated in the combustion space and the fuel rail of a diesel engine requires high-quality cost-effective sensors. A robust design with long-term stability is needed to achieve the 10^{10} 0–2,000 bar, operational lifetime load cycles required by large diesel engines. They need electronics capable of processing measured data quickly and demonstrating a high level of accuracy over a wide range of temperatures.

There are several different sensor technologies for detecting and transmitting high pressure but the most frequently used is the resistive measurement method because it references the measured pressure to the ambient conditions. This process makes use of two coating techniques. The first is to electrically insulate the active sensor components from the body of the device, achieved by a silicon dioxide layer being laid down using a chemical vapour deposition process. These are capable of withstanding 500V + AC. Secondly, resistors made of nickel and nickel chromium alloy are sputtered directly onto the silicon dioxide surface to form a Wheatstone bridge measurement circuit. In subsequent steps conventional lithographic techniques are used to create the meander-shaped conductor paths typically used in strain gauges.

The nickel-chromium alloy used to fabricate thin film devices has the additional advantages of having a very high manufacturing reproducibility and temperature stability up to 200°C. The latest high-pressure sensors utilise a non-welded design based on a monolithic body of precipitation-hardened 17–4PH nickel-chromium stainless steel. The sensor diaphragm, which must be of an exactly defined thickness, forms the bottom of a highly accurate deep bore machined into the centre of the device body, where the membrane is designed to operate at pressures in excess of 3,000 bar. The thin film structure is not directly located on the diaphragm but is deposited onto a tiny cantilever beam attached to it at the thinnest point, removing the inefficient need to create the sensor structures on diaphragms individually. To maintain low stress levels in relation to the elastic limit, which maximises the operational lifetime, a monolithic sensor body can be used instead of a welded structure. This offers enhanced security under fatigue stress conditions due to the seamless design and lower internal residual stress levels. A stable output signal is generated from the Wheatstone bridge, by the slight stretching of the diaphragm under pressure. However, the signal level is low, and with a k-factor of 2 for nickel–chromium strain gauges, outputs of 2 mV/V are measured. This disadvantage is reduced with the help of the latest application-specific integrated circuit (ASIC) technology. The circuit used in the sensor features an offset-free amplifier that operates in a closed control loop, thereby compensating for any offset drift. This allows even small output signals below 2 mV to be amplified and compensated with a high signal-to-noise ratio, at the 5V supply voltage. Sensors can be tailored to meet specific customer requirements. Analogue to digital conversion can be performed by a sigma-delta converter and error compensation effected by means of a 3D lookup table –

the error is evaluated and fed back via a fast pulse width modulated (PWM) signal. Temperature is measured either by the ASIC's on-chip sensor or directly by a nickel resistor on the strain gauge itself, the latter giving a good indication of the temperature of the pressure medium. The signal is measured with a bandwidth of 10 kHz, so that fast transient effects in the fuel injection process can be recorded. Electronics have to meet marine electromagnetic compatibility requirements and state the total error shown by a sensor.

The rapid, high-amplitude changes in pressure, as well as the easily recognised oscillations in the injector pressure tubing, are reproduced without error or delay. In contrast to piezoelectric measurement techniques, the static pressure is measured relative to the ambient value. The non-welded high-pressure sensor design described above allows reliable operation for the direct measurement of pulsating injection pressures in large engines. The fast signal-processing electronics with 10 kHz bandwidth permits optimisation of the fuel injection modulation process by the motor control system. By using the new sensors to monitor injector needle lift and cylinder pressure, the combustion process in each cylinder can be individually optimised, thereby allowing operators to reduce emissions, increase fuel efficiency and monitor engine health more closely.

Temperature sensors

The common technology employed in temperature sensors is the variation of electrical resistance in materials when the temperature of their surroundings changes with either a positive or negative coefficient. The sensors predominantly rely on contact with the measured medium but some measure the infrared radiation given off by a hot surface, for example. Some fire detectors work on this principle.

Sintered ceramic resistors made of heavy metal oxides are an example of semi-conductive materials which display an inverted exponential temperature curve. An example of an extreme precision sensor is a thin film metallic resistor integrated on a single substrate wafer. Temperature neutral 'trimming' resistors are used to give the high accuracy.

Force and torque sensors

Force sensors are either strain measuring or displacement measuring devices. The strain measuring devices might use the piezo-resistive of the magnetoelastic principle. The former being used to measure torque as well as force, they are the most widespread and are very reliable. Piezo-resistive sensors work on the fact that there is a change in resistance in the material due to its deformation.

Magnetoelastic devices use the principle that ferromagnetic materials change their magnetic properties under elongation or reduction in length due to the force on the component. This change in magnetic properties is detected and measured and an output signal is generated.

Mechatronics

Mechatronics is the technology that is starting to transform the world of control engineering. We have seen in Chapter 4 how the electronic control of diesel engines has taken them to new levels of performance. This has largely been due to the developments in manufacturing that have enabled the production of reliable and robust sensors.

These sensors can detect pressure and temperature to a much finer tolerance and act infinitely faster than any mechanical system. The power of microprocessors and the development of the algorithms required to interpret the signals from thousands of sensors has brought everything together at a time when the industry needs a step forward in efficiency.

Mechatronics is the synergistic integration of mechanical components, electronic devices, computer and software engineering along with embedded control features, coming together to keep machinery operating as close to its design condition as possible. It has been described as 'mechanical engineering for the 21st century'.

During the past 30 years or so, as we have seen, the technology consisted of pneumatic control equipment where precision-made mechanical instruments and other devices interpreted the condition of machinery and associated systems and transmitted pneumatic signals in an effort to control processes such as the temperature of cooling water or lubricating oil. These control systems were usually localised at the point of use, although some had remote indicators such as that for boiler water level. However, on the whole, control systems were independent of the main machinery alarm systems and the measurement of processes inside hostile environments, such as the engine combustion space, were just not possible.

As discussed, we are already seeing the electronic control of diesel engine combustion increasing engine flexibility and lowering emissions. Now, other marine machinery is also being produced with the ability to link into the central control and management system, and as algorithms become ever more sophisticated so will the efficient use of the equipment.

All communication systems need a common language and the focus for the development of mechatronics on board is now on the topology and protocols being used, with Modbus and CANBus being the top runners. However, some manufacturers are producing equipment that uses the protocol called Profibus.

The development of these standards into a unified system or one protocol is important so that manufacturers can move forward rather than having to cover different systems or having to change standards halfway through their production run, which is a possibility at present.

Mechatronics also has the potential to provide the industry with an increased quantity of information relating to the condition of machinery, thereby giving owners the evidence they require to ensure their vessels are complying with legislation. In addition, such information will enable managers to increase efficiency, thus decreasing fuel consumption and maximising profits. The capability of mechatronics to assist the industry with its move towards higher efficiencies, lower energy consumption and more ecofriendly vessels is on the horizon. Given the right development, more sophisticated and so-called intelligent systems will be capable of linking the control of machinery with the production of machinery condition data. This will drive efficient management systems, enabling ship and head office staff to work together by improving not only the vessel's voyage planning but also the efficient use of machinery. Progression will increasingly mean different types of fuel and/or propulsion types and will also include more efficient maintenance systems and changes in the general management of vessels. The integration of embedded systems into the control of marine machinery will once again bring to the fore the importance of crew competence and focused staff development.

We now have in place the 'Manila amendments' to the STCW convention. Specific outcomes from this update required a focus on the operation of pollution prevention equipment and, more generally, additional emphasis being given to environment management. There will inevitably be a concentration on professional development including diagnostic techniques for ship staff, since they will be in the front line in the event of any emergency for example, if the alarm monitoring system is indicating that the ship is not producing the correct exhaust emissions as the vessel is approaching an Emission Control Area (ECA).

This issue would clearly have to be resolved urgently so the vessel does not risk incurring a hefty fine. Integrated approach to design has potential for monitor efficiency gains. With sophisticated control systems in place, crew familiarisation will be essential, therefore continuity of staff, efficient handovers and the use of integrated management information systems will be required in running the modern fleet effectively.

Communication systems

The increased use of sensors and electronic control systems will undoubtedly bring a step change in both the accuracy and the amount of information that will be available to the engineers in charge of a marine power plant. The sensors are linked to Programmable Logic Controllers (PLC), microprocessors or full computer systems. However, as with any communication system they all have to be speaking the same language and using the same transportation system. A computer-controlled system employing a universal bus system instead of individually wired circuits is the modern trend for control engineering.

Compared to a control system with conventional wiring, a computer-controlled bus system has the following advantages:

- Cost, weight and construction savings due to the reduction in wiring (valves, pumps and actuators, etc are operated by data signals travelling along one communication bus instead of electrical signals travelling down individual circuits).
- Greater redundancy and operational reliability due to a much lower number of plug-in connectors and easy use of multiple transmission paths.
- Much simpler and reduced construction time.
- Reduced commissioning time due to much simpler connecting procedures.
- Multiple use of the sensor signals.
- Simple upgrade procedures.

Computer systems can be networked in a number of different ways. The most popular configurations are:

- Ring
- Star
- Bus

Volume 10 (*Instrumentation and Control Systems*) of the Reeds Marine Engineering Series explains the different communication protocols in more detail. However, students should know that Controller Area Networks (CAN) are the modern form of communication between devices linked together controlling the function of mechanical systems or equipment. Devices connected to a CAN bus network can send and receive 'packets' of data to/from the other devices or controllers connected to the system. Identification codes within the packets of instruction sets allow devices to follow or disregard the instructions contained within each packet.

One of the features of this system that promises to deliver so much is the production of real-time data. Machinery control, performance monitoring and maintenance systems can all be updated with real-time data. This will have substantial benefits for the efficiency of ship's machinery and for energy management. Remote monitoring will be made so much easier and therefore there will be knock-on effects for management systems, training and professional development of engineering officers.

The increased volume of data that can be recorded and transported about is many times the information that was in the older systems. This will lead to different ways of thinking; for example, humans can only process a limited amount of data in a very short space of time, therefore the control and information systems will need to filter out the 'unimportant' data. The challenge is to construct algorithms that ensure transmission of the correct and important information to the engineering officers. Supporting or additional (non-critical) information should be held back or placed to one side for inspection when requested. As the development of Artificial Intelligence progresses, so will the efficiency and use of these systems. Ships are already being designed and built that can operate with little or no intervention from people.

Standardisation has to be the key to progress in this field of engineering. Manufacturers will be designing and manufacturing their own equipment that will need to interface and communicate with other manufacturers' equipment. If there is not a common set of rules for the design and operation of the communication part of the equipment's operation, then nothing will work together as intended. The NMEA 2000 network standard sets out the uniform requirements for a serial based communications network. As mentioned the CAN system is a broadcast system where instructions from controllers are sent to all the components on the system. The instruction is only picked up by the component that it is intended for. However, students can see that if a malicious third party wanted to introduce their own component into the system then that component would have access to all the instructions broadcast. Therefore, one of the most important tasks of a standard is to outline the system for security and accuracy of data.

Unmanned Machinery Spaces (UMS)

In the past a UMS designated space may have been known under other names such as 'unattended' machinery space; however, IMO has chosen the term 'unmanned' and this is the terminology used in the STCW document as well as 'M' notices in the United Kingdom and by all participating Flag States.

Vessels that have a machinery space designated as unmanned does not mean that the engine room is unattended or unsupervised; in fact, quite the opposite is the case. STCW actually describes such an engine room as 'periodically unmanned engine-room', which is the key description to the system.

The watchkeeping engineering officer is just as much supervising, controlling, monitoring and working closely with the machinery as she/he would in a fully manned engine room. The difference is that some of the routine monitoring is undertaken by control systems that will activate a machinery alarm if any control condition goes outside a preset value. This means that the watchkeeper does not have to carry out routine adjustments to the machinery systems and his/her time is freed up to carry out other tasks. It also means, of course, that if the watchkeeper completes all the checks and duties, the machinery should operate for up to 8 h without the watchkeeper being physically present in the engine room.

The watchkeeper is, however, still on duty and responsible for the supervision of the machinery space, but to be effective she/he can monitor the alarm system remotely from the vessel's accommodation block. In the event of a machinery alarm the watchkeeper must be present in the machinery control room (MCR) and respond to the machinery alarm within 90 s of the alarm first sounding.

All modern ships are built with a sophisticated alarm and monitoring control system and they will all have the ability to run with UMS; however, not all will be operated in this way. Passenger ships, for example, have 15,000 alarm points but they still have an engineering watchkeeper in the MCR at all times due to the reassuring message that this sends to the customers.

Cargo ships on ocean passage will be able to operate the machinery space unmanned if the vessel has the appropriate certificate to do so. To get the approval, the vessel must have the following:

1. *Bridge control of propulsion machinery:* The bridge watchkeeper must be able to take engine control action in the event of a vessel emergency such as a navigational manoeuvre. Control and instrumentation must be as simple as possible for the bridge watchkeeper to use.
2. *Centralised control and instruments are required in machinery space:* Engineers may be called to the machinery space to answer a routine alarm or in the case of an emergency and controls must be easily reached and fully comprehensive.
3. *Automatic fire detection system:* Alarm and detection systems must operate very rapidly. Numerous well-sited and quick response detectors (sensors) must be fitted.

4. *Fire extinguishing system:* In addition to conventional hand extinguishers, a control fire station remote from the machinery space is essential. The station must give control of emergency pumps, generators, valves, ventilators, extinguishing media, etc.

5. *Alarm system:* A comprehensive machinery alarm system must be provided and repeater stations must be available in the accommodation areas.

6. *Automatic bilge high-level fluid alarms and pumping units:* Sensing devices in bilges with alarms and hand or automatic pump cut-in devices must be provided.

7. *Automatic start emergency generator:* Emergency generators must be situated outside the machinery space and connected to separate emergency bus bars. The primary function is to give protection from electrical blackout conditions.

8. *Local hand control of essential machinery.*

9. *Adequate settling tank storage capacity:* The watchkeeper will have to ensure that the engine has enough fuel to operate for the 8 hours that the engine room will be unmanned. If this is not done then the low-level alarms will sound and the watchkeeper will have to complete the task probably at a time that disturbs his/her rest.

10. *Regular testing and maintenance of instrumentation.*

7

DIESEL ENGINE SUPPORT SYSTEMS

Introduction

Small compression ignition (CI) and spark ignition (SI) engines can generally be described as having attached all the equipment that the engine needs to operate continuously. Equipment such as starting mechanisms, lubricating oil, fuel oil and cooling water pumps.

The larger engines used on ocean-going ships produce significant amounts of heat and therefore require support systems that are too large to fit on, or close to, the engine. Therefore, equipment such as pumps and heat exchangers need to be strategically located around the engine room, and the fluids piped to and from the engine.

This equipment is organised into different systems and requires careful monitoring. The temperatures and pressures within these systems are measured and reported back to the central machinery control room (MCR). Sophisticated control systems keep the systems performing optimally, which is important for the efficient running of the engine. It is, however, important that the Engineering Officer of the Watch (EOOW) understands the correct operation of these systems so that they know when something is not operating correctly.

For this to happen a detailed knowledge about the constructional details and operation of the machinery is vital to the EOOW's understanding.

Compressed Air

Compressed air is used for starting main and auxiliary diesels, operating whistles or typhons, testing pipe lines (e.g. CO_2 fire extinguishing system) and for workshop services. The latter could include pneumatic tools and cleaning lances as well as other hand tools and specialist tools such as engine exhaust valve or seat grinding wheels. The high-pressure compressed air for the starting of diesel engines will usually be stored in two large air receivers at around 30 kg/cm^2 (bar). The compressors will be low-volume, high-pressure machines, usually water-cooled. Classification societies require that the outlet temperature of the compressed air be kept at 98°C due to the risk of air start line fires. Class also require that at least two compressors are fitted to a vessel and that one must be propelled by an alternative power source (such as a diesel engine). Each compressor must have sufficient capacity to charge the starting air receiver from atmospheric pressure to full pressure in 1 h. The air receivers must have sufficient capacity to provide a minimum of twelve starts for a reversible engine and six starts for a non-reversible engine. It is important that air compressors do not carry over oil into the compressed air lines. If a fault develops in any of the air start valves, hot gases could start a serious fire or explosion in the air start line if oil is present. Due to the air being compressed there would be more oxygen than usual and the result could be violent. It is considered good watchkeeping practice to shut the isolating valves from the main air receivers while the engine is running to reduce the compressed air in the lines from the receivers to the engine. See Chapter 5 for more details of the starting air system.

The compressed air used for powering hand tools such as rotary wire brushes or needle guns is at a much lower pressure than the starting air, but the tools will require a considerable volume of air for them to operate properly and if there are two or three tools in operation at the same time then the air compressors will be working hard to keep up. The working air compressors operate at about 7 kg/cm^2 (bar) and are able to produce a high volume of air. The compressed air required for control engineering needs careful consideration as the air needs to be dry and carefully controlled if delicate controls and instruments are to work correctly. Instrument air compressors can be of the screw type, which may also be described as oil-injected or oil-free compressors. The compressors up to about 30 kW are generally air cooled and above 30 kW, fresh water cooling is available. The instrument air system must be free from both oil and water contamination for the instruments and controls to work properly, which is why the special 'oil-free' compressors are available. Instrument air can also be provided via a reducer/dryer combination fitted to the main air or working air system.

Air is composed of mainly 23% oxygen, 77% nitrogen by mass and since these are near perfect gases a mixture of them will behave as a near perfect gas, following Boyle's and Charles's laws (see Volume 3 of the Reeds series). When air is compressed its temperature and pressure will increase as its volume is reduced. There are several theoretical models for this process, which are as follows.

Isothermal compression

Isothermal compression of a gas is compression at constant temperature. This would mean in practice that as the gas is compressed heat would have to be taken from the gas at the same rate as it is being received. This would necessitate a very slow-moving piston in a well-cooled small-bore cylinder, which is not practical for an actual design.

Adiabatic compression

Adiabatic compression of a gas is compression under constant enthalpy conditions, that is, no heat is given to or taken from the gas through the cylinder walls and all the work done in compressing the gas is stored within it. Again, this is not easy to build as a practical solution.

In figure 7.1, the two compression curves show that there is extra work done by compressing adiabatically, therefore it would be better to compress as close to an isothermal compression as is possible. In practice, this presents a problem – if the compressor were slow running with a small-bore perfectly cooled cylinder and a long stroke piston, the air delivery rate would be very low.

Multi-stage compression

If we had an infinite number of stages of compression with coolers in between each stage returning the air to ambient temperature, we would be able to compress over the desired range close to isothermal conditions. However, this is still impracticable

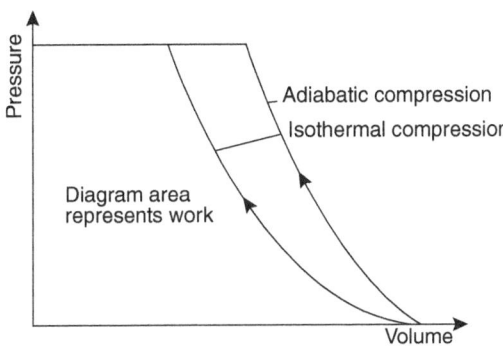

▲ Figure 7.1 *A comparison of compression processes*

and therefore a two- or three-stage compression with inter-stage and cylinder cooling is generally used when relatively high pressures have to be reached.

Figure 7.2 shows the work saved by using this method of air compression, but even with efficient cylinder cooling the compression curve is nearer the adiabatic than the isothermal and the faster the delivery rate the more this will be the situation.

To prevent overheating and consequential damage, cylinders have to be water- or air-cooled and clearance must be provided between the piston and the cylinder head. This clearance must be as small as practicable.

High-pressure air remaining in the cylinder after compression and delivery will expand on the return stroke of the piston. This expanding air must fall to a pressure below that in the suction manifold before a fresh air charge can be drawn in. Hence, part of the return or suction stroke of the piston is non-effective. This non-effective part of the suction stroke must be kept as small as possible in order to keep capacity to a maximum. This clearance is sometimes referred to as the 'bump clearance'; see the section 'Measuring the air compressor clearance'.

Volumetric efficiency

This is a measure of compressor capacity. It is the ratio of the actual volume of air drawn in each suction stroke to the stroke volume. Figure 7.3 shows what would happen to the compressor volumetric efficiency – and hence capacity – if the clearance volume were increased.

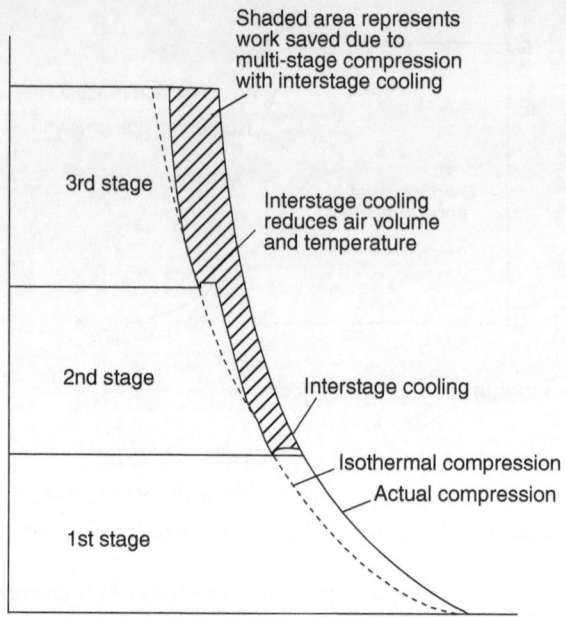

Shaded area represents
work saved due to
multi-stage compression
with interstage cooling

3rd stage

Interstage cooling
reduces air volume
and temperature

2nd stage

Interstage cooling

Isothermal compression
Actual compression

1st stage

▲ **Figure 7.2** *Three-stage compression*

Clearance volume can be calculated from an indicator card by taking any three points on the compression curve such that their pressures are in geometric progression, that is, $P_1/P_2 = P_2/P_3$ hence $P_2 = \sqrt{P_1 P_3}$ (figure 7.4). If V_c = clearance volume as a percentage of the readily calculable stroke volume and V_1, V_2, V_3 are also percentages of the stroke volume then:

$$P_1(V_1 + V_c)^n = P_2(V_2 + V_c)^n = P_3(V_3 + V_c)^n$$

i.e. $\dfrac{P_1}{P_2} = \left(\dfrac{V_2 + V_c}{V_1 + V_c}\right)^n$ and $\dfrac{P_2}{P_3} = \left(\dfrac{V_3 + V_c}{V_2 + V_c}\right)^n$

now $\dfrac{P_1}{P_2} = \dfrac{P_2}{P_3}$,

therefore $\left(\dfrac{V_2 + V_c}{V_1 + V_c}\right)^n = \left(\dfrac{V_3 + V_c}{V_2 + V_c}\right)^n$

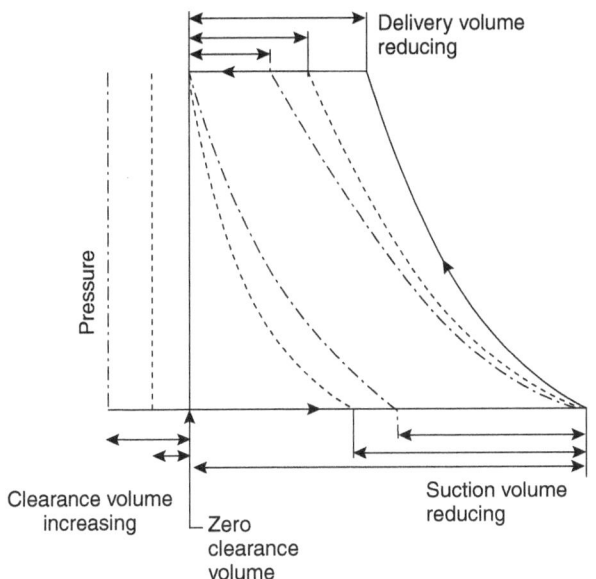

▲ **Figure 7.3** *Effects of increasing clearance volume*

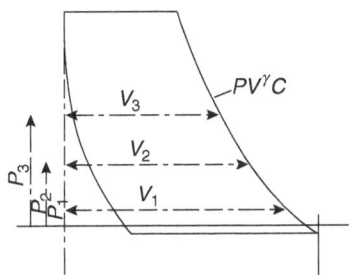

▲ **Figure 7.4** *Calculating clearance volume*

$$\text{hence}\quad \frac{V_2 + V_c}{V_1 + V_c} = \frac{V_3 + V_c}{V_2 + V_c}$$

$$V_c = \frac{V_2^2 - V_1 V_3}{V_1 + V_3 - 2V_2}.$$

Since V_1, V_2 and V_3 are known V_c can be calculated.

Measuring the air compressor clearance

Correct clearance must be maintained and this is usually done by checking the mechanical clearance between the top of the piston and the cylinder head (called the bump clearance) and adjusting it as required by altering the height of the piston relative to the cylinder and cylinder head. This is usually done by using inserts under the palm of the connecting rod. Bearing clearances should also be kept at recommended values because any wear in these bearings will also alter the bump clearance by moving the piston relative to the cylinder head. Two possible methods of ascertaining the mechanical clearance in an air compressor are:

1. Remove the suction or discharge valve assembly from the unit and place a small loose ball of lead wire on the piston edge, then rotate the flywheel by hand to take the piston over TDC. Remove and measure the thickness of the lead wire ball.

2. Put the crank on TDC, slacken or remove the bottom half of the bottom end bearing. Rig a clock gauge with one contact touching some underpart of the piston or piston assembly and the other on the crank web. Take a gauge reading. Then by using a suitable lever bump the piston, that is, raise it until it touches the cylinder cover. Take another gauge reading. The difference between the two readings gives the mechanical clearance.

EXTRA SPECIAL CARE MUST BE TAKEN TO ISOLATE THE COMPRESSOR BEFORE UNDERTAKING THIS WORK. This is a very important point and the Flag State examiner will be checking to ensure that candidates appreciate the importance of this procedure.

In practice, the effective volume of air drawn in per stroke is further reduced by the pressure in the cylinder, which on the suction stroke must fall sufficiently below the atmospheric pressure so that the inertia and spring force of the suction valve can be overcome and air under the force of atmospheric pressure will flow into the cylinder. Figure 7.5 shows this effect on the indicator card and also shows the excess pressure above the mean required upon delivery, to overcome delivery valve inertia and spring force and push the compressed air out of the cylinder.

Air compressors are either reciprocating or rotary types. The former are most commonly used at sea for the production of air for starting diesel engines or for driving power tools as outlined at the beginning of this chapter.

Reciprocating air compressors at sea are generally two- or three-stage types with inter-stage cooling. Figure 7.6 shows diagrammatically a tandem type of three-stage

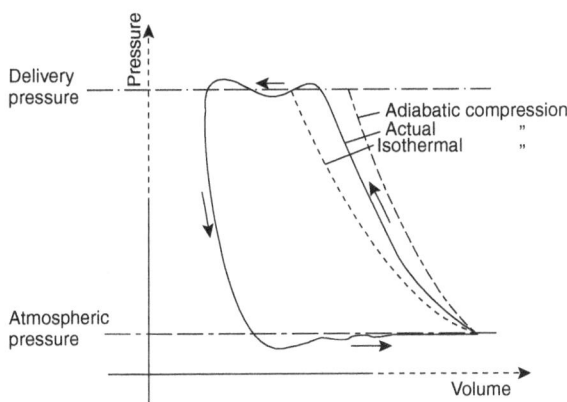

▲ **Figure 7.5** *Actual pressure–volume*

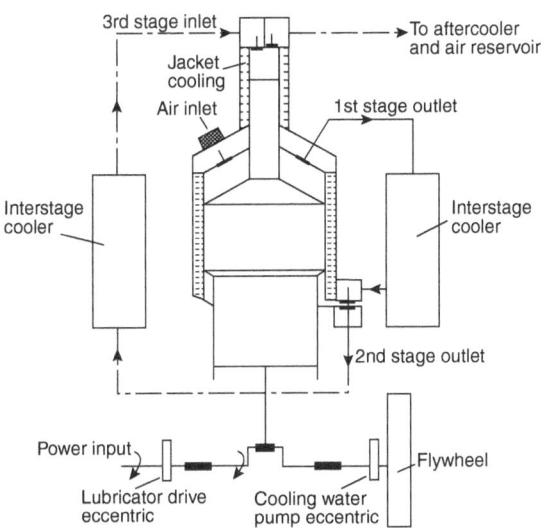

▲ **Figure 7.6** *Three-stage air compressor*

compressor. The pressures and temperatures at the various points would be roughly as follows:

	Delivery pressure	Air temperature	
		Before the coolers	After the coolers
First stage	4 bar	110°C	35°C
Second stage	16 bar	110°C	35°C
Third stage	40 bar	70°C	25°C

The above figures are for a salt water temperature of about 16°C. Final air temperature at exit from the after-cooler is generally at or below atmospheric temperature.

Drains

Fitted after each cooler is a drain valve; these are essential. To emphasise, if we consider 30 m³ of free air, relative humidity 75%, temperature 20°C being compressed every minute to about 10 bar, about ½ litre of water would be obtained each minute.

Drains and valves to the air storage unit must be open upon starting up the compressor in order to get rid of accumulated moisture. When the compressor is running, drains have to be opened and closed at regular intervals.

Filters

Air contains suspended foreign matter, much of which is abrasive. If this is allowed to enter the compressor it will combine with the lubricating oil to form an abrasive-like paste, which increases wear on piston rings, liners and valves. It can adhere to the valves and prevent them from closing properly, which in turn can lead to higher discharge temperatures and the formation of what appears to be a carbon deposit on the valves, etc. Strictly, the apparent carbon deposit on valves contains very little carbon from the oil, it is mainly solid matter from the atmosphere.

These carbon-like deposits can become extremely hot on valves that are not closing correctly and could act as ignition points for air–oil vapour mixtures, leading to possible fires and explosions in the compressor.

Hence air filters are extremely important. They must be regularly cleaned and where necessary renewed and the compressor must never be run with the air intake filter removed.

Relieving devices

After each stage of compression, a relief valve will normally be fitted. Regulations only require the fitting of a relieving device on the h.p. stage. Bursting discs or some other relieving device are fitted to the water side of coolers so that in the event of a compressed air carrying tube bursting, the sudden rise in pressure of the surrounding water will not fracture the cooler casing. In the event of a failure of a bursting disc a thicker one must not be used as a replacement.

Lubrication

Certain factors govern the choice of lubricant for the cylinders of an air compressor. These are: operating temperature, cylinder pressures and air condition. Students will recognise that it is the job of the Engineering Officer in charge of the watch to ensure that the correct grade of oil is used when topping up the air compressor. This will be detailed on the vessel's oil schedule, which is kept in the MCR and will be produced by the oil manufacturers. To comply with the Control of Substances Hazardous to Health (COSHH) regulations, the specification of the oil should also be available to the ship's crew.

Operating temperature

This affects oil viscosity and deposit formation. If the temperature is high, this results in low oil viscosity, very easy oil distribution, low film strength, poor sealing and increased wear. If the temperature is low, oil viscosity would be high. This causes poor distribution, increased fluid friction and power loss.

Cylinder pressures

If these are high, the oil requires to have a high film strength to ensure the maintenance of an adequate oil film between the piston rings and the cylinder walls.

Air condition

Air contains moisture that can condense out. Straight mineral oils would be washed off surfaces by the moisture and this could lead to excessive wear and possible rusting. To prevent this a compounded oil with a rust inhibitor additive would be used. Compounding agents may be from 5% to 25% of non-mineral oil, which is added to a mineral oil blend. Fatty oils are commonly added to lubricating oil that must lubricate in the presence of water; they form an emulsion, which adheres to the surface to be lubricated.

Two-stage air compressors

Most modern diesel engines use starting air at a pressure of about 26–30 bar and to achieve this, a two-stage type of compressor would be adequate. These compressors are generally of the reciprocating type, with various possible arrangements of the cylinders, or they could be a combination of a rotary first stage followed by a reciprocating high-pressure stage. This latter arrangement leads to a compact, high delivery rate compressor.

Figure 7.7 shows a typical two-stage reciprocating type of air compressor. The pressures and temperatures at the various points would be approximately as follows:

	Delivery pressure	Air temperature	
		Before the coolers	After the coolers
First stage	4 bar	130°C	35°C
Second stage	26 bar	130°C	35°C

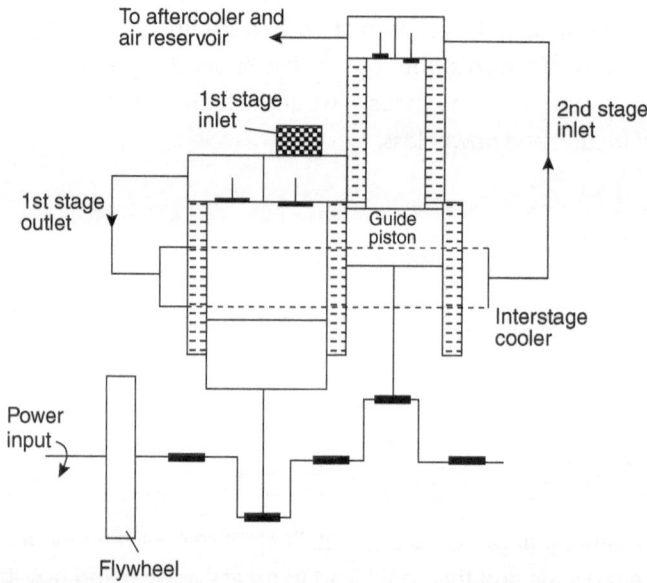

▲ **Figure 7.7** *Two-stage air compressor*

Compressor valves

An air compressor is a very simple device, but it is an essential piece of equipment for the correct operation of the machinery plant. The key to ensuring efficient operation of the compressor is in keeping the parts making up the 'compression' chamber in good working order. These are the pistons, piston rings and suction and discharge valves.

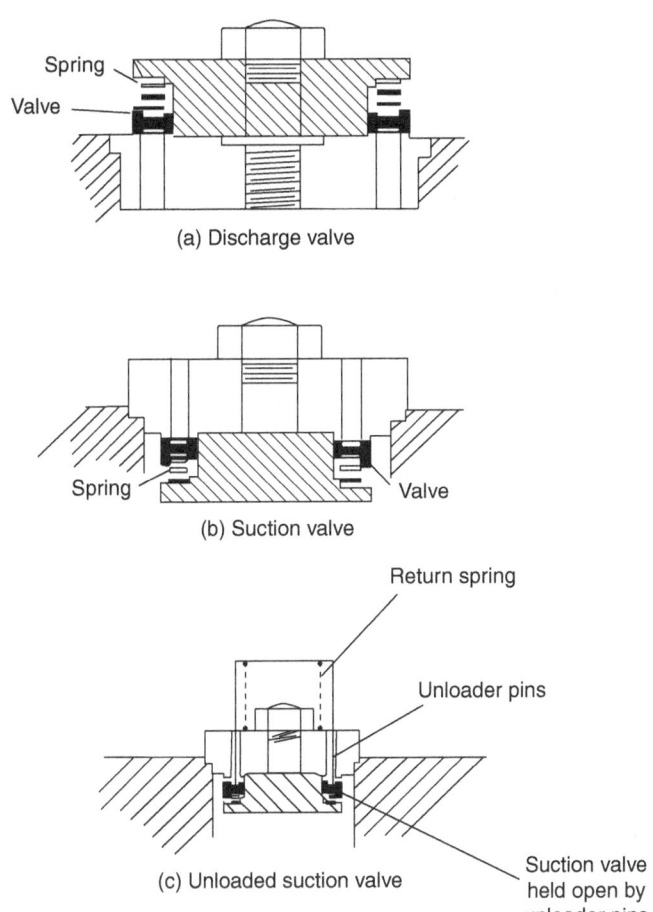

Spring

Valve

(a) Discharge valve

Spring   Valve

(b) Suction valve

Return spring

Unloader pins

Suction valve
held open by
unloader pins

(c) Unloaded suction valve

▲ **Figure 7.8** *Compressor valves*

The arrangement of the valves is shown in figure 7.8(a–c). Students should note that these would be suitable diagrams for reproduction in a Flag State administration's examination. Modern valves are somewhat more streamlined and lighter in order to reduce friction losses and valve inertia. Materials used in the construction are generally the following:

Valve seat

About 0.4% carbon steel hardened and polished working surfaces.

Valve

Nickel steel, chrome vanadium steel or stainless steel, hardened and ground, then finally polished to a mirror finish.

Spring

Hardened steel (*Note:* all hardened steel would be tempered).

Valve leakages do occur in practice and this leads to loss of efficiency and increase in running time.

Effects of leaking valves

1. *First-stage suction:* Reduced air delivery, increased running time and reduced pressure in the suction to the second stage. If the suction valve leaks badly it may completely unload the compressor.

2. *First-stage delivery:* With high-pressure air leaking back into the cylinder, less air can be drawn in. This means reduced delivery and increased discharge temperature.

3. *Second-stage suction:* High pressure and temperature in the second-stage suction line, reduced delivery and increased running time.

4. *Second-stage delivery:* Increased suction pressure in second stage, reduced air suction and delivery in second stage. Delivery pressure from first stage increased. Figure 7.9 shows the effect of a leaking second-stage delivery valve on the indicator cards of a compressor.

It must be remembered that it is not usual to find a facility for taking indicator cards from air compressors.

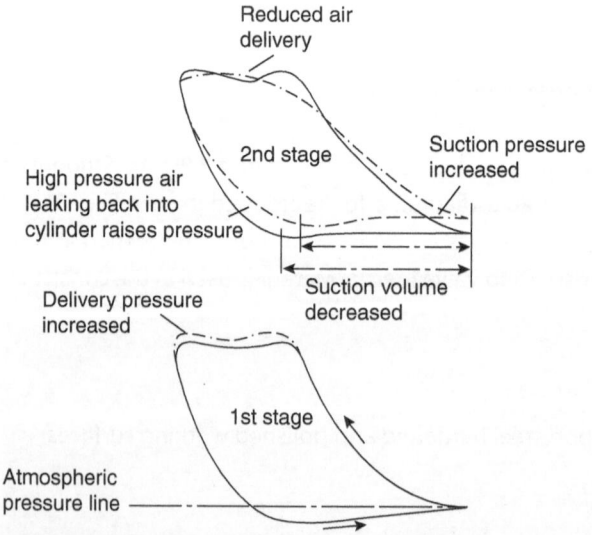

▲ **Figure 7.9** *Effect of leaking second-stage delivery valves*

Methods of regulating air compressors

We have seen that the role of the air compressor is to raise the pressure of air from a suction to a discharge by inputting energy into the medium. The compressed air has several applications and there are different types of compressor that have been developed to handle the different applications. These different types of compressor also have different methods of control depending upon the type and application of compressor. Therefore, to allow compressors to respond to fluctuations in system demand they are linked to an automatic pressure regulation controller and the controller will start a process to alter the output of the compressor. The most popular methods in use to date are given below.

Start stop control

A general observation would be that the torque required to drive a compressor increases with the speed of the machine. Also, the starting torque can be very high as is the case with reciprocating compressors. Some are fitted with star–delta starters but others are still direct online and for this reason the start–stop technology will only be suitable for electrically driven units. A pressure transducer attached to the air receiver set for desired max–min pressures would switch the current to the electric motor's starter either on or off. Drainage would have to be automatic and air receiver relatively large compared to the compressor unit requirements so that the number of starts per unit of time is not too great. It must be remembered that the starting current for an electric motor is about double the normal running current. During its operation the compressor does operate at its optimum efficiency and if the machine is stopped for long periods of time then the overall performance is acceptable.

Constant running control

This method of control is the one used most often for the higher volume machine running at a relatively low pressure. The compressor runs continuously at a constant speed and when the desired air pressure is reached the air compressor is unloaded in some way so that the air is NOT delivered and practically no work is done in the compressor cylinders.

The methods used for compressor unloading vary, but that most commonly used is to shut off the air to the suction side of the compressor. If the compressor receives no air then it cannot deliver any, or if the air taken in at the suction is returned to the suction, again no air will be delivered. In either case virtually no work would be done in the compressor cylinder or cylinders and this would provide an economy compared to discharging high-pressure air to the atmosphere through a relief valve.

Figure 7.10 shows diagrammatically a compressor unloading valve fitted to the compressor suction. When the discharge air pressure reaches a desired value it will act on the piston causing the spring-loaded valve to close, shutting off the supply of air to the compressor.

An alternative method of unloading the compressor, while continuing to run it, is to hold the suction valve open. When the compressor is unloaded the suction valve plates are held open by pins, which are operated by a relay valve and piston, not unlike that shown in figure 7.10. When the pressure in the air reservoir falls to a preset level, the piston's chamber is vented and return springs push out the holding pins, allowing the suction valve to operate normally (figure 7.8).

An alternative method of optimising the use of compressors is to have several smaller machines running in parallel. The number of machines running can be adjusted depending upon the demand. Some air compressors will employ an 'un-loading' system where a solenoid-operated unloading/drain valve can be used to reduce the energy required for starting/stopping the compressor. This has the added value of being able to drain any condensate from the compressor at regular intervals.

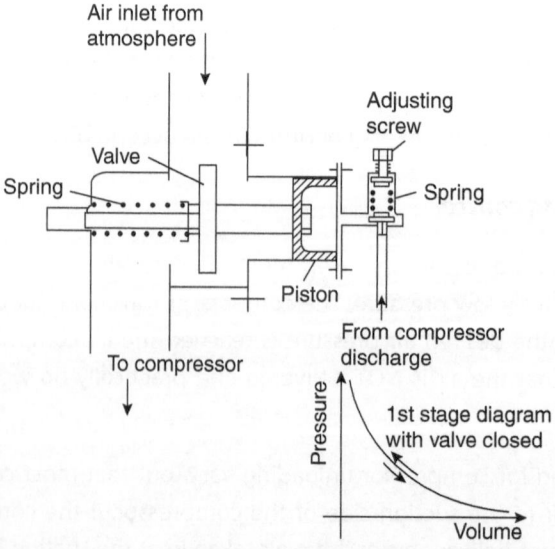

▲ **Figure 7.10** *Compressor unloading valve*

Variable-speed control

Modern electronics has allowed the development of thyristor control of a.c. synchronous motors and has added another dimension to the efficiency of machinery driven in this way. The ability to vary the speed of the compressor presents a number of major advantages over other methods of control. These are:

- A gradual start-up and increase in speed, meaning that there are no sudden peaks in the current supply to the motor. The stress of sudden acceleration on the mechanical components is reduced.
- The pressure can be controlled to a much finer tolerance because the speed and therefore the flow rate can be adjusted to match the demand. This reduces the range of the pressure fluctuations and also the stress on the pressure parts of the system. Initially, systems can be designed using smaller receivers.
- Due to the efficiency being optimised, so is the use of energy and therefore there will be a fuel saving for the vessel.
- Variable speed control is also suited to compressors operating in parallel. Here, one of the machines can be optimised by speed control while the others operate on an on–off basis.

Automatic drain

Figure 7.11 shows an automatic air drain trap, which functions in a near similar way to a steam trap. With water under pressure at the inlet the disc will lift, allowing

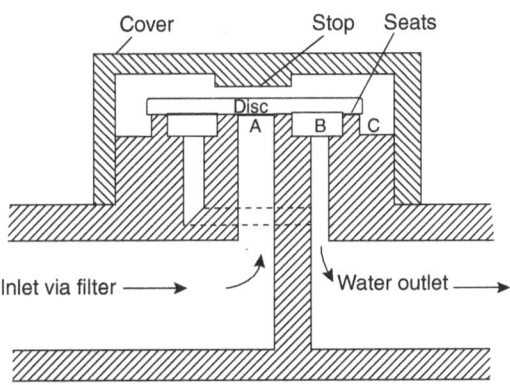

▲ **Figure 7.11** *Air drain trap*

the water to flow radially across the disc from A to the outlet B. When the water is discharged, the air now flows radially outwards from A across the disc. The air expands, increasing in velocity and ramming air into C and the space above the disc, causing the disc to close on the inlet. Because of the build-up of static pressure in the space above the disc in this way, and the differential area on which the pressures are acting, the disc is held firmly closed. It will remain so unless the pressure in the space above the disc falls.

In order that this pressure can fall, and the trap reopen, a small groove is cut across the face of the disc communicating B and C, through which the air slowly leaks to the outlet.

Obviously this gives an operational frequency to the opening and closing of the disc which is a function of various factors, for example, size of groove, disc thickness and volume of space above the disc. Therefore, it is essential that the correct trap be fitted to the drainage system to ensure efficient and effective operation. These traps should be checked by the watchkeeper by listening for their operation. After a while in operation, debris in the water can cause grooves to form across the disc and they stop working.

Air Vessels

Material used in the construction must be of good-quality low-carbon steel similar to that used for boilers, for example, 0.2% carbon (max.), 0.35% silicon (max.), 0.4% manganese, 0.05% sulphur (max.), 0.05% phosphorus (max.), u.t.s. 460 MN/m^2.

Welded construction has superseded the rivetted types and welding must be completed to class 1 or class 2 depending upon operating pressure. If above 35 bar approximately, then class 1 welding regulations apply.

Some of the main points relating to class 1 welding are that the welding must be radiographed, annealing must be carried out at a temperature of about 600°C and a test piece must be provided for bend, impact and tensile tests, together with micro-graphic and macro-graphic examination.

Mountings generally provided are shown in figure 7.12. If it is possible for the receiver to be isolated from the safety valve then it must have a fusible plug fitted, melting point approximately 150°C, and if carbon dioxide is used for fire fighting it is recommended

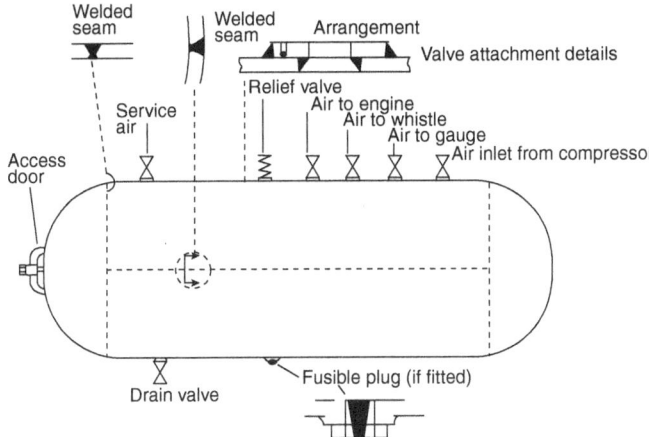

▲ **Figure 7.12** *Air reservoir*

that the discharge from the fusible plug be led to the deck. Stop valves on the receiver generally permit slow opening to avoid rapid pressure increases in the piping system, and piping for starting air has to be protected against the possible effects of explosion.

Drains for the removal of accumulated oil and water are fitted to the compressor, filters, separators, receivers and lower parts of pipelines. Before commencing to fill the air vessel after overhaul or examination, ensure that:

1. Nothing has been left inside the air vessel, for example cotton waste that could foul up drains or other outlets.
2. Check pressure gauge against a master gauge.
3. All doors are correctly centred on their joints.

Run the compressor with all drains open to clear the lines of any oil or water, and when filling open drains at regular intervals, observe pressure. After filling, close the air inlet to the bottle, check for leaks and follow up on the door joints. When emptying the receiver prior to overhaul, etc, ensure that it is isolated from any other interconnected receiver, which must, of course, be in a fully charged state.

Cleaning the air receiver internally must be done with caution. Any cleaner that gives off toxic, inflammable or noxious fumes should be avoided. A brush down and a coating on the internal surfaces of some protective, harmless to personnel, such as a graphite suspension in water, could be used.

Cooling Systems

These can conveniently be grouped into sections.

1. *Cylinder cooling or jacket cooling:* normally fresh or distilled water (figure 7.13). This may incorporate cooling of the turbine or turbines in a turbocharged engine and exhaust valve cooling.

2. *Fuel valve cooling:* This would be a separate system using fresh water or a fine mineral oil.

3. *Piston cooling:* This may be lubricating oil, distilled or fresh water. If it is oil, the system is generally common with the lubrication system. If water, a common storage tank with the jacket cooling system would generally be used.

4. *Charge air cooling:* This is normally sea water.

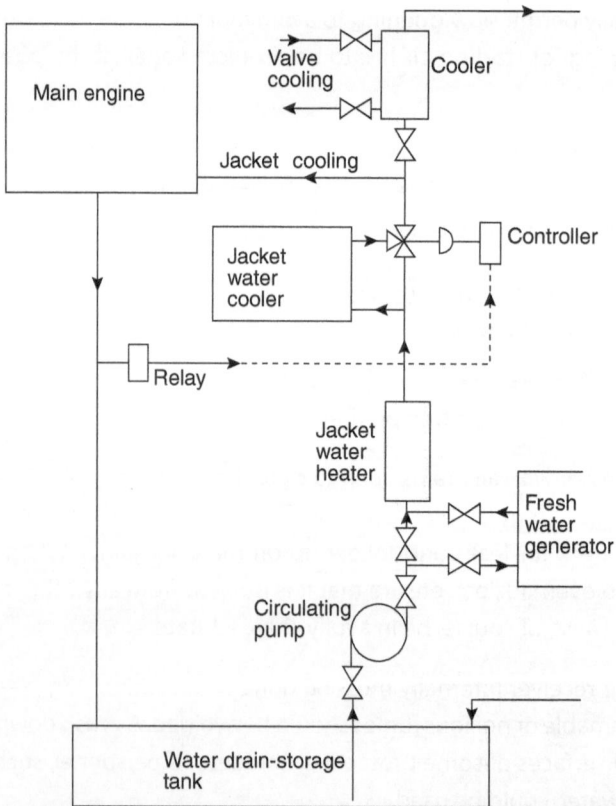

▲ **Figure 7.13** *Jacket cooling system*

Load-controlled cylinder cooling

In an effort to reduce the danger of local liner corrosion over the whole engine load, some manufacturers are employing cooling systems that are load dependent. In such a system, shown in figure 7.14, the cooling flow is split into a primary circuit, bypassing the liner, for cylinder head cooling. In the secondary circuit, uncooled water from the engine outlet is directed to cool the liner. To avoid vapour formation as a result of maintaining higher cooling temperatures, the system is pressurised to 4–6 bar.

The advantages claimed for such a system include:

1. Possible savings in cylinder lubrication oil feed rate.
2. Omission of cylinder bore insulation.
3. Reduced cylinder liner corrosion.

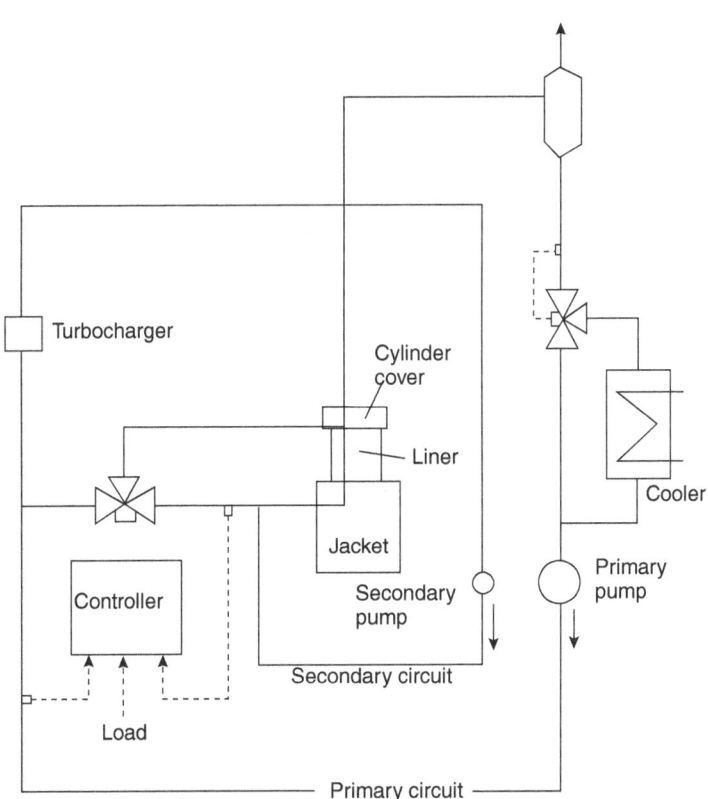

▲ **Figure 7.14** *Load-controlled cylinder cooling*

Comparison of coolants

Fresh water

Inexpensive, high specific heat, low viscosity. Contains salts that can deposit, obstruct flow and cause corrosion. Requires treatment. Leakages could contaminate lubricating oil system leading to loss of lubrication, possible overheating of bearings and bearing corrosion. Requires a separate pumping system.

It is important that water should not be changed very often as this can lead to increased deposits. Leakages from the system must be kept to an absolute minimum, so a regular check on the replenishing-expansion tank contents level is necessary.

If the engine has to stand inoperative for a long period and there is a danger of frost, (a) drain the coolant out of the system, (b) heat up the engine room or (c) circulate the system with heating on. It may become necessary to remove scale from the cooling spaces and the following method could be used. Circulate, with a pump, a dilute hydrochloric acid solution. A hose should be attached to the cooling water outlet pipe to remove gases. Gas emission can be checked by immersing the open end of the hose occasionally into a bucket of water. Keep the compartment well ventilated as the gases given off can be dangerous. Acid solution strength in the system can be tested from time to time by putting some onto a piece of lime. When the acid solution still has some strength and no more gas is being given off, the system is scale free. The system should now be drained and flushed out with fresh water, then neutralised with a soda solution and pressure tested to see that the seals do not leak.

Distilled water

More expensive than fresh water, high specific heat, low viscosity. If produced from evaporated salt water it would be acidic. No scale-forming salts. Requires separate pumping system. Leakages could contaminate the lubricating oil system, causing loss of lubrication and possible overheating and failure of bearings, etc.

Additives for cooling water

Those generally used are either anti-corrosion oils or inorganic inhibitors. If pistons are water cooled, an anti-corrosion oil is recommended as it lubricates parts that have sliding contact. The oil forms an emulsion and part of the oil builds up a thin unbroken film on metal surfaces. This prevents corrosion but is not thick enough to impair heat transfer. Inorganic inhibitors form protective layers on metal surfaces, guarding them against corrosion.

It is important that the additives used are not harmful if they find their way into drinking water – this is possible if the jacket cooling water is used as a heating medium in a fresh water generator. Emulsion oils and sodium nitrite are both approved additives, but the latter cannot be used if any pipes are galvanised or if any soldered joints exist. Chromates cannot be used if the cooling water is used in a fresh water generator as it is a chemical that must be handled with care.

Lubricating oil

This is expensive and generally there is no separate pumping system required (see Figure 7.15) since the same oil is normally used for lubrication and cooling. Leakages from the cooling system to the lubrication system are relatively unimportant provided they are not too large; otherwise one piston may be partly deprived of coolant with subsequent overheating.

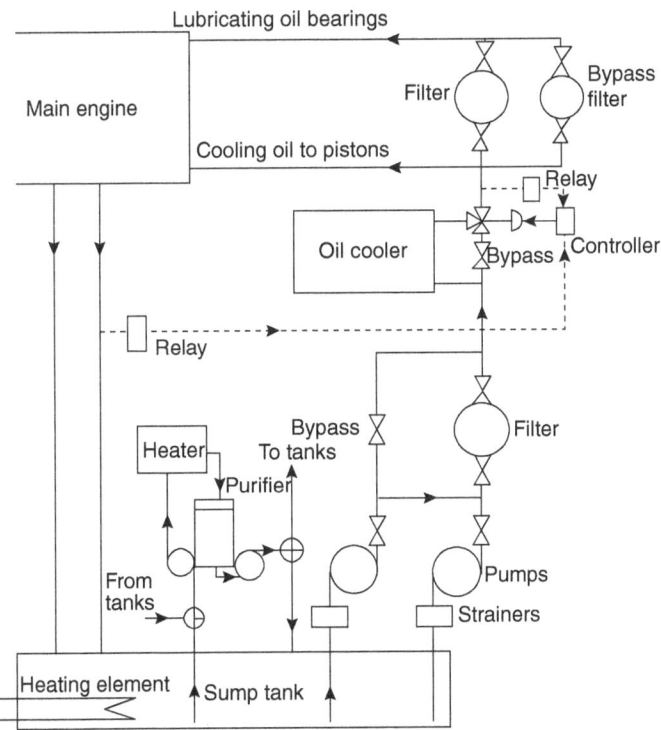

▲ **Figure 7.15** *Lubricating and cooling oil system*

Due to the reciprocating action of pistons, some relative motion between parts in contact with the coolant supply and return system must occur; oil will lubricate these parts more effectively than water. No chemical treatment required. Lower specific heat than water, hence a greater quantity of oil must be circulated per unit time to give the same cooling effect.

If the lubricating oil is subject to a high temperature it can burn, leaving carbon deposit as it does so. This deposit on the underside of a piston crown could lead to impairment of heat transfer, overheating and failure of the metal. Generally the only effective method of dealing with the carbon deposit is to dismantle the piston and physically remove it. Since oil can burn in this way, a lower mean outlet and inlet temperature of the oil has to be maintained. In order to achieve this, more oil must be circulated per unit time.

Some engines may use completely separate systems for oil cooling of pistons and bearing lubrication. The advantages gained by this method are as follows:

1. Different oils can be used for lubrication and cooling. A very low viscosity mineral oil would be better suited to cooling than lubrication.
2. Additives can be used in the lubricating oil that would be beneficial to lubrication, for example, oiliness agents, e.p. agents and V.I. improvers, etc.
3. Improved control over piston temperatures.
4. If oil loss occurs, then with separate systems the problem of detection is simplified and in the case of total oil loss in either system, the quality to be replaced would not be as great as for a common system.
5. Contamination of the oil in either system may take place. In the event, the problem of cleaning or renewal of the oil is not so great.
6. Oxidation of lubricating oil in contact with hot piston surfaces leads to rapid reduction in lubrication properties.

Disadvantages of having two separate systems are: greater initial cost due to separate storage, additional pipework and pumps. A sealing problem to prevent mixing of the two different oils is created and due to the increased complexity more maintenance would have to be carried out (figure 7.15).

Hydraulic Systems

Marine machinery is heavy and as a consequence some of the equipment cannot be operated or dismantled/assembled directly by people, without powerful assistance. The basic concepts of hydraulic power can be studied in the volume 2 (*Applied Mechanics*) of the Reeds Marine Engineering Series.

Hydraulic power is used in a number of applications onboard ships. One of the most important systems is the hydraulic steering gear (see volume 8 (*General Engineering Knowledge*) for more details). Another popular application is hydraulically-operated deck machinery. These could be hatch coverings or cargo handling equipment such as cranes. The windlasses and anchor winch could also be operated by hydraulic power.

A further common application is to see a hydraulic power pack situated in the machinery space and used for powering valves. Such systems can also be linked to the CANBus control systems for the main power plant.

8

MEDIUM-SPEED DIESELS

The term medium speed refers to diesels that operate within the approximate speed range of 300–800 rev/min. High speed is usually 1,000 rev/min and above.

The development of the medium-speed, usually four-stroke, engine has been considerable over the past 20 years and now it is a serious competitor for applications that were once only the domain of the large slow-speed two-stroke engines or the steam turbine.

The advantages and salient features of the medium-speed diesel are as follows:

1. Compact and space saving. The vessel can have reduced height and broader beam, which can give the vessel's owners the flexibility to offer a service to ports where shallow draught is of importance. The considerable reduction in engine height compared to direct drive engines and the reduced weight of components means that lifting tackle, such as the engine room crane, is reduced in size as it will have lighter loads to lift through smaller distances. More cargo space is made available and because of the higher power to weight ratio of the engine a greater weight of cargo can be carried.

2. Through using a reduction gear, a useful marriage between ideal engine speed and ideal propeller speed can be achieved. For optimum propeller speed, hull form and rudder have to be considered; the result is usually a slow-turning propeller (for large vessels this can be as low as 50–60 rev/min). Gearing enables the naval architect to design the best possible propeller for the vessels without having to consider any dictates of the engine. Engine designers can ignore completely propeller speed and concentrate solely upon producing an engine that will give the best possible power weight ratio.

3. Modern tendency is to utilise unidirectional medium-speed geared diesels coupled to either a reverse reduction gear, controllable pitch propeller (CPP) or electric generator. The second two of these methods are the ones primarily used in a number of new buildings and the advantages to be gained are considerable. They include:

   a. Less starting torque required, clutch disengaged or CPP in neutral.

   b. Reduced number of engine starts, hence starting air capacity can be greatly reduced and compressor running time minimised. Classification society requirements are six consecutive starts without air replenishment for non-reversible engines and twelve for reversible engines. Cylinder liner wear rate increases during starting.

   c. Engines can be tested at full speed with the vessel alongside a quay without having to take any special precautions.

   d. With the mechanical drive arrangement and the engine or engines running continuously, power can be taken off via a clutch or clutch/gear drive for the operation of electric generators or cargo pumps, etc. Hence the main engine has become a multi-purpose 'power pack'. This is especially the case when there are several engines being used to power generators in the 'power station' setup (see figure 9.1).

   e. Improved manoeuvrability, vessel can be brought to rest within a shorter distance by intelligent use of the engines and CPP.

   f. Staff workload during 'stand-by' periods is reduced and the system lends itself ideally to simple bridge control.

4. With two engines coupled via gearing, one may be disengaged and overhauled while the other supplies the motive power. This reduces off hire time as the voyage is continued at slightly reduced speed with a fuel saving.

5. Spare parts are easier to store and manhandle, therefore unit overhaul time will be greatly reduced.

Engine Couplings, Clutches and Gearing

Various arrangements of geared engines coupled together are possible. The basic arrangement depends upon the services the engine has to supply; for example, a high electrical load in port may have to be catered for with the alternator being driven at a higher speed than the engine. Hence, a step-up gearbox would be required along with some form of clutch. Large-capacity cargo pumps operating at high speed would require a similar arrangement. Figure 8.1 shows different types of arrangements with different types of clutches or couplings being used.

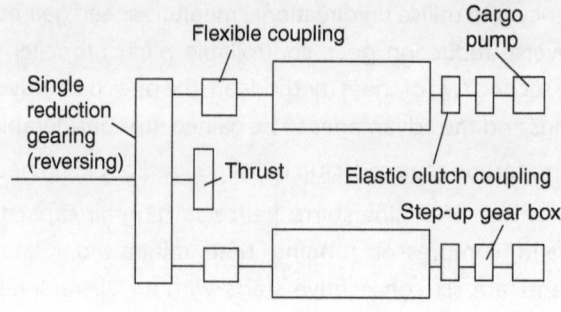

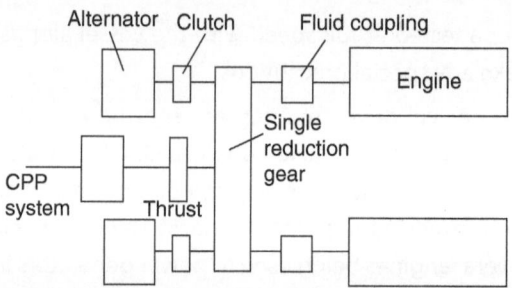

▲ **Figure 8.1** *Engine arrangements*

Fluid couplings

These are completely self-contained; apart from a cooling water supply, they require no external auxiliary pump or oil feed tank. A scoop tube when lowered picks up oil from the rotating casing reservoir and supplies it to the vanes for coupling and power transmission; withdrawal of the scoop tube from the oil stops the flow of oil to the vane; which then drains to the reservoir. During power transmission a flow of oil takes place continuously through the cooler and clutch.

Fluid clutches operate smoothly and effectively. They use a fine mineral lubricating oil and have no contact and hence no wear between driving and driven members. Torsional vibrations are dampened out to some extent by the clutch and transmitted speeds can be considerably less than engine speed if required by suitable adjustment of the scoop tube. It is possible to have a dual entry scoop tube for reversible engines; this obviates the use of CP propellers or reversible reduction gears but the control problem is considerably more complex with reversible engines, which have to be stopped and started, and if four-stroke engines are used, camshafts have to be moved, etc (figure 8.2).

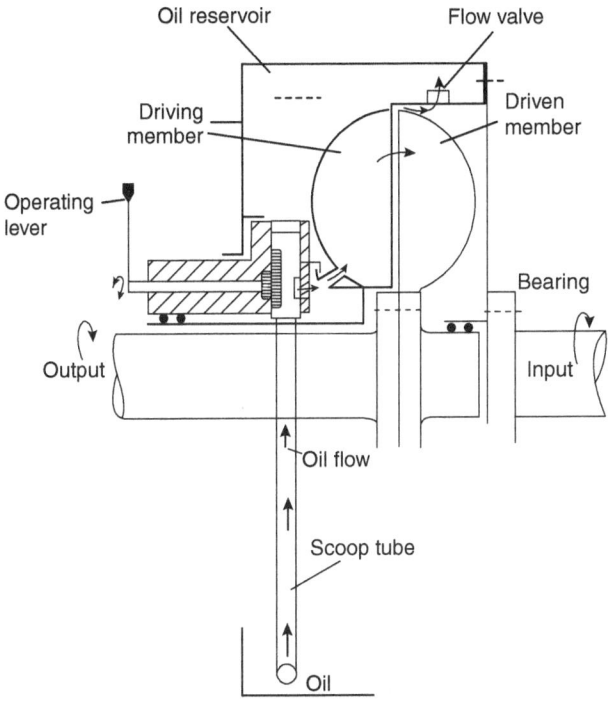

▲ **Figure 8.2** *Fluid coupling (vulcan)*

Reverse reduction gear

These gear systems are mainly restricted, at present, to powers of up to about 4800 kW for twin-engined single-screw installations. Their obvious advantages are as follows:

1. Unidirectional engine.
2. No CP propeller required.
3. Ability to engage or disengage either engine of a twin-engine installation from the bridge by a relatively simple remote control.
4. Improved manoeuvrability, etc.

When dealing with higher powers, the friction clutches used in the system can become excessively large, great heat generation during engagement may require a cooling system, the overall arrangement becomes more expensive and it may be cheaper to use direct reversing engines – however, it may also, for reasons previously outlined, be prudent to use a CP propeller.

Two systems of reverse reduction gear are shown in figures 8.3 and 8.4. In figure 8.3, the engine drives a steel drum that has two inflatable synthetic rubber tubes bonded to its inner surface. These tubes have friction material, like brake lining, on their inner surface. Air is supplied through the centrally arranged tube, or the annulus formed by the tube and shaft hole to one or other of the inflatable tubes. Two flanged wheels are connected via hollow shafts and gears to the main gearwheel and shaft.

For operation ahead, air would be supplied to inflatable tube A, which would then by friction on flanged wheel B bring gears 1 and 2 up to speed; gears 3, 4 and 5 together with flanged wheel D would be idling.

For astern operation, air would be supplied to inflatable tube C (A evacuated) and by friction on flanged wheel D gears 3, 4, 5 and 2 would be brought up to speed; gear 1 and drum B would be idling. For single reduction, gears 3 and 4 would be the same size and so would gears 1 and 5.

An alternative system, either single or double reduction but probably the latter, is shown in figure 8.4. Friction clutches A and B are pneumatically controlled from some remote position. Gears 1, 2, 3 and 4 would have to be the same size if the gear were to be single reduction – but this is most unlikely.

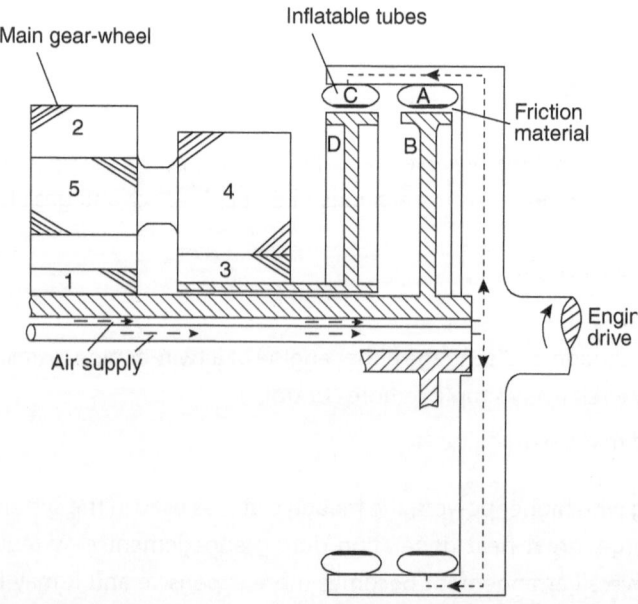

▲ **Figure 8.3** *Friction clutch*

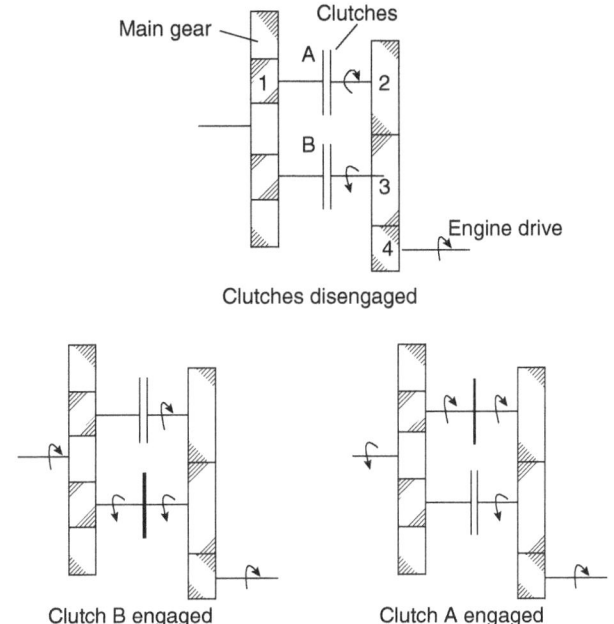

▲ Figure 8.4 *Reversible reduction gear*

Flexible couplings

These are used between the engine and gearbox to dampen down torque fluctuations, reduce the effects of shock loading on the gears and engine and cater for slight misalignments. They are also used in conjunction with clutches for power take-off when required. In construction they may be similar to the well-known multi-tooth type to be found in turbine installations or employ diaphragms or rubber blocks. Those types that use rubber or synthetic rubber, such as nitrile, give electrical insulation between driving and driven members, but all types will minimise vibration and reduce noise level.

Figure 8.5 shows a combination of flexible couplings and pneumatically operated friction clutch, the arrangement of which gives a smooth transition of speed and torque during engagement; it could be typical of an arrangement for the take-off for electrical power or cargo pumps, etc. The rubber blocks would be synthetic if oil is likely to be present as natural rubber is attacked by oil.

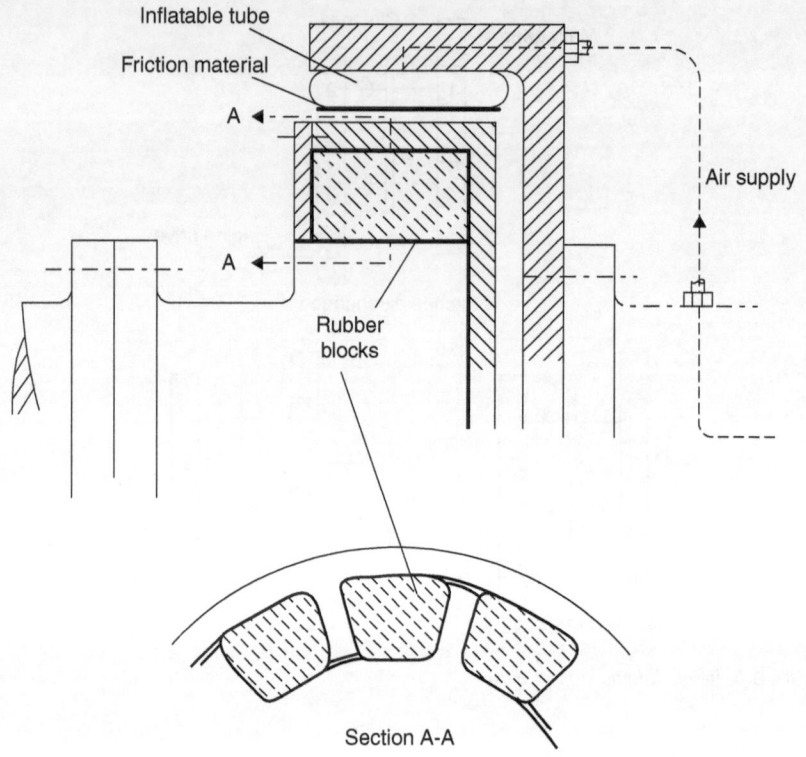

Inflatable tube

Friction material

A

Air supply

A

Rubber blocks

Section A-A

▲ **Figure 8.5** *Flexible clutch coupling*

The Geislinger coupling

The main function of a Geislinger coupling is to assist in the damping out of torsional vibrations. This is accomplished by connecting the engine crankshaft to the load via flexible steel leaf springs arranged radially in the coupling, which is also filled with oil. As torsional fluctuations occur they are absorbed by the leaf springs, which deflect and displace oil to adjacent chambers, slowing down the relative movement between the inner and outer components of the coupling. The makers claim that this effective damping is achieved without problems of wear because of the absence of friction (figure 8.6).

Damping oil is supplied from the engine oil system through the centre of the coupling. It is returned to the engine through hollow coupling bolts. Maintenance is limited to cleaning, inspection and the replacement of 'O' rings.

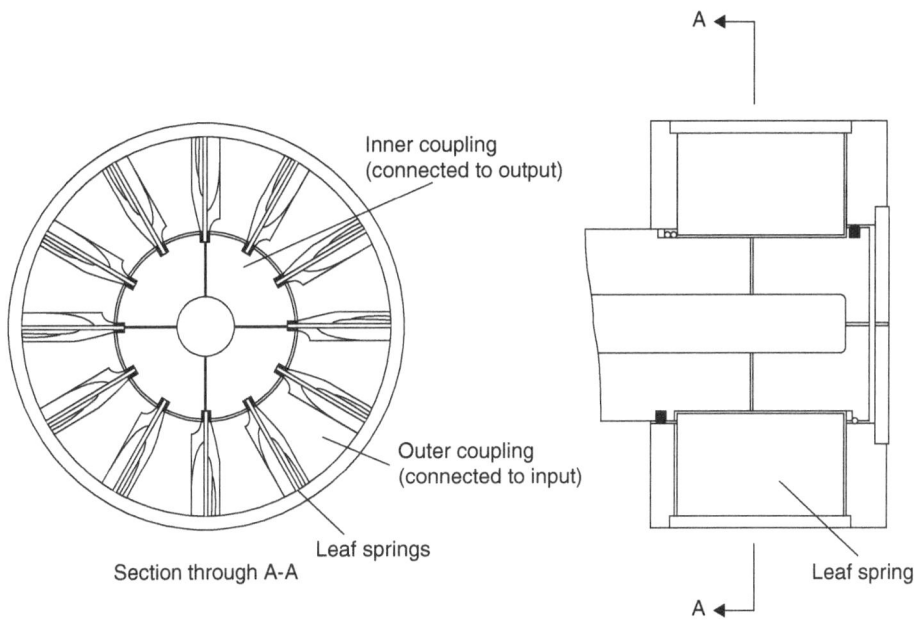

Inner coupling
(connected to output)

Outer coupling
(connected to input)

Leaf springs

Section through A-A

Leaf spring

▲ **Figure 8.6** *Geislinger torsional vibration damping coupling*

Gearboxes, Thrust Blocks, Shafting and Controllable Pitch Propeller (CPP)

Shafting and CPPs are covered in more detail in Chapter 6 of Volume 8 of the Reeds series. However, the use of CPP operating through a gearbox and coupled with medium-speed diesels is a fairly common arrangement for higher power ocean-going vessels such as product-carrying tankers and therefore needs to be described as part of this chapter.

Gearboxes are very interesting and need a great deal of care in both manufacture and ongoing care. They have to transmit sometimes large forces through relatively small areas of contact. The metal obviously transmits that power but metal-to-metal contact would mean that the component parts would not last for long. This means that the quality of the oil and the oil supply is vital to the ongoing success of the gearbox.

Although the gears are said to be meshing, they are actually sliding over one another and the oil needs to be in the right place to ensure that the gear teeth perform correctly. Another major consideration for the ship's engineering staff is to ensure that no metal, tools or other foreign bodies are allowed to enter the casing of the gearbox. The effect is catastrophic if metal gets between the teeth on the gearwheel due to the small clearance between the gears.

Gearboxes transmit power through a drive train. The gears can be arranged to increase or decrease the speed of rotation of input and output shafts or they can be used to transmit power through an angle so that the output shaft is pointing in a different direction to the original shaft.

The basic arrangement is to have straight gears around the outside of a wheel, fixed to the end of an input shaft, linking with a second wheel, of a different diameter, fixed to the start of an output shaft. The dimensions of the gear wheels will determine the different input and output speeds to and from the gearbox. Any combination of speeds can be chosen to suit the designer's needs.

Straight-cut gears present a problem in so much as they have the minimum surface-to-surface contact area through which to transmit the power. Therefore they have to be sized accordingly. If the teeth are set at an angle across the end of the wheel to form a helical gear then the area for transmitting power is increased and the gearwheel can be more compact than a gear transmitting the same power using straight teeth.

The problem here is that because the teeth are set at an angle, there will be forces transmitted at different angles. One component of the force will be transmitted through the gear as required and another will be transmitted along the shaft as a vector component of the total force from the input shaft.

This will result in a lateral thrust being transmitted along the shaft. The value of the thrust will depend upon the angle of the teeth and the total power from the input shaft. This means that with a single helical gearwheel a thrust block of some sort will be needed to counteract the thrust from the gearwheel.

Another answer to this problem is to arrange for half the width of the gearwheel to have helical teeth set in one direction and the other half of the wheel to have teeth set in the opposite direction. This means that the thrust from one set of teeth is offset by the thrust from the other set of teeth and the need of a thrust block has been overcome.

The profile of the teeth is very important to the smooth operation of the gearbox because for the teeth at the end of the input shaft to mesh and transmit power, they have to slide into the space in-between the gears on the output shaft. It is also important for the tip of the gear not to make any contact with the root of the opposing gearwheel as this will also impose forces on the gears, resulting in gearbox failure.

Smaller gearboxes and low-power gearboxes might be lubricated by relying on the oil splashing onto the gears as they operate. However, this means that at start-up the gears are not so well protected and wear can occur during this time. Larger or more powerful arrangements will have the oil pumped into the gearbox where it is arranged to spray directly onto the meshing gears, ensuring that even at the start-up stage the gear teeth are well lubricated.

Gearboxes do need to be checked and looked after. Any unusual noises must be investigated and routine inspections must be made at the appropriate intervals. It is not a good idea to make frequent visual inspections because there is more chance of introducing foreign materials inside the casing.

When an inspection is made, the engineer should be looking out for the following:

- Broken teeth on the gearwheels
- Discolouration anywhere on the gearwheel of teeth (indicating overheating)
- Excessive wear on the faces of the gear teeth (indicating a lack of lubrication)
- Condition of the oil.

A sample of oil can be sent away for further analysis. This will be checked for any metal or water content and from this analysis a picture of the condition of the gearbox can be formed. This process obviously takes some time; therefore some companies, such as Kittywake International, are now supplying analysis kits that can be used on board. The next step is to offer online real-time testing of lubricating oil. This will then start to move the industry towards a CBM approach, which is described in more detail in Chapter 12 of Volume 8 of the Reeds series.

Propellers

Although normally described under naval architecture, propellers have become the focus of efficiency gains in recent years and therefore will come under engineering knowledge as it could be an area where fuel savings could be made by retrofitting an updated system not available when the vessel was built.

Exhaust Valves

The arrangement for four-stroke medium-speed diesels is to incorporate two exhaust valves per cylinder and if we consider a moderate-sized installation consisting of two

12-cylinder V engines, this gives a total of 48 exhaust valves. A not inconsiderable quantity, and if the plant is to burn fuel of high viscosity, the maintenance problem for these valves could be considerable.

In order to minimise maintenance and to prolong valve life, bearing in mind that burning of high viscosity oil is essential due to the higher cost of light diesel oil, certain design parameters and operating procedures must be followed. These are:

1. Separately caged exhaust valves are preferred even though they increase the initial cost. If they are made integral with the cylinder head and used with poor-quality fuel then there will be an increased frequency of valve replacement and overhaul. Cylinder head removal each time becomes a tedious time-consuming operation and the caged valves save a lot of time. However, part-load or short-trip operation can be a problem as the exhaust valves could be running at a temperature where the dew point of the gases is reached. Some cross-channel operators have in the past had problems with acid erosion of exhaust valve spindles on uprated Pielstick PC2.5 because they had water-cooled exhaust valve cages. The previous version of the engine running on the short voyages did not have the same problem.

2. All connections to the valves, cooling, exhaust, etc, should be capable of easy disconnection and reassembly.

3. Materials that have to operate at elevated temperatures must be capable of withstanding the erosive and corrosive effects of the exhaust gas. When burning oils of high viscosity that contain sodium and vanadium, deposits can form on the valve seats, which, at high temperatures (in excess of 530°C at the valve seat), become strongly corrosive sticky compounds that lead to burned valves. Hence the need for materials that can withstand the corrosion and for intense cooling arrangements for valve seats.

4. Stellite valve seats have started the quest for improved durability of exhaust valves. Stellite is a mixture of cobalt, chromium and tungsten, extremely hard and corrosion-resistant, that is fused onto the operating surfaces.

Low-temperature corrosion due to sulphur compounds can occur during prolonged periods of running under low load conditions. The valve spindle and guide, which would be at a relatively low temperature, are the principal places of attack due to the effective cooling in this region. Ideally, valve cooling should be a function of engine load with the valve being maintained at a uniform temperature at all times. As stated, this could prove complicated and expensive to arrange for part-load and low-load conditions.

Further to the use of Stellite, a nickel-chromium alloy, strengthened by additions of titanium, aluminium and carbon, called Nimonic 80A, has gained favour for use in exhaust valve construction. Recently, MAN have found that welding a high-temperature resilient Ni-Cr alloy onto a stainless steel spindle would dramatically improve the hardness and ductility of the valve seat as well as its resistance to cracking when compared to chromium- and nickel-based hard facings, including Nimonic 80A.

In the first stage of the process, the stainless steel DuraSpindle is placed through a new robotic welding procedure where Inconel, an alloy traditionally used in gas turbines, is welded into the groove of an exhaust spindle valve seat.

Once the alloy has been welded in place, the DuraSpindle is then machined, after which more than 10 tonnes of force is used during the special rolling process to work harden the Inconel weld to 500 HV. While the spindle is being rolled and rotated, three or four concentric grooves, depending on the spindle size, are etched into the seat at a depth of several millimetres. This further hardens a relatively ductile material.

The rolling process provides compressive stresses into the component, as opposed to tensile stresses, which may cause cracking in the seat area. Compressive stressing significantly reduces the probability of cracking even in the advent of welding defects.

The hard facing on the spindle seat is further hardened by heating the material up to 600–700°C. The metallurgical reaction, called precipitation hardening, further hardens the seat to 600 HV.

Compared with an Alloy 50-type hard facing material, DuraSpindle is 20% harder and 50% harder if compared to a spindle with Stellite hard facing or Nimonic 80A.

5. Effective lubrication of the valve spindle is necessary to avoid risk of seizure and possible mechanical damage due to a valve 'hanging up'. In order to minimise lubricating oil usage the lubrication system for the valves would be similar to that used for cylinder lubrication and since the amount of oil used would therefore be in small quantities, any contamination of the oil by combustion products and water, etc would be minimal, and this would also increase the life of crankcase lubricating oil.

Rotocap

This simple device when fitted to exhaust valves causes rotation of the valve spindle during valve opening. Wear of the valve seat is reduced, seat deposits are loosened and valve operation life is extended. Figure 8.7 shows the Rotocap, which operates

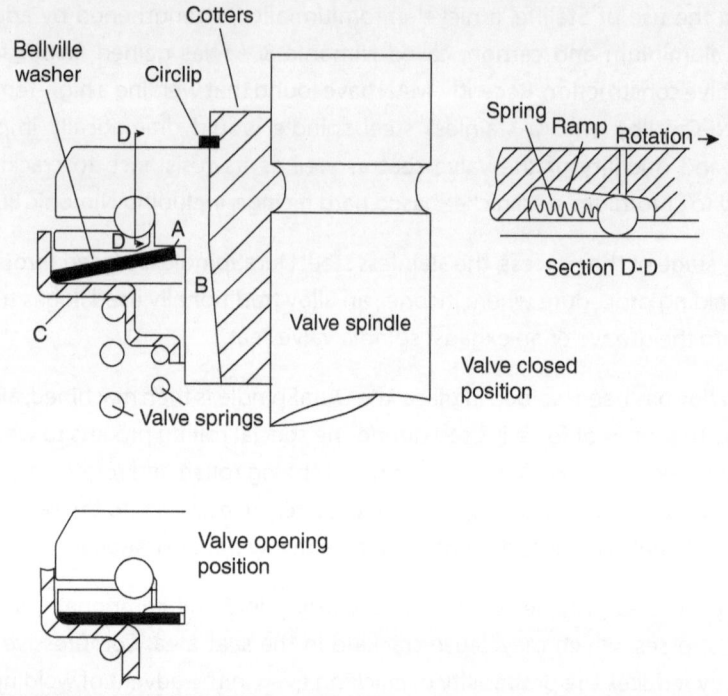

▲ **Figure 8.7** *Rotocap*

as follows: an increase in spring force on the valve as it opens flattens the belleville washer so that it no longer bears on the bearing housing B at A. This removes the frictional holding force between B and C, the spring cover. Further increase in spring force causes the balls to move down the ramps in the retainer, imparting as they move a torque that rotates the valve spindle. As the valve closes, load from the belleville washer is removed from the balls and they return to the position shown in section D–D.

Figure 8.8 shows an exhaust valve with welded stellited seat around which cooling water flows, keeping the metal temperature at full load conditions well below 500°C and minimising the risk of attack by sodium-vanadium compounds. The valve is housed in a 'cage', which can be easily removed for maintenance without disturbing the cylinder cover.

As stated in Chapter 2, modern medium-speed four-stroke engines usually have four valves per cylinder head to maximise the CSA of the ports and thus improve gas flow through the engine. The gas flow of a typical four-valve cylinder head is shown in figure 8.9.

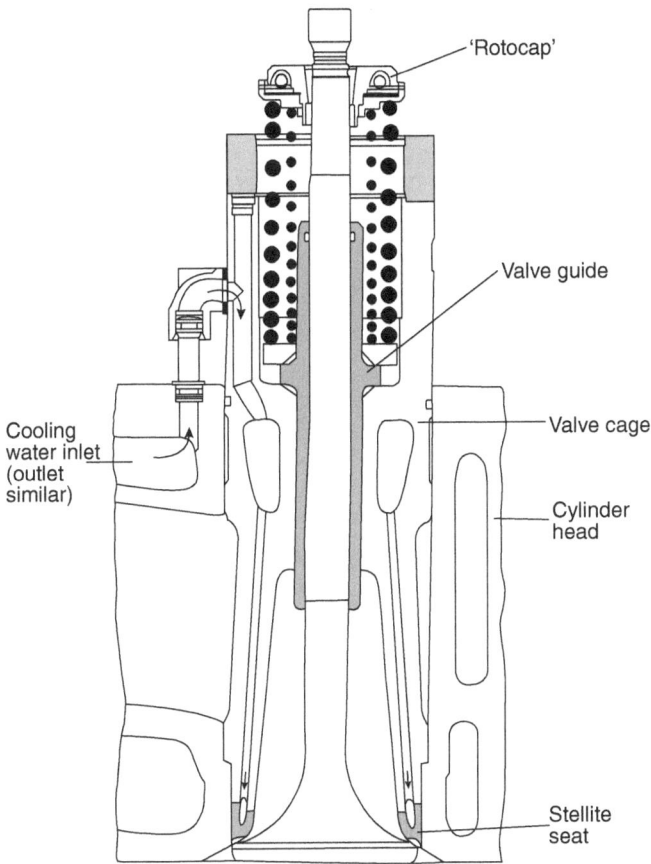

'Rotocap'

Valve guide

Valve cage

Cylinder head

Cooling water inlet (outlet similar)

Stellite seat

▲ **Figure 8.8** *Exhaust valve*

Engine Design

The principal design parameters for medium-speed diesel engines are:

1. High power to weight ratio.
2. Simple, strong, compact and space saving.
3. High reliability.
4. Able to burn a wide range of fuels.
5. Easy to maintain; the fact that components are smaller and lighter than those for slow-speed diesels makes for easier handling, but accessibility and simple to

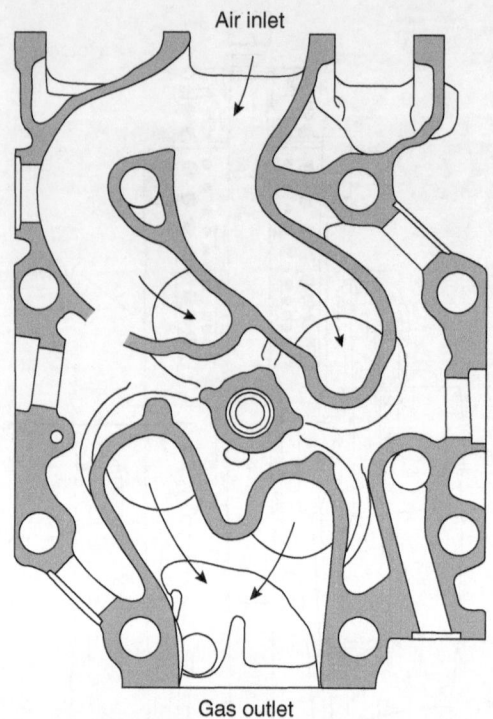

Air inlet

Gas outlet

▲ **Figure 8.9** *Gas flow of typical four-valve cylinder head*

understand arrangements are inherent features of good design (figures 8.10 and 8.12).

6. Easily capable of adaption to unmanned operation.

7. Low fuel and lubricating oil consumption.

8. High thermal efficiency.

9. Low cost and simple to install.

10. Four-stroke design leads to electronic control and use of advanced environmental techniques such as the Miller cycle (see page 13 for further information about the Miller cycle).

Types of engine configuration

Either two- or four-stroke cycle single acting turbocharged with 'in line' or 'V' cylinder configuration. The main choice is, certainly at present, for the four-stroke engine and there are various reasons for this.

1. They are capable of operating satisfactorily on the same heavy oils as slow-speed two-stroke engines.

2. Effective scavenging is relatively easy to achieve in slow-speed two-stroke engines but it becomes more difficult with an increase in mean piston speed. Modern medium-speed engines are generally, but not exclusively, of the four-stroke configuration. With large inlet and exhaust valve overlap, effective scavenging can be accomplished. Scavenging is further improved by utilising high turbocharger pressure ratios. The current versions of turbochargers using single-stage aluminium compressors achieve pressure ratios of 4.5–5.0. However, two-stage turbocharging is required for engines using the extreme Miller cycle. Using the Diesel or Otto cycle, good scavenging and high turbocharger pressure ratios result in engines producing high BMEP figures. The use of the Miller cycle reduces the maximum possible but also reduces the maximum temperature and the NOx produced.

3. The mean piston speed is calculated by multiplying twice the stroke times the rev/s. For medium-speed diesels it would be approximately 9–10 m/s and for slow-speed diesels 7–9 m/s would be an average figure. The latest MAN engines have been type approved with a mean piston speed of 8.97 m/s for the S80ME-C9.2 and 8.49 m/s for the G80ME-C9.2 as can be seen from the ultra-long-stroke engine, which has a reduced piston speed over the super-long-stroke engine. However, the cyclic stresses involved are greater for the medium-speed engine. In order that greater power can be developed in the cylinder, the working fluid must be passed through the engine faster, hence the higher the mean piston speed for a given unit the greater the power. Practical limitations govern the piston speed, such as the relation between cylinder CSA and areas of exhaust and air inlet, method of turbocharging and inertia forces are the main limitations. To reduce inertia forces designers have in the past utilised aluminium alloy for piston skirts and in some cases entire pistons. However, as the output of medium-speed engines has increased the limitations of aluminium have become apparent. Designers of high-output engines now specify cast or forged steel for piston crowns and nodular cast iron for piston skirts. The greater mass of this type of piston means that higher inertia forces result and cognisance of this must be made when designing the connecting rod and bottom end arrangements. Inertia forces must be taken into account for bearing loads – important in trunk piston engines (ie the majority of medium- and high-speed diesels) where the guide surface is the cylinder liner; a smaller side thrust means less friction and cylinder liner; and piston wear.

4. Engine can operate with the turbocharger out of commission; this would present a considerable problem with a two-stroke engine of the medium-speed type.

5. Turbocharger size and power can be reduced.
6. Specific fuel consumption is comparable with the two-stroke engines.

Typical 'V' type engine

The following is a brief description of a medium-speed diesel engine currently in use:

- Cylinder bore = 400 mm
- Stroke = 560 mm
- BMEP = 23 bar
- Maximum cylinder pressure = 160 bar
- Four-stroke turbocharged with up to 18 cylinders developing approximately 700 kW (MCR) per cylinder at approximately 600 rev/min.

Overall dimensions of a 18 cylinder 'V' type

- Length = 10.25 m
- Height = 5.0 m
- Width = 4.0 m
- Dry weight = 145 tonnes.
- Specific fuel consumption = 175 g/kWh.

Bedplate and cylinder blocks are of heavy section cast iron; this gives a strong, compact arrangement with good properties for damping out vibrations.

The crankshaft, of an 'underslung' design, is a solid forging. The connecting rod is also forged but is of the 'marine-type' bottom end and is two pieces. Pistons are of a composite design with forged steel crown and a cast iron skirt. Piston crown is bore cooled. Liners are of good-quality grey cast iron alloy and are bore cooled in the vicinity of the combustion space.

Future development

The trend in the field of the medium-speed engine is towards higher power outputs per cylinder, with high reliability when operating on cheaper high-viscosity fuels. Much development work is being carried out by manufacturers to improve the combustion process. This work focuses on the timing and duration of fuel injection to achieve reliable combustion and manufacturers are now testing engines operating with firing pressures in excess of 210 bar.

- Cylinder bore = 580 mm
- Stroke = 600 mm
- Speed = 450 rev/min
- Power per cylinder = 1,250 kW.

Typical lubrication and piston cooling system

A pump, which could be main engine driven, supplies oil to a main feeder pipe wherein oil pressure is maintained at approximately 6 bar. Individual pipes supply oil to the main bearings from the feeder; the oil then passes through the drilled crankshaft to the crankpin bearing then flows up the drilled connecting rod to lubricate the small end bush. It then flows around the cooling tubes cast in the piston crown then back down the connecting rod to the engine sump. Oil would also be taken from the main feeder to lubricate camshaft gear drive, camshaft bearings, pump bearings, etc (figure 8.10).

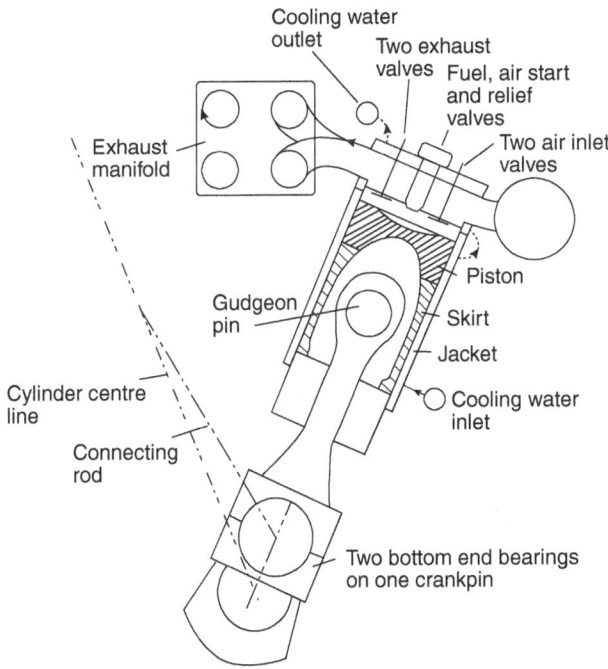

▲ **Figure 8.10** *V-type engine*

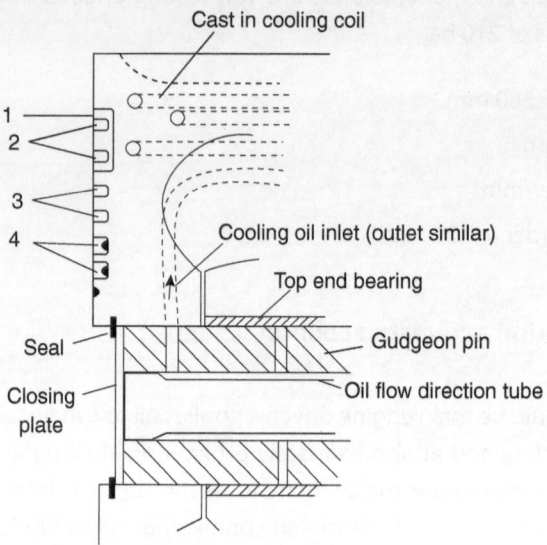

▲ **Figure 8.11** *Piston cooling*

Figure 8.11 shows in simplified form a typical cooling system for alloy pistons. Cast in the piston is a cooling coil and a cast iron ring carrier (marked (1) in the diagram); (2) are two chromium-plated compression rings; (3) two copper-plated compression rings; (4) two spring-backed downward-scraping scraper rings of low inertia type. They are spring backed to give effective outward radial pressure since the gas pressure behind the ring would be very small. The oil flow direction tube is expanded at each end into the gudgeon pin and it is so passaged to direct oil flow and return to their respective places without mixing.

Due to complex vibration problems that can arise in medium-speed engines of the 'V' type, it would appear important to have a very strong and compact arrangement of bedplate, etc. Excessive vibration of the structure can lead to increased cylinder liner wear and considerable amounts of lubricating oil being consumed.

Alkaline lubricating oil of the type used in these engines is expensive and because the engines are mainly trunk type, consumption rates can be high. Positioning, and type, of oil scraper ring is important. With some engines they have been moved from a position below the gudgeon pin to above since considerable end leakage sometimes occurred from the gudgeon bearing. The rings should scrape downwards and there may be two scraper rings fitted, each with two downward-scraping edges, spring backed and of low inertia (figure 8.12).

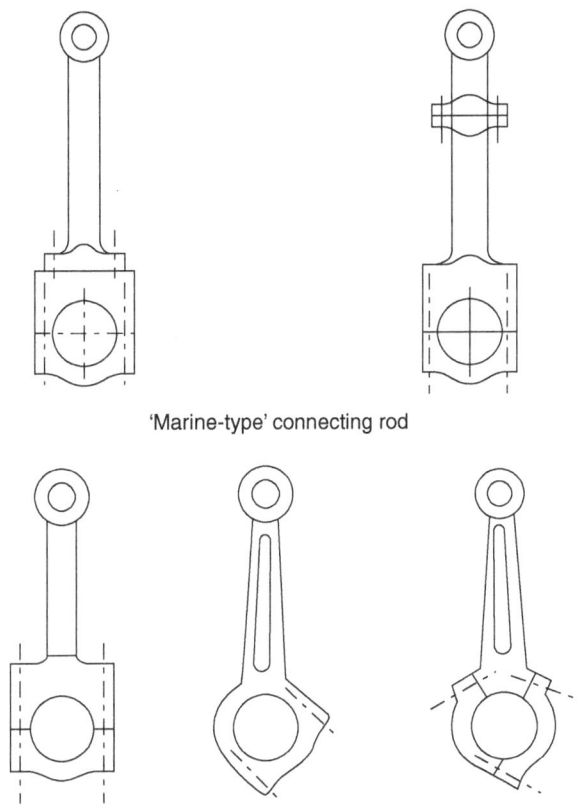

'Marine-type' connecting rod

Connecting rods may be round or 'H' section

▲ **Figure 8.12** *Variations of connecting rod design*

Future trends for four-stroke engines

LNG as a fuel

One view for the future development of the four-stroke medium-speed engine is to use LNG fuel as a solution for lower exhaust emissions. Rolls-Royce developed the Bergen gas engine range to span powers from 1,460 kW to 7,800 kW. Bergen Diesels were then sold to Langley Holdings and now contribute to their drive for reduced emissions from IC engines.

Engine manufacturers are facing up to the challenge of increasingly strict requirements for exhaust emissions. There is a growing pressure to reduce CO_2 and IMO Tier II regulations on NOx emissions has been superseded by much tougher Tier III limits.

Both Bergen diesel engines and Bergen gas engines are attractive for marine propulsion and auxiliary duties. The B32:40 and C25:33 diesel engines with Clean Design notation comfortably meet IMO Tier II requirements without additional off-engine clean-up. The B35:40 and C26:33 gas engines have NOx emissions lower than the strict Tier III limits and net CO_2 equivalent emissions (which also take into account methane slip) and are about 22% less than an engine burning diesel fuel, with negligible SOx.

For many applications the gas engine is a natural choice but acceptance was retarded by complexity of safety rules and lack of LNG bunkering infrastructure. An acceptable regulatory structure is now in place and the infrastructure is being filled out. With the price difference between liquid fuel and LNG increasing, the case for gas is becoming even stronger.

The market for gas engines is advancing. Bergen gas engines in marine applications have now accrued more operating hours in vessels as diverse as cargo ships, feed supply vessels, ferries and offshore supply vessels that are now equipped with Bergen gas engines. Gas tanks and the gas supply system to the engine are now established technology.

The C26:33 series combines well-proven lean burn gas engine technology with the main mechanical components of the compact C25:33 diesel engine range. The first-generation engines will be produced with six, eight or nine cylinders in line, and an introductory power range from 1,469 to 2,430 kW at 900/1,000 rev/min for generator and mechanical drive applications.

CO_2 equivalent emissions are reduced by 22% compared with engines burning liquid fuel, NOx emissions are cut by 92% while emissions of SOx and particulates are negligible. The design of the C26:33 cuts methane slip, which has been seen as a disadvantage of gas engines, to very low levels. The engine meets both IMO Tier III and the forthcoming emission limits for SOx.

With the BV35:40 and C26:33 gas engines in service, an inline version of the B35:40 was developed to complete a seamless range of Bergen marine gas engines spanning power requirements from 1,460 kW to 7,800 kW. The new C26:33 takes over from the K-series gas engine, which proved highly successful both on land and in pioneering marine applications, going through four generations before reaching its limit of development.

There is a growing awareness and discussion about the feature of IC engines known as methane slip. Driven by the fact that methane is more than 20 times more effective at global warming than CO_2 the subject has the potential to be fuelled by sentiment rather than a study of the facts.

Methane (CH_4) slip is the pheromone where some of the methane from the fuel moves through the engine and out of the exhaust without being burned. Some manufacturers are keen to point out that this occurs more on the engines that operate on the Otto cycle rather than the Diesel cycle. However, the industry is confident that as the mechanics of the methane slip become better understood, so changes in combustion design will reduce the problem. Some suggestions for how methane can bypass the combustion process include being injected early or late in the combustion cycle and the gas is therefore caught in the scavenge port and gets sucked through during the overlap period. Another possibility is that the air/gas mix in the Otto cycle can be caught just above the piston ring where it remains unburned and escapes with the exhaust.

It then follows that older, fuel oil, combustion space designs could be more prone to these imperfections than would new engines that are designed with methane slip in mind. It also follows that any reduction in fuel injection performance could make the situation worse.

As the engine design improves so will the combustion efficiency and therefore less unburned fuel will pass through the process, making the modern purpose-built 'gas' engine less and less prone to methane slip. The wider industry view is that methane slip is a real issue but is only part of the issue CIMAC discussions focused on reducing all engine emissions and not looking at any one part in isolation.

Using gas as a fuel reduces the CO_2 considerably, cuts the NOx by 90% and reduces the SOx emissions to practically zero according to the in-service experience of Rolls-Royce, who now have in excess of 30,000 h operational experience from which to draw upon. In the face of so much saving of emissions, a temporary small amount of methane is a good transient solution.

Variable Valve Timing

The great advantage of using modern computer-controlled engine management systems is that the control actions can be freed from the constraints of mechanical control mechanisms.

For example, the older 'fuel injector' (figure 3.3) relies on the build-up of pressure in the fuel to lift the needle that allows the fuel to flow. To us this is an instantaneous action but to a fast-moving piston, waiting for its next injection of energy, any delay will mean a loss of efficiency.

Delays could easily be caused by any one of a number of components being slightly worn and therefore not allowing the correct rise in pressure due to some 'leakage'.

A much better system would be to have a very fast opening, electronically operated fuel valve that allows high-pressure fuel to be injected at exactly the correct time at every time of asking – which of course is the basis of the 'common rail' fuel injection system.

The point is that the CR system has different components and methods of working from the older systems. This is the same if variable valve timing (VVT) is to be used in the design of an engine.

VVT is a very useful tool to use in extending the operating envelope of the engine; however, it is very difficult to achieve when only mechanical control solutions are available.

The most flexible systems are the ones that replace the engine's camshaft with an electromagnetic or electro-hydraulic valve operating system. This arrangement would allow the software in the management systems to switch between different theoretical combustion cycles (see page 9) to suit the current operating conditions.

9

HYBRID AND ELECTRIC PROPULSION

Introduction

Electric propulsion systems are gaining popularity as the flexible propulsion system for merchant vessels. However, it is now required that much more knowledge about these systems is gained by the operational engineering staff required to look after the machinery on a daily basis.

IMO has now agreed to the details of the knowledge and skills requirement of an Electro Technical Officer due to the growing sophistication and complexity of the electrical and electronics used on modern ships. There is also a requirement under the Manila Amendments of STCW for the Chief Engineering Officer to know more about high-voltage distribution systems and especially the safety side of such systems. Therefore, this new chapter is designed to give the engineering officer an overview of the different systems, how they are arranged and why they are being used. The in-depth understanding and calculations involved with the current and voltage waveforms will be covered in Volume 7 of the Reeds series.

Additional features of the Hybrid Plant

The energy density contained in a tank full of fuel oil is difficult to replicate with the technology available today. However, improving the overall efficiency of a traditional power plant is one way to make the maritime industry more environmentally friendly. This is achieved with the use of technology that assists the main power plant to overcome its worse areas of inefficiency.

For example, a diesel engine's design will be optimised around an operating requirement for the duty that it will be performing. If this is to be driving a ship on long ocean voyages for most of its operating life, then the owner will need that engine to be at its most efficient during this time. It then follows that at other times, such as when changing speed/load during manoeuvring, the engine will not be at its optimum efficiency. Helping the engine during these times with other energy sources can improve the overall efficiency of the plant. This will save fuel and reduce cost and improve the environmental footprint of the vessel.

Chapter 10 and 11 explore ideas about how a ship's environmental footprint can be improved. However, here we can focus on the central power plant. As already mentioned, a diesel engine is not very efficient when asked to change load, especially at short notice. The load increase is requested, and more fuel is sent to each cylinder. However, the turbocharger needs to speed up to supply the additional air required to burn the fuel. In order to accomplish this task, it needs more exhaust gas, which comes at the end of the combustion of the higher amount of fuel.

A battery on the other hand can react quickly and with energy that has been stored at a time of efficient plant operation. Therefore, overall, there is an efficiency gain for the plant. The limitation for batteries accomplishing this task is when they are asked to supply a very high power in a short timescale. Here the 'supercapacitor' is used to bridge the gap as they can charge and discharge in much shorter timescales.

Hybrid power plant using fuel cells, batteries, supercapacitors, diesel engines and waste heat recovery systems can be arranged into an efficient propulsion package. The challenge comes due to all of the different components working to different types and levels of electricity. Some work on producing direct current (DC) while others produce and consume electricity as alternating current (AC).

One answer is to have a direct current (DC) electrical energy transmission system and have the different components connected into the DC distribution system via their own interface. Therefore, each power source or consumer can work at their most optimum and connect to each other through the DC system and not have to share a common frequency on an AC system.

Using this system also has advantages for the future when other forms of energy generation/storage become available; they can also be plugged into the system without disturbing the original components (the so-called plug and play approach). Possible new sources could include solar and wind power.

General Arrangements

There have been diesel electric and turbo-electric marine propulsion systems in the past. The P&O passenger liner *Canberra*, for example, had a turbo-electric drive system fitted as new when the vessel was built in 1961 and the *Queen Elizabeth II* was re-engined in 1987, and a diesel electric drive system replaced the original conventional steam turbine shaft line drive.

The initial driving force behind fitting an electric drive system to a ship was the flexibility in propulsion plant layout. This was particularly relevant for the passenger liner/cruise ship part of the business as the propulsion plant could be arranged so that additional 'revenue earning' passenger space could be accommodated for a given power output.

The second consideration, again initially for cruise ships, was the adoption of the 'power station' principle of operation. The electrical load of the passenger ship is considerable, even without electric drive motors. Therefore, it is often more economical for a vessel to be able to call upon a number of smaller generators, that can be matched to the load, than it is to have a smaller number of larger generators running at part load. When electric propulsion motors are added to the design of the vessel then the advantages of the 'power station' principle are even greater (figure 9.1).

Refinements to the different systems are happening all the time as the advancement in technology takes place. The developments in power electronics have allowed considerable efficiency gains and as different energy sources start to become more effective the idea of the direct current (d.c.) bus transmission system allowing a 'plug-n-play' style of system is becoming more relevant.

In the past, d.c. motors and control systems provided excellent speed control with electrical drive systems. However, d.c. motors are complicated, heavy and need more power than the equivalent-sized a.c. machine. The problem is that a.c. motors rely on the frequency of the supply system to operate. The rotor of the a.c. motor follows the sinusoidal waveform of the supply, which is determined by the frequency of

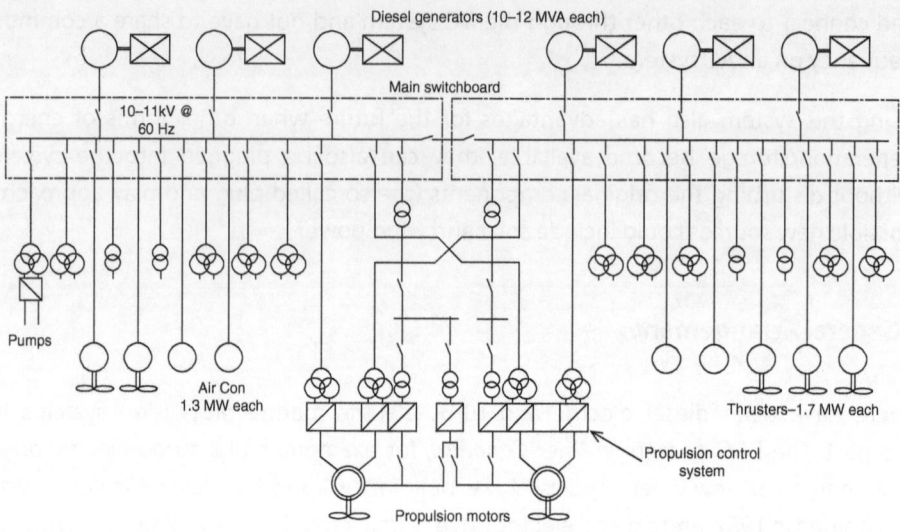

▲ **Figure 9.1** *Traditional a.c. electric drive propulsion drive system*

the a.c. system. This means that until recently a.c. motors have been single-speed machines.

This might not be so much of a problem if an a.c. propulsion motor is coupled to a CPP, which would then be used to provide the variable propulsion required to manoeuvre the vessel. There is, however, considerable energy saving potential in reducing the speed of the motor when the full speed is not required.

To enable a reduction in the speed of an a.c. motor the frequency of the electrical supply must also be reduced and there are now a number of different methods that are being used to control the speed of marine propulsion motors. The method that is currently the most popular in the systems is called the PWM. Here, the voltage between the different phases of the supply is switched on and off, or modulated, at high speed. This switching changes the waveform of the flux density, which in effect changes the magnetic field setup within the motor and alters the speed. The PWM control comes from a variable frequency inverter.

Harmonics is the term used to describe a distortion in the behaviour of rapidly changing physical quantities such as noise and electricity. Pure notes are noises vibrating at a given frequency but when we talk about 'harmonics' we are describing the overtones of the pure note produced by some interference or distortion. In electrical systems the distortion can be extensive and accumulative and therefore the fundamental frequency of 60 Hz would have a second harmonic of 120 Hz and a third harmonic of 180 Hz. The values can keep on rising where the number of the harmonic distortion is multiplied by the fundamental frequency to give the distortion value, for example, the tenth harmonic will be 600 Hz.

In speed control circuits of a.c. propulsion motors, the high-speed switching action of the electronic components, in the power converters, will cause a harmonic distortion of the original 'pure' waveform of the original supply from the generators.

In marine electrical installations, electric variable speed drives are the main load on the system and therefore the harmonic disturbance of the fundamental frequency does, in turn, have an effect on all the connected loads regardless of their position in the system. Symptoms of harmonic distortion in the electrical system are as follows:

- Occasional unexplained occurrences, such as:
  o flickering lights
  o alarms sounding
  o fuses, circuit breakers and earth leakage devices tripping for no apparent reason
  o cables running hot
  o hot switchboards
  o overheating motors
  o frequent need to replace your motor's bearings and insulation.

Some of the common and unpredictable effects of excessive harmonic distortion on marine installations include:

- Overheating and sustained damage to bearings, laminations and winding insulation on generators, transformers and induction motors causing early life failure, which could potentially result in fire.
- Overheating of the stator and rotor of fixed speed electric motors; risk of bearing collapse due to hot rotors. This is especially problematic on explosion-proof motors with increased risk of explosion, more especially with ExN (non-sparking motors).
- Overheating of cables and additional risk of failure due to resonance. Harmonics also decrease the ability to carry rated current due to 'skin effect', which reduces a cable's effective CSA.
- Disruption in the operation of uninterruptible power supplies (UPS).
- Spurious tripping or failure of sensitive electronic and computer equipment, measurement and protection relays.
- Voltage resonances leading to transient overvoltage and overcurrent failures in the electrical network.

- Electromagnetic interference (EMI) resulting in disruption to communication equipment.
- Malfunction of circuit breakers and fuses.

Total Harmonic Distortion (THD)

Harmonic distortion can be multiples of either the voltage or current waveforms and the THD is a term used to describe the contribution of all the harmonic waveforms in the electrical power generation and distribution system. It is expressed as a percentage of the ratio of the root mean square (RMS) value of the total harmonic content to the RMS value of the fundamental frequency.

Lloyd's Register rules on harmonic distortion of voltage state are as follows. Unless specified otherwise, the THD of the voltage waveform at any a.c. switchboard or section board is not to exceed 8% of the fundamental for all frequencies up to 50 times the supply frequency and no voltage at a frequency above 25 times supply frequency is to exceed 1.5% of the fundamental of the supply voltage. All other classification societies place a limit of 5% on THD of voltage (THDv). The Institute of Electrical and Electronic Engineers' (IEEE) Recommended Practice for Electrical Installations on Shipboard (IEEE Standard 45–2002) states:

A dedicated propulsion bus should normally have a voltage total harmonic distortion of no more than 8%. If this limit is exceeded in the dedicated propulsion bus, it should be verified by documentation or testing that malfunction or overheating of components does not occur. A non-dedicated main generation/distribution bus should not exceed a voltage total harmonic distortion of 5%, and no single voltage harmonic should exceed 3%.

IEC 60034–1, 2004, Rotating Electrical Machines – Part 1: Rating and Performance requires that the THDv for synchronous motors above 300 kW output should not exceed 5%. It does not specify distortion levels for individual harmonics. However, keeping low THD values on a system will further ensure proper operation of equipment and a longer equipment lifespan.

Importance of mitigating THD

There are several methods used to counter the effects of harmonic distortion in marine power systems, including:

- Active or passive filters.

- Increasing the number of pulses in power converters by using multiple-phase shifted secondary windings in propulsion motor supply transformers.
- Installing generators with a large sub-transient reactance.

The predominant harmonics that are expected to occur in the electrical power conversion systems are calculated at the design stage.

Keeping low THD values on a system will further ensure proper operation of equipment and a longer equipment lifespan.

Power Quality Measurement

Land-based utilities monitor their power quality as a matter of routine. In a marine vessel where harmonic distortion has the potential to disrupt its electrical network, possibly leading to a blackout and loss of control in restricted waters, the need for power quality surveillance is even more significant.

Regular monitoring of power quality, using a predetermined pattern of propulsion motor loading, with a complete record of operational parameters, would help ensure that the harmonic distortion levels on board are closely monitored as the vessel and its equipment age and operating configurations change.

An online monitoring system that records all the parameters and can be triggered to make specific recordings of transient voltage spikes or resonances would be invaluable in assessing the ongoing quality of power. It would also be a very useful tool to investigate the root cause of accidents caused by anomalies in the electrical network and to identify incipient faults in these systems.

PWM control strategies were introduced in the early 1980s to overcome the heating and torque pulsations of the then 'square wave drives' (also known as 'quasi-square wave' or 'six-step drives'). The purpose was to reduce the output harmonics, especially the low-order harmonics, to the motor. Since that time, the various PWM strategies have been improved significantly such that present series of drives usually have output current waveforms (ie not the output voltage) that are relatively sinusoidal. This was achieved due to a combination of PWM techniques and advances in fast power semi-conductors such as insulated gate bipolar transistors (IGBTs). One example of this system is included in figure 9.2.

Such systems are termed active front end (AFE) systems. Compared to similarly rated conventional six-pulse a.c. PWM drives, the AFE drive has significantly higher conducted and radiated EMI emissions, and therefore special precautions and installation techniques may be necessary when designing such a system. The Rolls-Royce 'hybrid system' is an adaptation of the AFE system and the general arrangement is described in the following.

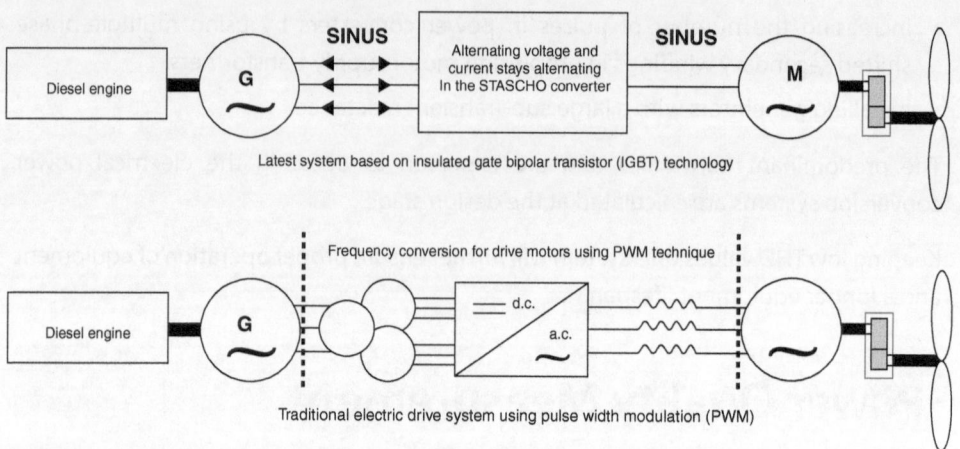

▲ Figure 9.2 *IGBT technology a.c. electric propulsion drive system*

Advantages of the electric and hybrid drive systems

Rolls-Royce has designed and equipped many vessels with various combinations of diesel-electric, gas-electric and hybrid propulsion solutions for offshore support vessels and related multi-role vessels that demonstrate substantial fuel savings and reduced emissions compared with mechanical systems. This system is particularly effective where that is a requirement for the sudden acceleration of a diesel engine: vessels such as tug boats, vessels fitted with dynamic positioning and vessels working extensively in ice. The advantages of the electric and hybrid drive systems include:

- Up to 50% reduction in fuel consumption resulting in reduced NOx/CO_2 emissions compared to diesel-mechanical propulsion.
- High levels of flexibility and redundancy in the configuration of the propulsion system and electrical plant.
- Fast AFE frequency control allows rapid and easy manoeuvring of the vessel.
- Up to 30% reduced maintenance due to fewer running hours and less mechanical stress when using frequency control drives.
- Reduced noise and vibration giving greater crew comfort.
- No need for heavy, space-consuming transformers.

A modern electric propulsion system today uses AFE technology. The system consists normally of two main propulsion shaft lines or thrusters each driven by an electric motor controlled by an AFE frequency converter. Depending on the type of the vessel there will also be two to four thrusters with AFE frequency control for manoeuvring and position keeping. Electric power is produced by three to six generators driven by diesel or gas engines. When using AFE converter technology in an electric propulsion system there is no need for heavy, space-consuming transformers as in a traditional 12- or 24-pulse system and the THD will be below 2%.

Low-voltage electric drive systems have their limitations, with a maximum of power generation of approximately 20,000 kW. This is due to limitation of short-circuit levels on the switchboards. When more power is needed, or when the vessel requires a very large bollard pull, a hybrid propulsion system should be considered. Efficient hybrid propulsion systems combine mechanical and diesel/gas electric transmissions. Individual systems are tailored to the vessel and its operating profile in terms of total installed power and how much of this power travels the diesel electric route.

A typical vessel with the Rolls-Royce system will have a twin-screw layout with CPP. Each shaft line comprises a medium-speed engine, a reduction gearbox and a clutch between gearbox and engine. At the forward end of each main engine are a second clutch and a large shaft generator. A frequency-controlled variable speed electric motor is connected to a power take-in (PTI) drive in the gearbox. In addition to the two main shaft lines there are two or more auxiliary gensets and there are also tunnel and azimuth thrusters to assist in manoeuvring and positioning.

Efficiency under all operating conditions is the defining principle, achieved by running only the number of engines actually required, and avoiding having powerful engines operating at low part loads with a resulting high specific fuel consumption. Further, utilisation of frequency-controlled electric motors eliminates the zero-pitch propeller losses, which may become significant for long periods of operations at low load. At the same time the energy losses associated with electric transmissions are reduced at higher powers by routing all or most of the propulsion power through a low-loss mechanical transmission.

Cycloconverters are a common form of electrical variable speed drive in the higher power range and, as such, are used for main propulsion drives. Unlike other forms of a.c. drives, such as a.c. PWM drives and load commutated inverters (LCI), both of which have an intermediate stage (ie d.c. bus) to facilitate dual conversion (a.c. to d.c. and d.c. to a.c.), the cycloconverter is a direct conversion drive converting one frequency to another without the need for an intermediate stage. Cycloconverters have been in service, in different industries, for a long time but in the past they have been based on mercury arc rectifiers and they have operational constraints by having a maximum output frequency of 33% of the input frequency.

Developments of speed control of synchronous induction motors, known as the Static Kramer drive, uses a cycloconverter to further enhance the system under the new term of 'static Scherbius' drive. When used in power conversion systems the operation of a cycloconverter is complex, with both positive and negative bridges necessary for each motor phase. To briefly describe their operation it is necessary to consider the operation of a single-phase-to-single-phase device with full-wave rectifiers and a resistive load.

Cycloconverter input current characteristics and associated harmonic content are complex and dependent upon a number of factors, including:

- The pulse number of the cycloconverters
- The relative magnitude of the output fundamental voltage
- The ratio of the input and output frequencies
- The displacement power factor of the load
- The firing control strategy.

In applications with large drives, six-pulse drives are not common. Multi-pulse drives, including 12-pulse, are the norm to minimise the input harmonic currents and associated disruption of the power supply system. One development of the a.c. motor control is based upon IGBT transistors, thyristors and bypass switching.

The basic reason for the adoption of electric drive systems is to improve the efficiency of converting the energy from the fuel into useful propulsive and power generation. The more energy conversion steps in the line, the more potential there is for losses. The use of a d.c. grid or distribution system is an effort to reduce the number of changes in energy between the fuel and the use of the power output.

For example, converting the fuel directly into thrust is not yet possible. Using mechanical components only is inefficient and as we have seen earlier in this chapter there are a large number of changes involved in the a.c. drive system using the PWM or AFE system where we have fuel converted into mechanical force and then through the magnetic field to an a.c. output; next, it is changed to another a.c. frequency then to d.c. and back to a.c. before being converted once again to mechanical power and finally trust. Much of these changes can be removed with the use of the d.c. grid, as shown in figure 9.3.

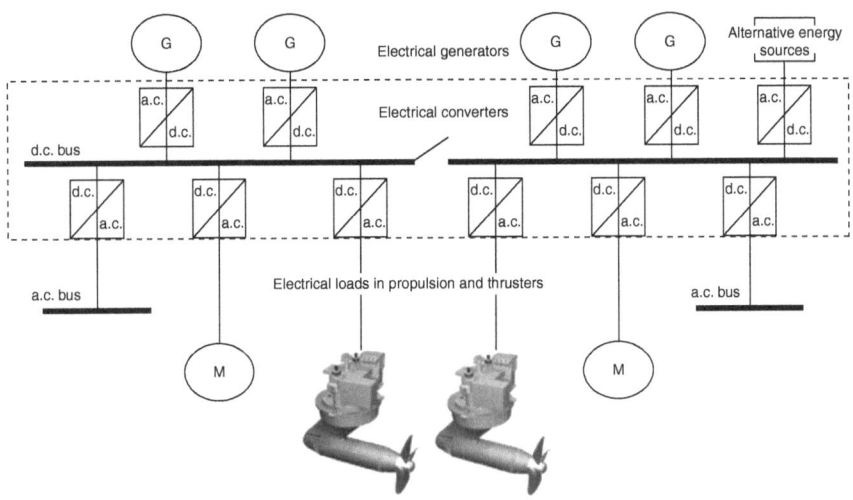

▲ **Figure 9.3** On-board d.c. grid (electric propulsion drive system)

The eSiPOD system from Siemens offers owners two propellers at either end of a single-podded drive. The aim is to allow maximum flexibility in vessel design thus maximising the 'cargo carrying' ability of the vessel. The twin or mono propeller systems give maximum choice to optimise the underwater hull design.

10

WASTE HEAT RECOVERY SYSTEMS

General Details

Reference should be made to Chapter 1 for general comments relating to heat balance. Figure 1.2 details an approximate heat balance for an IC engine showing significant losses to the exhaust and cooling. Every attempt made to utilise energy in WHR from both exhaust and coolant is established practice. Sufficient energy potential can be available in exhaust gas at full engine power to generate sufficient steam, in a waste heat boiler, to supply total electrical load and heating services for the ship. The amount of heat actually recovered from the exhaust gases depends upon various factors such as steam pressure, temperature, evaporation rate required, mass flow of gas, condition of heating surfaces, etc. Waste heat boilers can recover up to about 60% of the loss to atmosphere in exhaust gases. Heat recovery from jacket cooling water systems at a temperature of 70–80°C is generally restricted to supplying heat to the fresh water generator.

Combustion equipment

Most modern ships have boiler arrangements for raising steam. A thermal fluid system alternative is available but the preferred system is steam auxiliary boilers. During low engine power conditions or when the main engine is not in use, the boiler has to combust fuel to provide the heat source. It is therefore appropriate to repeat some very general remarks on combustion with details of the typical boiler equipment in use on board ship. A more detailed explanation of boilers appears in Chapter 3 of Volume 8 of the Reeds series.

Good combustion is essential for the efficient running of the boiler as it gives the best possible heat release and the minimum amount of deposits upon the heating surfaces. To ascertain if the combustion is good we measure the % CO_2 content (and in some installations the % O_2 content) and observe the appearance of the gases.

If the % CO_2 content is high (or the % O_2 content is low) and the gases are in a non-smokey condition then the combustion of the fuel is correct. With a high % CO_2 content the % excess air required for combustion will be low and this results in improved boiler efficiency since less heat is taken from the burning fuel by the small amount of excess air. If the excess air supply is increased then the % CO_2 content of the gases will fall.

Condition of burners, oil condition pressure and temperature, condition of air registers, air supply pressure and temperature are all factors that can influence combustion.

Burners

There are two basic types of burners, the pressure jet and the rotary cup. The pressure jet as its name suggests relies on the fuel oil supply pressure to force the fuel through a series of small nozzles in the end of a long tube. The holes are set at an angle and will therefore give a spin to the fuel as it exits from the burner. This spin or swirl gives the fuel the right action to mix thoroughly with the air delivered by the air register and therefore when the mixture hits the flame front it is ignited. The rotary or spinning cup type of burner does not rely on the fuel pressure to give atomisation. The low-pressure fuel oil is released into the centre of the rotary cup that is spinning at about 5,000 rev/min. As the fuel follows the conical cup it eventually comes to the rim where centrifugal force makes it fly from the edge into the path of the primary airflow. The air hits the fuel, atomising it as it does so. The primary and secondary air provides the oxygen for combustion as well as the shape of the flame propagation. If any of these components are dirty or the sprayer plates damaged then effective atomisation will not be achieved, resulting in poor combustion.

Oil

If the oil is dirty it can foul up the burners. (Filters are provided in the oil supply lines to remove most of the dirt particles but filters can get damaged. Ideally the mesh in the last filter should be smaller than the holes in the burner sprayer plate.)

Water in the oil can also affect combustion; it could lead to the burners being extinguished and a dangerous situation arising. It could also produce panting (unstable combustion leading to pressure fluctuations), which can result in structural defects.

If the oil temperature is too low the oil does not readily atomise since its viscosity will be high; this could cause flame impingement, overheating, tube and refractory failure. If the oil temperature is too high the burner tip, becomes too hot and excessive carbon deposits can then be formed on the tip causing spray defects. These could again lead to flame impingement on adjacent refractory and damage could also occur to the air swirlers. Oil pressure is also important since it affects atomisation and lengths of spray jets.

Air register

Good mixing of the fuel particles with the air is essential, hence the condition of the air registers and their swirling devices is important. If they are damaged mechanically or by corrosion then the airflow will be affected.

Air

The combustion air supply is governed by the combustion controller fuel/air ratio setting. If this is set too low then insufficient air will be supplied, resulting in incomplete combustion and the generation of black smoke. If the fuel/air ratio is set too high then too much air will be supplied for combustion, resulting in a greater percentage of free oxygen in the uptakes than is desirable, causing the boiler efficiency to fall.

It is generally considered that the appearance of the boiler uptake gases will give an accurate indication of the effectiveness of combustion. While this is undoubtedly true, it should be noted that clear uptake gases can be achieved while supplying excess air, resulting in a reduction in boiler efficiency. To achieve maximum boiler efficiency the fuel/air ratio setting should be reduced until the setting for optimum combustion, commensurate with clear uptake gases, is reached.

Boiler operation

Boilers are potentially one of the most dangerous places in the engine room of a ship. For this reason the Flag State examiner issuing a certificate of competency to a motor ship marine engineer will not do so unless she/he is sure that the marine engineer can cope with the dangers of the steam raising plant.

Auxiliary boilers on modern ships are usually fire tube boilers operating with a working steam pressure of about 7 bar. This is enough pressure to supply all the necessary heating required on board the vessel. However, because the fire tube boiler has a relatively large amount of water for the size of boiler, it also has a greater potential for causing a lot of damage if there was a structural failure.

It must be remembered that if the steam is at 7 bar pressure then any parts in contact with the steam are also at 7 bar pressure. If the pressure on the water within the boiler was suddenly reduced to atmospheric pressure, due to some form of structural failure, then the water would flash off into steam.

Steam requires 1,600 times the volume of water; therefore, when the pressure is released and the water flashes into steam a considerable force is released and large sections of the boiler can be moved at considerable speed.

One of the most important dangers to guard against is loss of water. The metal furnace close to the burner relies upon the cooling effect of water on the other side of the furnace to ensure that it does not overheat and fail. There are a number of reasons for a loss of feed water and the motor engineer will need to understand his/her system and be able to explain to the examiner how to guard against a loss of water in the auxiliary boiler.

The gauge glass is the primary source of information about the water level in the boiler. There are always two gauge glasses in case one becomes blocked; however, on a marine boiler they are situated on opposite sides of the boiler. The reason for this is that if the vessel is rolling and one glass is empty then the other should be showing a high level and vice versa. More details can be found in Volume 8 of the Reeds series.

Fire is another major concern with boilers. Fuel oil is led to the boiler where it is burned after passing through the burner and associated filters and pumps. As the burners are sometimes temperamental in their operation they have to be opened, cleaned and adjusted from time to time. Oil builds up around the furnace front and can be the cause of a fire if engineers are not careful.

The watchkeeper will be responsible for the safe operation of the auxiliary boiler during his/her watchkeeping duty period; therefore, it is essential that she/he understands the following safety-related start-up and operating procedure. Safe start-up procedure involves a purge cycle. This means that the boiler will run the forced draught fan for a few seconds before trying to light the boiler. This is to ensure that any unburned hydrocarbons from the previous cycle are taken away from the burner that will be lighting up soon.

With the air operating correctly, the next step is to introduce the heat source. This is usually in the form of a high voltage passing from one electrode to another via an air gap (a bit like the spark in a car). When the heat and the oxygen are both in place, it will then be okay to introduce the fuel and start the burn. Feedback to the boiler controls saying that the boiler is alight is provided by a photoelectric cell. Using this sequence there is very little chance of unburned gases accumulating in the furnace and causing a violent start-up or an explosion.

The watchkeeping engineer could be called upon to start a boiler that has 'locked out'. It would only do this if it had failed to light for some reason. It is very important that the watchkeeper carries out some basic checks before trying to relight the boiler.

The first and most important check is to look at the water level in the boiler and make sure that the boiler has not 'locked out' due to low water level. Make sure that you look at both gauge glasses and if you are unsure you need to follow the gauge glass check procedure described in Chapter 3 of Volume 8 of the Reeds series. If a low water level is suspected, it is very important that you DO NOT start the burner until the correct level is restored. The problems here could range from failure of the feed water pump to deliver a sufficient quantity of water to a malfunction of the feed water control float that dictates when and how much water is sent to the boiler.

The feed water pump operates under difficult conditions because not only does it have to pump water that is close to its boiling condition but it also has to pump it at sufficient pressure to overcome the boiler's working pressure. Sometimes these pumps 'gas up', in other words, the water's vapourisation condition is met and the feed water starts to turn to steam inside the pump, stopping the flow of water.

The other, not uncommon, problem is with the float sticking in the feed water controller. I have seen some of these corroded so much that the float has broken away from the activating arm and therefore was not capable of working. If the water level has been the problem then the boiler can be started again following the restoration of the feed water to the boiler.

The watchkeeper should monitor the boiler as it works through its safety purge cycle and then the starting sequence described earlier. Some boilers are fitted with a burner viewing port, which is to be treated with great respect. These should never be used when the boiler is starting up as serious injury has been caused in the past due to 'blowback' as some unburned fuel has caught alight.

A malfunction in the burner starting sequence is another reason for the boiler 'locking out'. If the boiler fails to light – some control systems may allow two cycles before giving an alarm – then the watchkeeper will need to identify the reason and rectify the fault.

If the 'lock out' is due to the burner then there are three conditions to check:

1. Boiler fan working to give the correct amount of air to the burner
2. Igniter working and in the correct position
3. Fuel supply to the burner nozzles or spinning cup.

Conditions 1 and 3 are relatively easy to check because the fan not working or the air register blocked will be easy to spot, as will be the lack of fuel. Igniter faults and subsequent setup, however, is generally more difficult to deal with but is a more common fault than the other two.

Sometimes there could be quite a bit of pressure to get the boiler back online, for example. If the heavy fuel oil to the generator was cooling down due to lack of steam, there might not be much time in which to have the boiler up and running again or to change the generator over to a light distillate fuel.

The problem with the igniter is that it has to sit right in the most turbulent area of the air/fuel mix and the forces involved are enough to knock it out of alignment occasionally and therefore it stops working. It has to be reset to the correct position, which is just where the air/fuel ratio is correct for combustion. If it is placed too close to the burner the mixture is too rich for ignition and if it is placed too far away the mixture could be too lean or it might miss the spray of fuel entering the combustion space. Then, the burner will cut out due to the length of time trying to light the fuel.

Generally, the engineers will know the settings and be able to work away at getting the boiler going again but late at night, by yourself and under pressure makes the job much more difficult.

Package Boilers

Although such boilers are not necessarily involved with waste heat systems, it is considered appropriate to include them at this stage. These boilers are often fitted on motorships for auxiliary use and the principles and practice are a good lead into general boiler practice. Two types of design involving modern principles will now be considered.

Sunrod vertical boiler

The design sketched in figure 10.1 is the Sunrod Marine Boiler. This boiler utilises a water-cooled furnace incorporating membrane-walled construction. The membrane water wall is backed by low temperature insulation (figure 10.2a). The water wall tubes

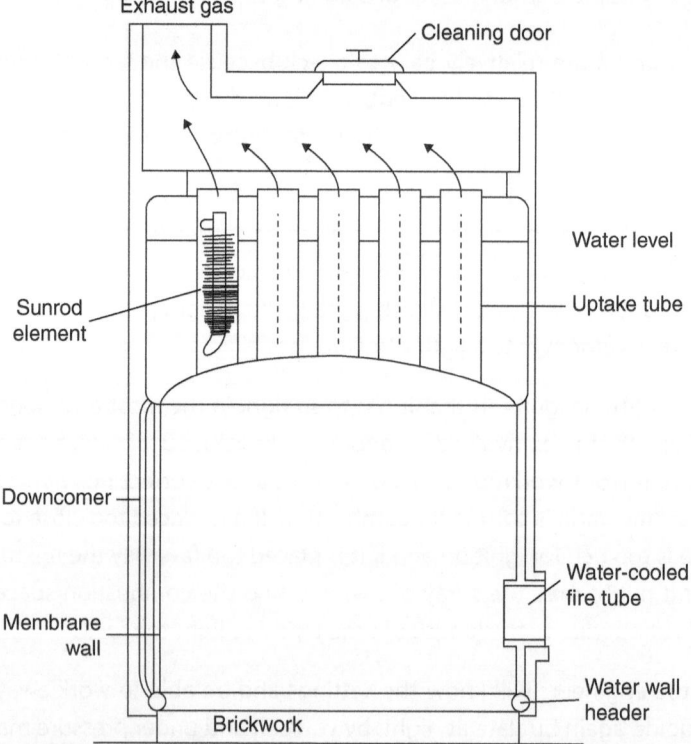

▲ **Figure 10.1** *Sunrod marine boiler*

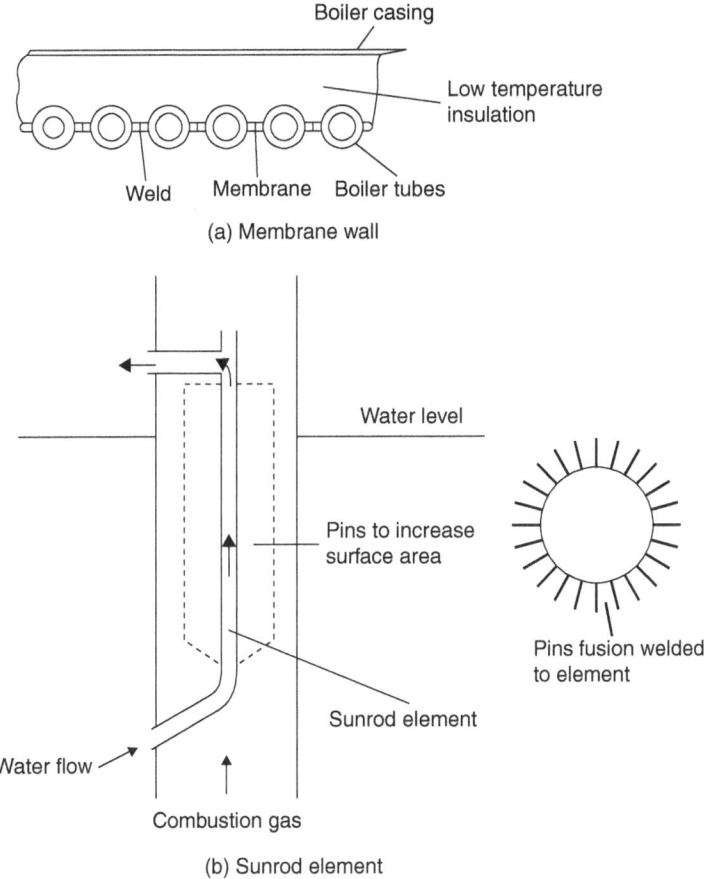

▲ Figure 10.2 *Sunrod boiler detail*

are joined at the lower end to a circular header and at their upper ends to the steam chamber. Good circulation is assured by the arrangement of a number of downcomers as shown in the diagram. The steam chamber has a number of smoke tubes each fitted with a 'Sunrod element'. The purpose of the Sunrod element is to increase the heating surface area of the boiler. This is accomplished by welding pins onto the element as shown in figure 10.2b. In some Sunrod designs the fire tube is also water-cooled. This design is manufactured in sizes ranging from 700 kg/h to 35,000 kg/h with pressures up to 18 bar. The boiler is usually fitted with automatic start-up/shutdown and combustion control.

Due to the absence of furnace refractory lining, this type of boiler is extremely robust and easy to operate. Cleaning the boiler is also relatively easy and is accomplished, when the boiler is shut down, by simply removing the cleaning doors, opening the drain and spraying with high-pressure fresh water.

Pressure control of the steam is accomplished by flashing the boiler when pressure drops below a preset level during periods of high steam load and dumping steam to the condenser when the pressure rises due to low steam load.

Vapour vertical boiler (coiled-tube)

Figure 10.3 shows in a simplified diagrammatic form a coiled-tube boiler of the stone-vapour type. It is compact, space saving, designed for UMS operation, and is supplied ready for connecting to the ship's services. A power supply, depicted here by a motor, is required for the feed pump, fuel pump (if fitted), fan and controls.

Feed water is force-circulated through the generation coil wherein about 90% is evaporated. The un-evaporated water travelling at high velocity carries sludge and scale into the separator, which can be blown out at intervals manually or automatically. Steam at about 99% dry is taken from the separator for shipboard use.

The boiler is completely automatic in operation. If, for example, the steam demand is increased, the pressure drop in the separator is sensed and a signal, transmitted to the feed controller, demands increased feed, which in turn increases air and fuel supply.

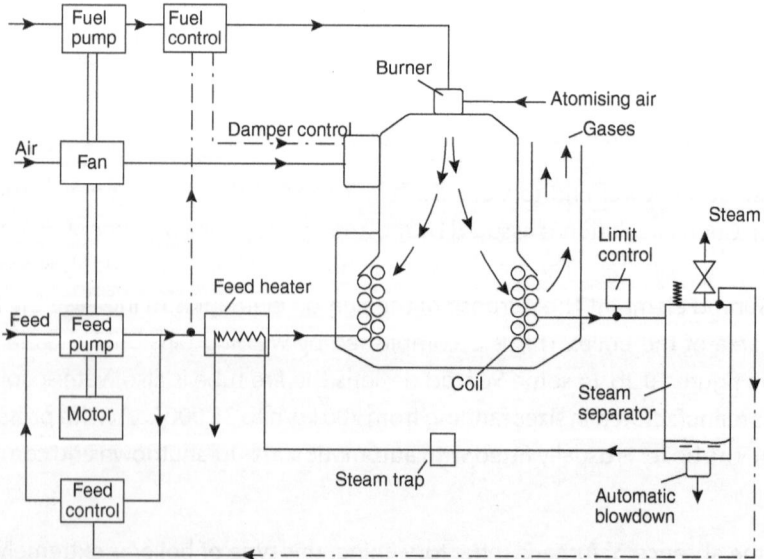

▲ **Figure 10.3** *Package coil-type boiler*

With such a small water content, explosion due to coil failure is virtually impossible and a steam temperature limit control protects the coil against abnormally high temperatures. In addition the servo-fuel control protects the boiler in the event of failure of water supply. Performance of a typical unit could be:

- Steam pressure = 10 bar
- Evaporation = 3,000 kg/h
- Thermal efficiency = 80%
- Full steam output in about 3–4 min.

Note: Atomising air for the fuel may be required at a pressure of about 5 bar.

Steam-to-steam generation

In vessels that are fitted with water tube boilers, a protection system of steam-to-steam generation may be used instead of desuperheaters and reducing valves, etc (see later).

Hybrid and Power Take-Off and Take-In Systems

New electric drives now offer economy with flexibility, which is an improvement over the simple use of shaft alternators that were and still are, widely used to provide electric power on ships of many types. By taking power from the main engine to drive the generator, current is supplied more economically, and with fewer running hours on the auxiliary generator sets. But there have been limitations. The ship's electrical system normally requires a fixed frequency, and this means that the engine speed has to be constant, though some vessels are designed to accept variations from 60 Hz down to 50 Hz allowing some speed change. There have also been electrical and mechanical systems for holding generator speed constant despite variations in engine speed. Medium-speed engines are usually constant speed engines and they lend themselves more to the use of shaft alternators than do the slow-speed engines (figure 10.4).

The introduction of the HSG system by Rolls-Royce was one manufacturer's answer to improve the efficiency of main engine-driven alternators. HSG stands for hybrid shaft generator, and it is actually an advanced power electric system for conditioning the power coming from a shaft alternator so that the switchboard sees a constant voltage

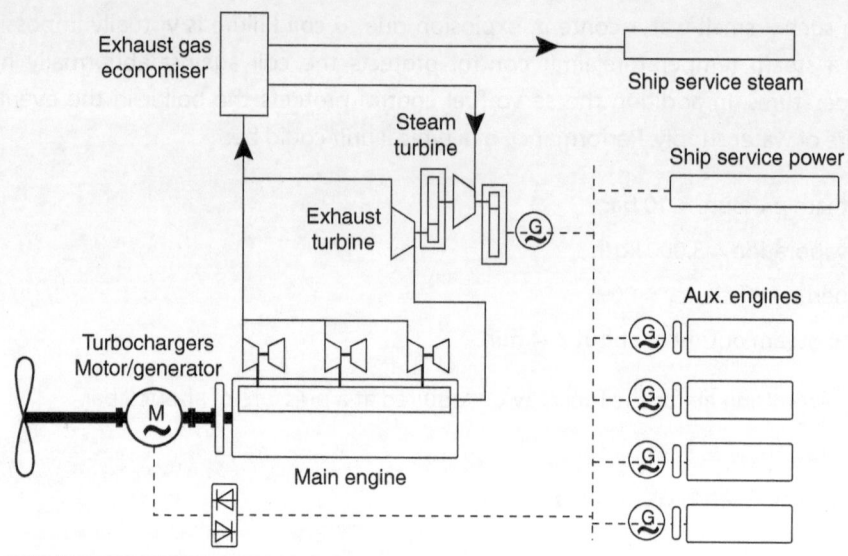

▲ **Figure 10.4** *Wärtsilä WHR system*

and frequency, and the correct phase angle to match other generator sets running in parallel. This opens the way for much more flexible use of engine and propeller speed variations to maximise both propeller and engine efficiencies by running them at their design points. HSG gets away from the straitjacket of fixed speed. Hence, for some applications the HSG concept can give remarkable fuel reductions over a comparable diesel mechanical installation. The system also helps to reduce exhaust CO_2 and NOx emissions.

HSG can in addition control the shaft generator to enable it to act as a motor, feeding in power to the propeller. This would add flexibility to the power plant by allowing the generated power from the switchboard to move the vessel in a very limited way without running the main propulsion motor.

An example of the savings possible is that one 6,500 kW CPP will have about 900 kW loss at fixed nominal RPM and with zero pitch. The HSG gives the opportunity of reducing the engine RPM, and hence the shaft line RPM, down to idling speed, still with fixed nominal frequency and voltage on the electrical network. Consequently, the zero pitch losses are reduced by 800 kW. In addition, such reductions of engine RPM will give up to 5–8% additional direct fuel saving based on higher efficiency of the diesel engine.

Even at higher load there is a fuel saving by reducing RPM. For instance, with the same example propeller as above on a vessel with a maximum speed of 20 knots, the normal procedure is to sail (eg at 14 knots) with fixed nominal engine and propeller RPM, and reduced pitch. But with the HSG system installed, pitch may be increased to 100%, engine and shaft RPM reduced by 30%, the same 14 knots maintained, but power consumption may be cut by approximately 20% from 1,900 kW to 1,500 kW. Further, the ability to use the same HSG concept to employ the shaft generator as a power take-in motor gives further potential for optimising the energy efficiency.

Vessels with medium-speed engines driving CPP or main thrusters through reduction gears are prime candidates for HSG. Typically, the vessel's speed is controlled by varying propeller pitch. Alternatively, combinator control can be used, allowing some variation in engine speed as well as pitch. As the case quoted above shows, running the engine(s) at full revolutions is often inefficient, particularly because a propeller turning at full speed but low pitch has high losses.

HSG can transform the situation. Engine and shaft speed can be optimised to allow power production at its most economical, while the propeller operates at its maximum efficiency speed and pitch for the given conditions. The shaft generator continues to function down to very low shaft speeds, feeding the main switchboard and supplying the ship's electrical load, avoiding the need to run auxiliary generator sets.

Using the shaft generator as a PTI motor with the HSG concept can be attractive where a vessel may have to steam very slowly, or loiter waiting, for a place at the quay. The main engine can be shut off instead of idling, and power generated instead by a genset operating at an efficient load.

The drive is also applicable to merchant ships with direct-coupled low-speed diesel engines and fixed pitch propellers. The traditional problem with shaft generators in this type of propulsion system is that all ships' speed control is by altering engine revolutions. Even when on passage, weather conditions cause small engine speed variations, enough to cause problems with fixed frequency shaft generator systems. HSG concept allows the generator to follow the speed changes, but the electrical consumers still receive the normal voltage and frequency. Much more use can therefore be made of a shaft generator deriving its power from a big diesel engine operating with a high thermal efficiency, eliminating the need to have auxiliary gensets running continually.

Refrigerated cargo vessels can save even more, because a shaft generator can supply the power for cooling down prior to loading, probably reducing the number and size of generator sets. Offshore vessels use shaft generators extensively. For vessels with hybrid electrical/mechanical transmission, the cost and complication of dedicated PTI motors can be saved, and HSG adds another level of flexibility to multi-mode operation. Changing from mode to mode requires less work from the crew in setting up systems, and can be automatic – simply select the desired mode for the next planned operation and issue the command.

Other vessel types with operation modes requiring widely different amounts of power for propulsion, such as yachts and fishing boats, can also reap the benefits of hybrid propulsion using the HSG system, in terms of reduced fuel burn and less noise and vibration.

The HSG concept works on the basis of a two-step electrical conversion. It uses power electronics and AFE technology developed by Rolls-Royce, avoiding the need for bulky transformers. The first step is from variable frequency a.c. to d.c. The second step is from d.c. to fixed frequency a.c., with the added feature of a 'speed droop' characteristic, which makes the system appear to the switchboard and other connected power suppliers as if it were a standard generator set running in parallel and sharing load in a stable way. The system is housed in a standard Rolls-Royce cabinet. The drive is suitable for 440 V or 690 V systems, and the power ranges from 100 kW to 5,000 kW.

This concept is not restricted to new vessels. Converting ships in service can be cost effective and a significant help in reducing emissions. Where an existing shaft generator is driven by a secondary PTO from the gearbox, conversion can be simple. The HSG concept can handle both synchronous and asynchronous electrical machines, so the existing generator might be retained, conversion involving installation of the drive cabinet and some switchboard alterations. Other types of vessel should be evaluated on a case-by-case basis.

Turbogenerators

Such turbines are fairly standard l.p. steam practice and reference, where necessary, could be made to Volume 9 of the Reeds series. Steam from the exhaust gas boiler can be used to drive a turbine. Detailed instructions are provided on board ship for personnel unfamiliar with turbine practice. For the purposes of this chapter, the short extract description given below should be typical and adequate.

Turbine

A single cylinder, single axial flow, multistage (say 5) impulse turbine provided with steam through nozzles at 10 bar and 300°C, preferably with superheat to limit exhaust moisture to 12%. Axial adjustment of rotor position is usually arranged at the thrust block and protection for overspeed, low oil pressure and low vacuum is provided. Materials and construction for the turbine unit and single reduction gearing are standard modern practice.

Electrical

The turbine at 100–166 rev/s drives the alternator and exciter through a reduction of about 6:1 to produce typically 450–600 kW at 440 V, 3 ph., 60 Hz. A centrifugal shaft-driven motorised governor arranged for local or switchboard operation would operate the throttle valve via a hydraulic servo. Straight line electrical characteristics normally incorporate a speed droop adjustment to allow ready load sharing with auxiliary diesel generators or an extra turbo unit.

Ancillary plant

This is normally provided as a package unit with condenser, air ejector, auto gland seals, gland condenser, motorised and worm-driven oil pumps, etc. A feed system is provided, either integral or divorced from the turbine-gearbox-alternator unit. Exhaust can be arranged to a combined condenser incorporating cargo exhaust. Control utilises gas bypass, dumping steam, etc.

Silencers

Normally, waste heat boilers act as spark arresters and silencers at all times. The silencer sketched in figure 10.5 would not usually be fitted if such boilers were used but a short description of the silencer may be useful.

Three designs have been utilised. The tank type has a reservoir of volume about 30 times the cylinder volume. Baffles are arranged to give about four gas reversals. The diffuser type has a central perforated discharge pipe surrounded by a number of chambers of varying volume. The orifice type is sketched in figure 10.5 and the construction should be clear. Energy pulsations and sound waves are dissipated by repeated throttling and expansion.

Gas Analysis

A number of factors have been stated that affect the design and operation of the plant and some salient points will now be briefly considered.

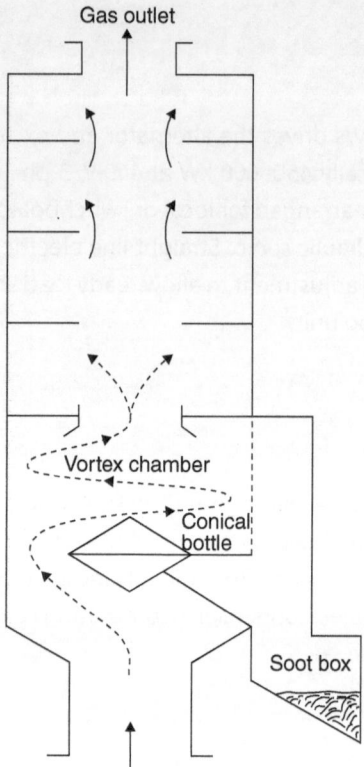

▲ **Figure 10.5** *Silencer and spark arrester*

Optimum pressure

This depends on the system adopted but in general the range is from 6 bar to 11 bar. The lower pressures give a cheaper unit with near maximum heat recovery. However, higher pressures allow more flexibility in supply with perhaps more useful steam for certain auxiliary functions together with reserve steam capacity to meet variations in demand. Low feed inlet temperatures reduce pressure and evaporative rate.

Temperature

There must be a temperature differential for heat to transfer from the exhaust gas to the water circulating in the tubes of the economiser (exhaust gas boiler). The size of the temperature difference will be affected by fouling, gas velocity, gas distribution, metal surface resistance, etc, as well as the load on the engine. However, a reduction in the service engine revolutions will cause a reduced mass of gas but a temperature increase as long as the power is maintained constant. A similar effect will be apparent under operation in tropical conditions. The effect of increased back pressure will be to raise the gas temperature for a given air inlet temperature. Figure 10.6 illustrates: (a) typical heat transfer diagram, and (b) gas temperature/mass–power curves. A common temperature differential is about 40°C, that is, water inlet 120°C and gas exit 160°C.

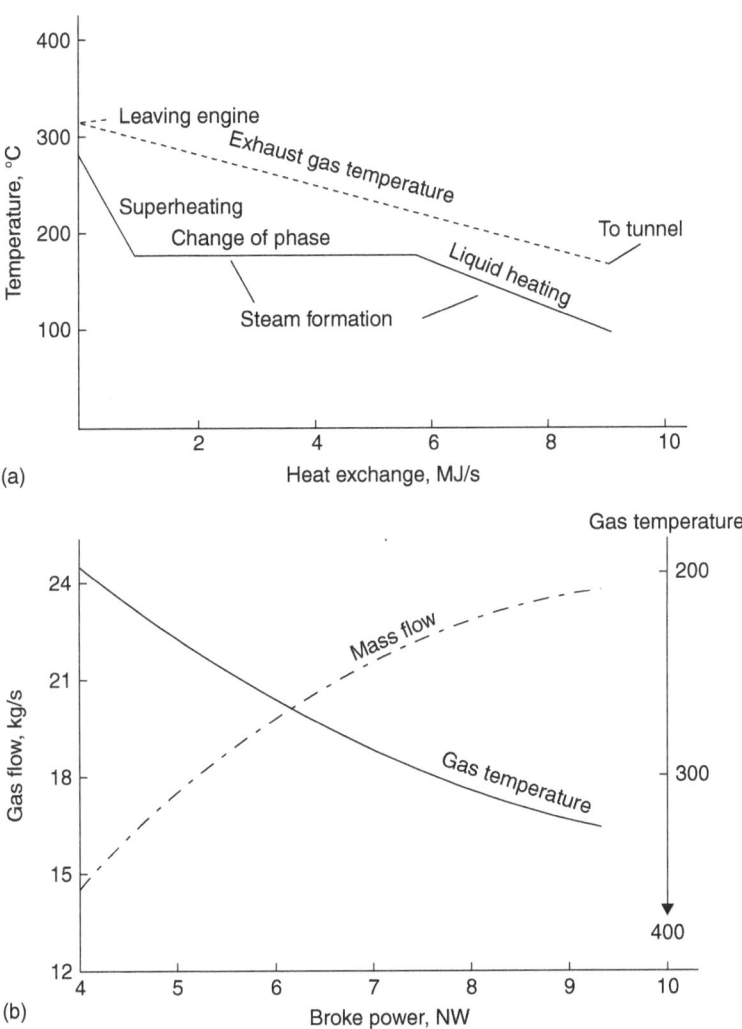

▲ **Figure 10.6** *Exhaust gas conditions*

Corrosion

The acid dew point expected is about 110°C with a 3% sulphur fuel and a high rate of conversion from SO_2 to SO_3 is possible. Minimum metal temperatures of 120°C for mild steel are required.

Exhaust system

The arrangement must offer unrestricted flow for gases so that back pressure is not increased. Good access is required for inspection and cleaning. On designs with alternate gas-oil firing, provision must be made for quick and foolproof changeover with no possibility of closure to atmosphere and waste heat system at the same time.

Gas/Water Heat Exchangers

Waste heat economisers

Such units are well proven in steamship practice and similar all-welded units are reliable and have low maintenance costs in motorships. Gas path can be staggered or straight through with extended surface element construction. Large flat casings usually require good stiffening against vibration. Water wash and soot blowing fittings may be provided.

Waste heat boilers

These boilers have a simple construction and fairly low cost. At this stage a single natural circulation boiler will be considered and these normally classify into three types, namely: simple, alternate and composite.

Simple

These boilers are not very common as they operate on waste heat only. Single- or two-pass types are available, the latter being the most efficient. Small units of this type have been fitted to auxiliary oil engine exhaust systems, operating mainly as economisers, in conjunction with another boiler. A gas change valve to direct flow to the boiler or atmosphere is usually fitted as described below.

Alternate

This type is a compromise between the other two. It is arranged to give alternate gas and oil firing with either single- or double-pass gas flow. It is particularly important to arrange the piping system so that oil fuel firing is prevented when exhaust gas is passing through the boiler. A large butterfly type of changeover valve is fitted before the boiler so as to direct exhaust gas to the boiler or to the atmosphere. The valve is so arranged that gas flow will not be obstructed in that as the valve is closing one outlet, the other outlet is being opened. The operating mechanism, usually a large external square thread, should be arranged so that with the valve directed to the boiler, fuel oil is shut off. A mechanical system using an extension piece can be arranged to push a fork lever into the operating handwheel of the oil fuel supply valve. When the exhaust valve is fully operated to direct the gas to atmosphere, the fork lever then clears the oil fuel valve handwheel after changeover travel is completed. It is also very important to ensure full fan venting and proper fuel heating-circulation procedures before lighting the oil fuel burners.

Composite

Such boilers are arranged for simultaneous operation on waste heat and oil fuel. The oil fuel section is usually only single pass. Early designs utilised Scotch boilers, with, say, a three-furnace boiler, it may mean retaining the centre or the wing furnaces for oil fuel firing. The gas unit would often have a lower tube bank in place of the furnace, with access to the chamber from the boiler back, thus giving double pass. Alternative single pass could be arranged with gas entry at the boiler back. Exhaust and oil fuel sections would have separate uptakes and an inlet changeover valve was required. In general Scotch boilers as described are nearly obsolete and vertical boilers are used. As good representative, and more up-to-date, common practice, two types of such boilers will be considered.

Cochran boiler

The Cochran boiler, whose working pressure is normally of the order of 8 bar, is available in various types and arrangements, some of which are as follows: single-pass composite, that is, one pass for the exhaust gases and two uptakes, one for the oil fired system and another for the exhaust system; double-pass composite, that is, two passes for the exhaust (figure 10.7) gases and two uptakes, one for the oil fired system and one for the exhaust system. (Double-pass exhaust gas, no oil fired furnace and a single uptake, is

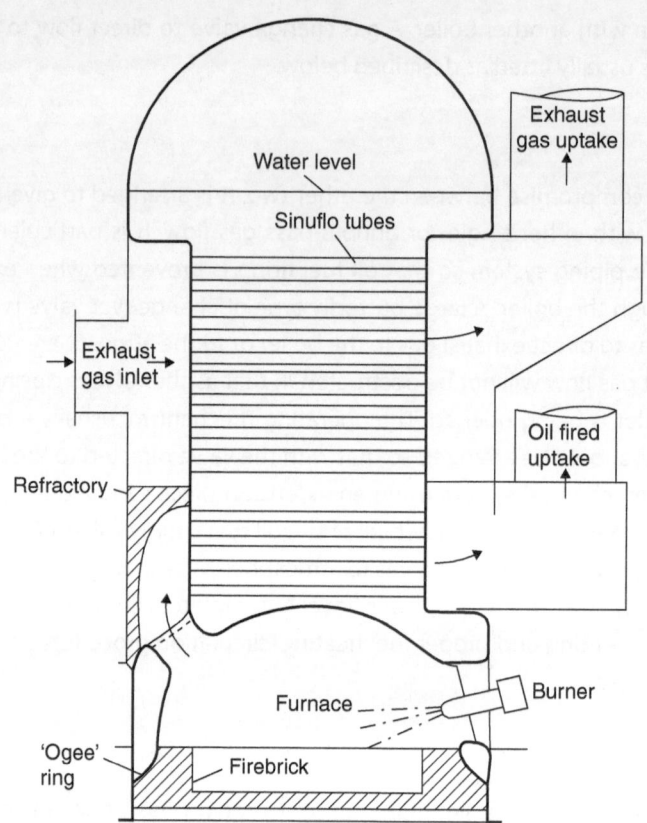

▲ **Figure 10.7** *Diagrammatic arrangement of a single-pass composite Cochran boiler*

available as a simple type. Or, double-pass alternatively fired, that is, two passes from the furnace for either exhaust gases or oil fired system with one common uptake.)

The boiler is made from good-quality, low carbon, open hearth, mild steel plate. The furnace is pressed out of a single plate and is therefore seamless. Connecting the bottom of the furnace to the boiler shell plating is a seamless 'Ogee' ring. This ring is pressed out of thicker plating than the furnace; greater thickness is necessary since circulation in its vicinity is not as good as elsewhere in the boiler and deposits can accumulate between it and the boiler shell plating. Hand-hole cleaning doors are provided around the circumference of the boiler in the region of the 'Ogee' ring.

The tube plates are supported by means of tube stays and by gusset stays, the gusset stays supporting the flat top of the tube plating. Tubes fitted are usually of special design (Sinuflo), being smoothly sinuous in order to increase heat transfer by promoting turbulence. The wave formation of the tubes lies in a horizontal plane when the tubes are fitted; this ensures that no troughs are available for the collection of dirt or moisture. This wave formation does not in any way affect cleaning or fitting of the tubes.

Thimble tube boiler

There are various designs of thimble tube boiler, including: oil fired, exhaust gas, alternatively fired and composite types.

The basic principle with which the thimble tube operates was discovered by Thomas Clarkson. He found that a horizontally arranged tapered thimble tube, when heated externally, could cause rapid ebullitions of a spasmodic nature to occur to water within the tube, with subsequent steam generation. Figure 10.8 shows diagrammatically an alternatively fired boiler of the Clarkson thimble tube type capable of generating steam with a working pressure of 8 bar. The cylindrical outer shell encloses a cylindrical combustion chamber, from which radially arranged thimble tubes project inwards. The combustion chamber is attached to the bottom of the shell by an 'Ogee' ring and to the top of the shell by a cylindrical uptake. Centrally arranged in the combustion chamber is an adjustable gas baffle tube.

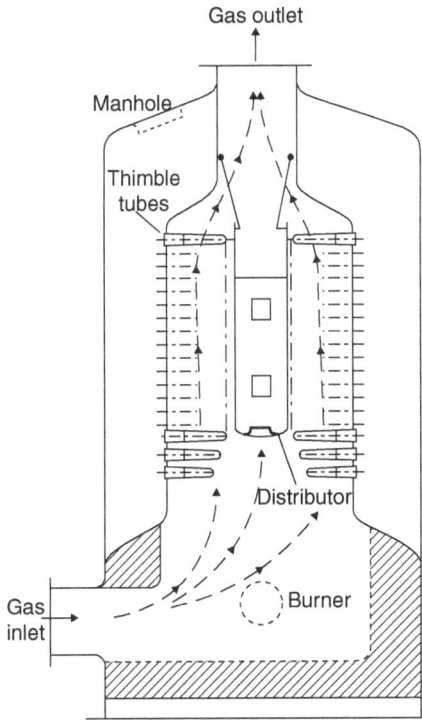

▲ **Figure 10.8** *Alternatively fired thimble-tube boiler*

Exhaust Gas Heat Recovery Circuits

Many circuits are possible and a few arrangements will now be considered. Single boiler units as discussed, while cheap, are not flexible and have relatively small steam generating capacity. The systems now considered are based on multi-boiler installations.

Natural circulation multi-boiler system

It is possible to have a single-exhaust gas boiler located high up in the funnel, operating on natural circulation whereby a limited amount of steam is available for power supply while the vessel is at sea. In port or during excessive load conditions, the main boiler or boilers are brought into operation to supply steam to the same steam range by suitable cross-connecting steam stop valves (figure 10.9). In port, the exhaust gas boiler is

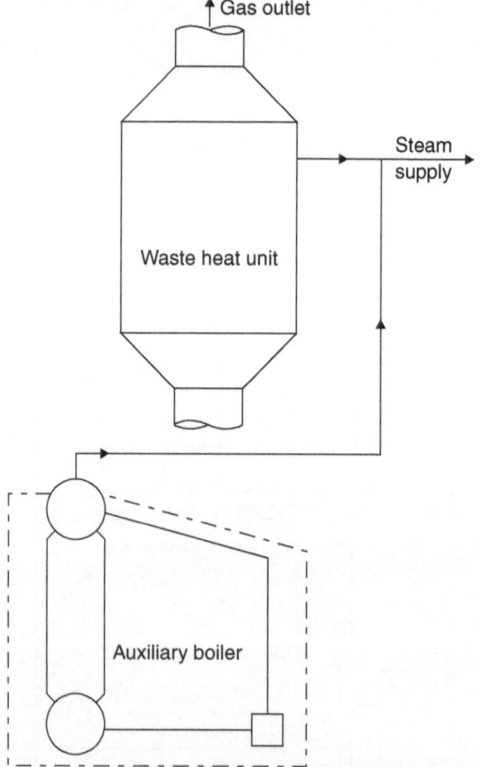

▲ **Figure 10.9** *Natural circulation/waste heat plant and W.T. boiler*

secured and all steam is supplied by the oil-fired main boilers. This system is suitable for use on vessels such as tankers where a comparatively large port steaming capacity may be required for operation of cargo pumps, but suffers from the disadvantage that the main boilers must either be warmed through at regular intervals or must be warmed through prior to reaching port. Further to this, the main boilers are not immediately ready for use in the event of an emergency stop at sea unless the continuous warming through procedure has been followed.

Forced circulation multi-boiler system

In order to improve the heat transfer efficiency and to overcome the shortcomings of the previous example, a simple forced circulation system may be employed. The exhaust gas boiler is arranged to be a drowned heat exchanger, which, due to the action of a circulating pump, discharges its steam and water emulsion to the steam drum of a water-tube boiler. The forced circulation pump draws from near the bottom of the main boiler water drum and circulates water at almost 10 times the steam production rate, thus giving good heat transfer. The steam/water emulsion on being discharged into the water space of the main boiler drum separates out exactly in the same way as if the boiler were being oil fired. This arrangement ensures that the main boiler is always warm and capable of being immediately fired by manual operation or supplementary pilot operated automatic fuel burning equipment (figure 10.10). Feed passes to the main boiler and becomes neutralised by chemical water treatment. Surface scaling is thus largely precluded and settled-out impurities can be removed at the main boiler blowdown. If feed flow only is passed through an economiser type unit, parallel flow reduces risks of vapour locking. Unsteady feed flow at normal gas conditions can result in water flash over to steam and rapid metal temperature variations. Steam, hot water and cold water conditions can cause thermal shock and water hammer.

Contra flow designs are generally more efficient from a heat transfer viewpoint, giving gas temperatures nearer steam temperature, and are certainly preferred for economisers if circulation rate is a multiple of feed flow. The generation section is normally parallel flow and the superheat section is contra flow. Output control could be arranged by output valves at two different levels, thus varying the effective heat transfer surface utilised. In addition a circulating pump bypass arrangement gives an effective control method (figure 10.11).

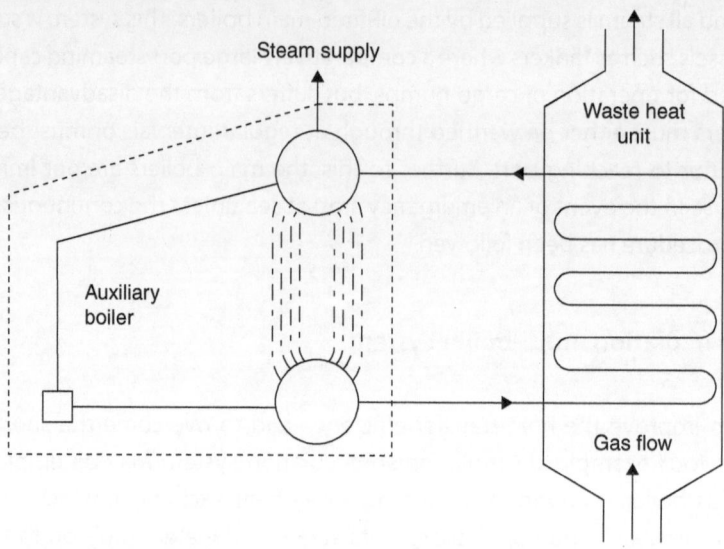

▲ **Figure 10.10** *Natural circulation*

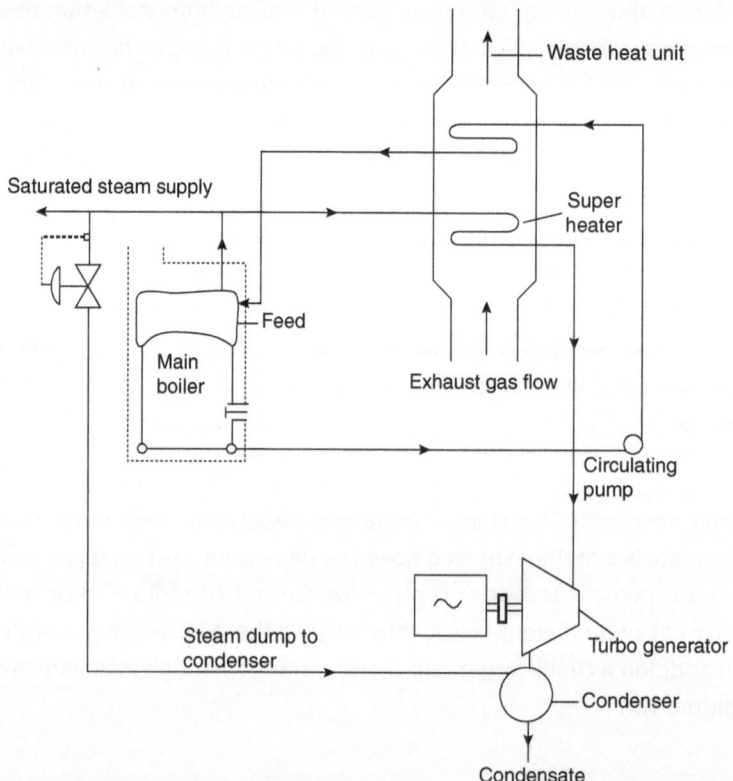

▲ **Figure 10.11** *Forced circulation waste heat plant and main boiler*

Dual pressure forced circulation multi-boiler system

This concept has been incorporated in the latest waste heat circuits and the sketch illustrates how the general principle can be applied in conjunction with a waste heat exchanger to supply superheated steam. By this means every precaution has been taken to minimise the effect of contamination of the water-tube boiler.

Steam generated in the water-tube boiler by either oil firing or waste heat exchanger passes through a submerged tube nest in the steam/steam generator to give lower grade steam, which is subsequently passed to the superheater.

A water-tube boiler, steam/steam generator and feed heater may be designed as a packaged unit with the feed heater incorporated in the steam/steam generator. The high-pressure high-temperature system at, say, 10 bar will supply a turbogenerator for all electrical services while the low-pressure system at, say, 21/2 bar would provide all heating services. Obviously the dual system is more costly. Numerous designs are possible, including separate low-pressure and high-pressure boilers, either natural or forced circulation, indirect systems with single- or double-feed heating, etc (figure 10.12).

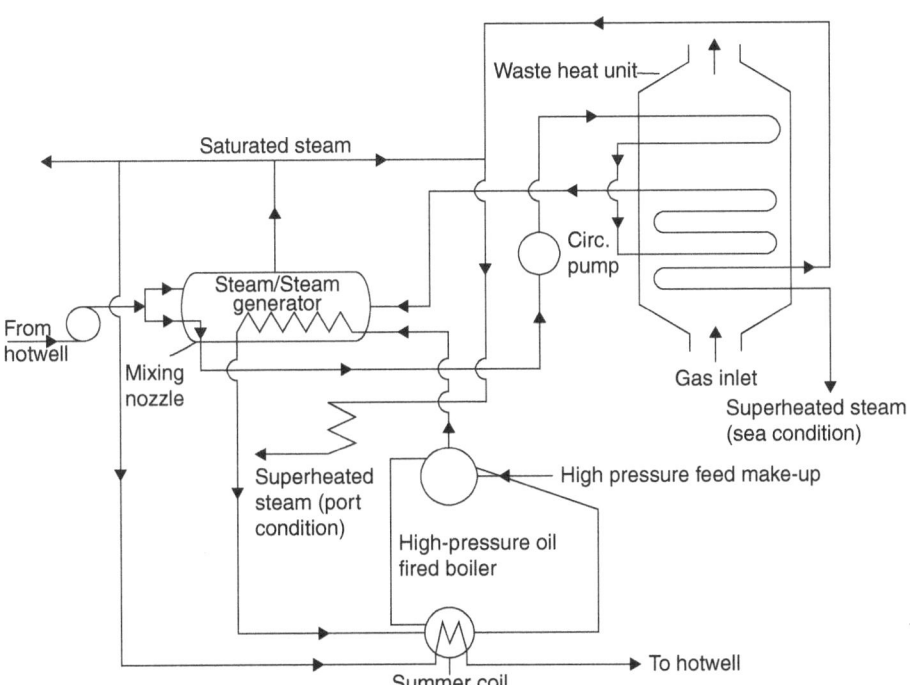

▲ **Figure 10.12** *Dual pressure system*

It is very important that students are very familiar with the current thinking in boiler design and operational practice. This is an important safety issue and will be considered very carefully by the Flag State examiner. Students must understand the watchkeeping issues with boilers identified at the start of this chapter.

Water/Water Heat Exchangers

Evaporators

The basic information given on evaporators in Volume 8 of the Reeds series should *first* be considered. In motor ship practice, efficient single effect units incorporating flexible elements and controlled water level are in service. Evaporators utilising jacket cooling water as the heating medium producing an output of 20–25 tonnes per day are common. Flash evaporators have increasingly been fitted on large vessels utilising multi-stage units. Also multiple effect evaporators of conventional form are used. The steam circuit of many modern motor ships has developed in complexity to approach successful steam ship practice. The reverse osmosis (RO) plant is very popular for making large amounts of water, mostly for drinking purposes.

Feed heating

The advantages of pre-heating feed water are obvious. Three methods will be considered, namely: economisers, mixture and indirect. Economiser types have been included in previous discussion and sketches. It is sufficient to repeat that such systems require a careful design to cope with fluctuations of steam demand and that particular attention is necessary to ensure protection against corrosive attack. Mixture systems employ parallel feeding with circulating pump and feed pump to the economiser inlet. Such circuits require careful matching of the two pumps and control has to be very effective to prevent cold water surges leading to reducing metal temperatures and causing corrosion. Indirect systems require a water/water exchanger feed heater.

This design reduces the risk of solid deposit in the economiser and maintains steady conditions of economiser water flow, thus protecting the economiser against corrosive attack. A typical system is shown in figure 10.13. If boiler pressure tends to rise too high the circulation bypass will be opened.

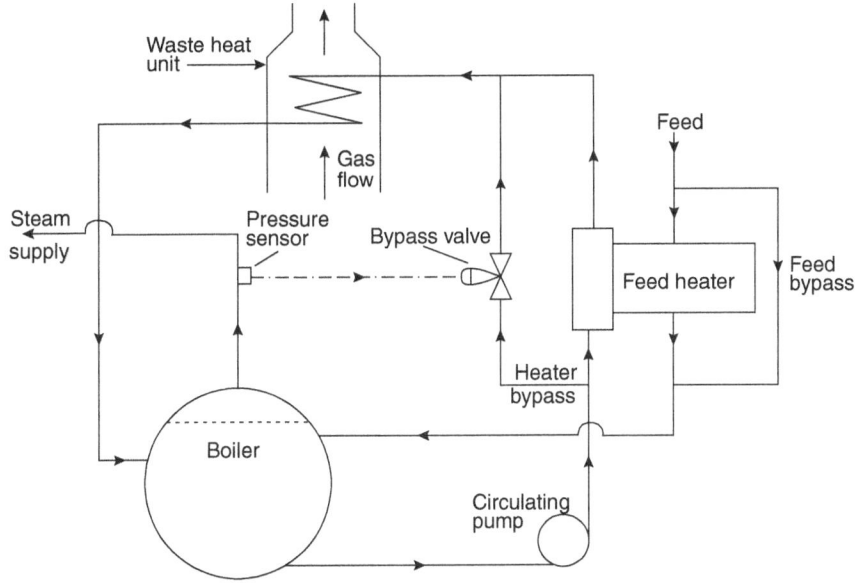

▲ **Figure 10.13** *Indirect feed heating*

The effect will be twofold, that is, feed water will enter the boiler at a lower temperature and water temperature entering the economiser is at a higher temperature. These two effects serve to reduce boiler pressure and thus control the system. Obviously this system is more costly but is very flexible.

Combined Heat Recovery Circuits

The low-grade heat of engine coolant systems restricts the heat recovery in such secondary circuits to temperatures near 7–8°C. As such, it is normally restricted to use with distillation plants. Combined or compound units involving combination between engine coolant and exhaust gas systems are complicated by the need to prevent contamination and utilise the large volume of low-temperature coolant in circulation. Jacket water coolant temperatures have increased in recent motorship practice but even if the engine design can be modified to suit even higher temperatures, there is always a problem of high radiation heat loss from jackets to confined engine rooms.

Exhaust Gas Power Turbine

In an effort to improve overall plant efficiency, the turbocharger manufacturer ABB has developed a system that exploits surplus exhaust gas in a power turbine where the turbine's power output is fed either to the engine crankshaft or to an auxiliary diesel or steam turbine generator (figure 10.14). The latter is only feasible if the demand for electricity is greater than the output of the power turbine. This development has been made possible by the improved turbocharger efficiencies achieved in recent years, resulting in surplus energy being available in the exhaust gas.

The power turbine can be brought in and out of service, as conditions require, by operating a flap in the exhaust line. The turbine part of the power turbine is similar to those of turbochargers. The drive from the turbine is via epicyclic gearing and a clutch to the chosen mode of power input.

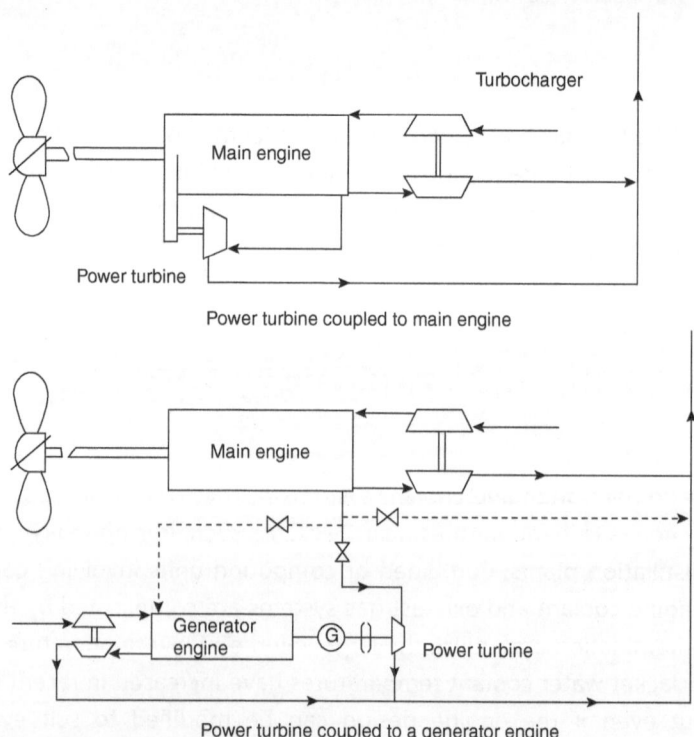

Power turbine coupled to main engine

Power turbine coupled to a generator engine

▲ **Figure 10.14** *Recovery of surplus exhaust gas*

Gas Turbines

The gas turbine theoretical cycle and simple circuit diagram have been considered in Chapter 1. Marine development of gas turbines stemmed from the aero industry in the 1940s. Apart from an early stage of rapid progress, the application to marine use has been relatively slow until recently. Consideration can best be applied in two sections, namely, industrial gas turbines and aero-derived types. In general this can largely be considered as a 'marinisation' of equipment originally designed for other duties such as powering aircraft or driving land-based electrical generators.

Energy in a Power Turbine

Industrial Gas Turbines

The simplest design is a single-shaft unit, which has low volume and light weight (5 kg/kW at 20,000 kW). Fuel consumption (specific) may be about 0.36 kg/kWh on residual fuels. This consumption is not normally acceptable for direct propulsion and initial usage was as emergency generators in MN practice and the RN for small vessels or as boost units in larger warships. Compared to steam turbines (32% output, 58% condenser loss), the simple gas turbine (24% output, 73% exhaust loss) is less efficient but the addition of exhaust gas regeneration gives 31% output (specific fuel consumption = 0.28 kg/kWh) and combined RN units 36% output. Normally a two-shaft arrangement was preferred in MN practice, in which load shaft and compressor shaft are independent (figure 10.15).

A design was available by 1955 for main propulsion with maximum turbine inlet conditions of 6 bar, 650°C and specific fuel consumption approaching 0.3 kg/kWh. Starting of the twin-shaft unit was by electric motor, power variation by control of gas flow, conventional gear reduction and propeller drive by hydraulic clutch with astern torque converter (more modern practice uses variable pitch propeller). Turbine and turbo-compressor design utilised standard theory and simple module construction utilising horizontally split casings, diffusers, etc, and easily accessible nozzles.

To improve efficiency even further it is necessary to use much higher inlet gas temperatures (1,200°C would give a specific fuel consumption of about 0.2 kg/kWh).

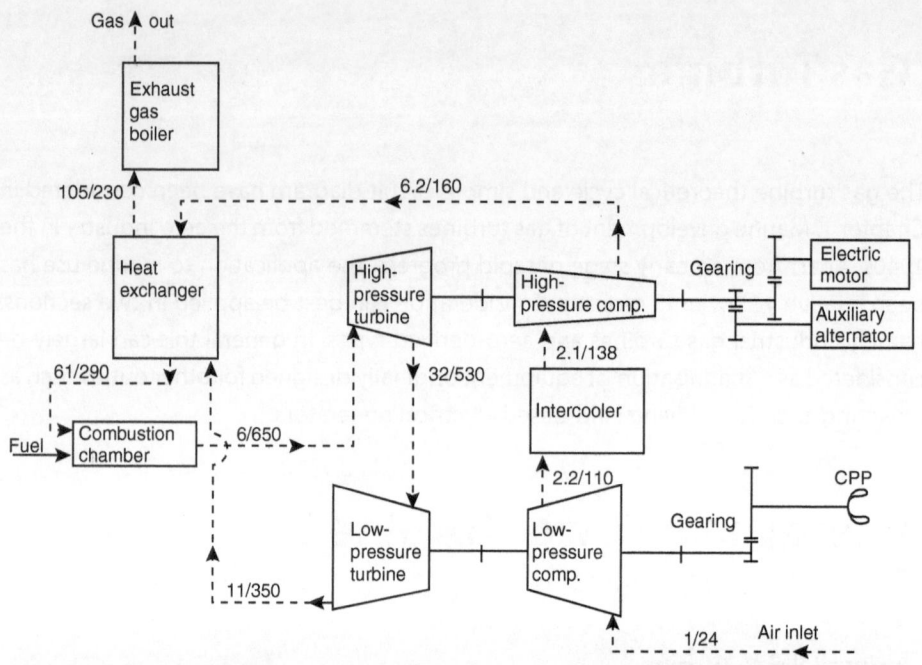

▲ **Figure 10.15** *Open cycle marine gas turbine*

The limiting factor is suitable materials. Experiments have been, and still are, being carried out with ceramic blades and with cooled metallic blades. Essentially the problem is the same for steam turbine plant and there has been no marked incentive for the shipowner to install gas turbine plant in preference to equally economic and established steam systems. During the 1960s experience was established in the vessels *Auris, John Sergeant* and *William Paterson*. It may well be that direct gas-cooled reactors in conjunction with closed cycle gas turbines in electric power generation may be an attractive possibility in nuclear technology. GEC. produce a wide range (4,000–50,000 kW) of industrial gas turbines, now effectively marinised for marine propulsion. In addition to reliability, easy maintenance, low volume, etc, the very easy application to electric drives and to automation make the units attractive. Geared drive usually utilises locked train helical gears or alternatively epicyclic gearing.

CPP development has also broadened the possibilities of various propulsion systems, including geared diesel – gas turbine systems. Marine gas turbines do run with a high noise level and they require to be water washed at regular intervals, the latter depending upon the type of fuel being used. The recently changing design of ships has meant that the owner, or operator, needs to analyse propulsion systems carefully for all economic factors, which vary greatly for VLCC, Rolls-Royce, LNG, container vessels, etc. Gas turbines have been exclusively adopted for RN surface vessels.

Aero-derived

Apart from RN units so derived from aero gas turbines, the first British MN vessel so engined was the g.t.s. *Euroliner* in 1970. Turbo Power and Marine Systems Inc. twin gas turbines, 22,500 kW each at 3,600 rev/min, drive separate screw shafts at 135 rev/min through double reduction locked train gears, with CPP. Main electrical alternators are driven from the gearbox.

11

SAFETY AND ENVIRONMENTAL

When people decided to take their trade to the open ocean, they knew that they were working in a harsh and difficult environment. Despite this, the lore of the sea was compelling and many have succumbed to its power. Modern life demands that we endeavour to make all our places of work and leisure as safe as possible. This is accomplished by using a combination of rules and the creation of a safety culture where working safety is part of being a 'professional seafarer'.

As ships ventured beyond national borders there had to be some mechanism for extending the rule of law to the vessel operating across the ocean. Therefore, ships were registered, usually in the country of the owners or where the ship traded to/from the most.

Ship owners were (and still are) required to register their ships in the chosen country. The vessels were then subject to the rules and regulation, and the domestic law, of the country of registry. For example, ships registered in Holland were governed by Dutch law and flew the Dutch national flag and ships registered in the UK came under UK law and flew the 'Red Ensign', which is the flag of the British Merchant Navy.

With each nation having their own set of rules, regulations and laws, this meant that ships operating under different administrations (flags) might be slightly different in their requirements for things such as the standards of construction or the requirements for the skills and education of the ship's staff.

Most ships were registered with the 'traditional' seafaring nations where the rules and standards had been built up over many years, leading to a safe way of operating vessels in some of the most challenging environments of earth.

However, during the 1960s, 1970s and 1980s ship owners became increasingly frustrated by what they saw as the overly bureaucratic and inflexible rules required by the different administrations where ships were registered. In addition, the competition from owners who were operating vessels under 'alternative' administrations, where costs were not as high, started proving too much and many of the owners of the time could no longer stay in business.

Therefore, during this time there was a big move for owners to register their vessels with countries that did not have much experience of setting the rules and regulations for operating safe ships. As a consequence, these administrations gained the name 'Flags of Convenience' and the reputation of allowing safety standards and minimum skill requirements for ships staff to drop.

The industry needed to address the decline in standards but how do you impose and police a common set of rules right across the world? All nations capable of regulating ships would need to agree to adhere to a minimum set of rules and the organisation capable of setting the rules is the 'Maritime' department of the United Nations (UN).

The UN established the Inter-Governmental Maritime Consultative Organization (IMCO) in 1948 so that seafaring nations that were regulating international shipping had a place to discuss the various aspects of shipping and agree among themselves about future action. In 1948 the emphasis was on establishing an environment where commercial trade could flourish and removing unfair and restrictive practices.

The name of the UN assembly was change to the International Maritime Organization (IMO) in 1982 and its aim was expanded to include a mechanism for participating governments to discuss and agree on regulation and best practices relating to the technicality and safe practices of operating international shipping.

It should be recognised that the IMO does not make 'laws'. The IMO provides a secretariate and facilities where member states can discuss and agree on issues of maritime importance, such as pollution prevention and the safe manning of ships. IMO treaties are formed following extensive discussions within one of the five main committees or sub committees.

Member states must then incorporate the IMO treaties into their own domestic law. The subject concerned can then be applied to the ships that operate under the administration of each member state. In addition, member states are required to inspect ships that are visiting ports within their jurisdiction to ensure that they are conforming to the IMO criteria. Ships that do not meet the requirements can be held in a port (for a serious defect) until the issue is rectified.

This system of 'policing' the agreed IMO regulations is referred to as 'Port State Control' (PSC). However, this does present a slight problem as a ship might call at different ports that are close to each other, but which are in different countries. The ships could then be subjected to multiple inspections in close succession, placing unnecessary cost on and causing disruption to the vessel's trading routine.

In addition to this there could be slight differences in the inspections as one member state focuses on one part of legislation in detail and another member state concentrates on another part of the legislation.

Therefore, IMO has grouped together the PSC states into a number of 'regimes' that are linked by their location in the world. The PSCs agree to work together, and each have a memorandum of understanding (MoU). For example, the Paris MoU includes the PSC from members that are situated in and around Europe.

This means that if a ship has been successfully inspected at the port of a member within the Paris MoU area, then that ship will not be inspected again for a period of time that is determined by that ship's risk profile. Usually, low risk can be up to three years between inspections, standard risk every year and high risk ships can be inspected every six months. If a vessel fails an inspection several times in a row it can be banned from entering the ports or anchorages within the region.

This system is important for the Marine Engineering Officer (MEO) to understand as it has a direct bearing on the responsibilities that he or she is expected to undertake. For example, any pollution from ships needs to be prevented and the special equipment used to complete this function, such as the oily water separator (OWS), needs to be kept fully operational. If not and this is spotted by the PSC inspector then the ship will be given a penalty, the severity of which will depend upon the type of defect.

The training of MEOs will need to include ensuring that they are fully familiar with the key IMO conventions, which are:

- International Convention for the Safety of Life at Sea (SOLAS), 1974, as amended.
- International Convention for the Prevention of Pollution from Ships, 1973, as modified by the Protocol of 1978 relating thereto and by the Protocol of 1997 (MARPOL).
- International Convention on Standards of Training, Certification and Watchkeeping for Seafarers (STCW) as amended, including the 1995 and Manila Amendments and further amendments.

A more in-depth description about the mechanisms of regulating shipping is covered in volume 8 (*General Engineering Knowledge*) of the Reeds Marine Engineering Series, chapter 12, "Management and Leadership". However, it is appropriate to cover the latest efficiency regulations in this chapter. An explanation about the two new regulations of the Energy Efficiency Existing Ship Index (EEXI) and carbon intensity indicator (CII) appear on page 343.

Seafarer Safety in the Engine Room (crankcase explosions)

Introduction

Venturing out to sea in ships is still a hazardous occupation and therefore seafarers must take extra care for their own safety as well as the safety of fellow shipmates.

The person presenting themselves to a Flag State administration for the Engineering Officer of the Watch qualification needs to be very familiar with all aspects of engine room safety and safe working practices.

The UK Government has been publishing versions of the code of safe working practices for merchant seafarers for many years. This document is a fantastic resource full of information about the safe working practices on board ships.

The document has been updated to include 'Managing Occupational Health and Safety', which gives detailed advice about developing a 'safety culture' as well as information about a seafarer's personal health.

Ships' staff working regularly in the engine room will need to guard against excessive:

- noise (ear protection is ESSENTIAL at all times)
- heat (intake plenty of water)
- irritant substances (use barrier cream or surgical-style gloves)

Fire hazard

To gain useful energy from fuel, we need to bring together the three conditions required to start a fire or controlled explosion. Obviously, this is completed under carefully controlled conditions and the aim of converting the chemical energy into useful work is achieved.

It is equally obvious that deliberately bringing together heat/fuel/air could be hazardous because if they come together in the wrong place then a fire could ensue. This makes the engine room potentially the most hazardous place on the ship.

There are several places in the engine room that deserve special note and where the watchkeeper must be especially diligent in monitoring for any potential danger. These are:

- Oil fired boiler fronts
- Diesel exhausts
- Oil that has been allowed to accumulate in the bilges
- Any welding or gas cutting operations
- Swarf and oily rags placed in rubbish bins.

The first line of defence is 'good housekeeping', which is a general term for being tidy, cleaning up spilt oil, clearing away rubbish carefully, posting a fire watch when welding or burning and generally being alert to any possible source of ignition and taking steps to reduce or eliminate the risk of fire.

When running the machinery, however, there will be other things to guard against. Some of the more significant hazards are described over the next few pages.

Crankcase explosions

The student should first refer to Volume 8 of the Reeds series for a consideration of spontaneous ignition temperatures and also limits of inflammability of gases and vapours in air. Crankcase explosions have occurred steadily over the years, with perhaps that of the *Reina del Pacifico* in 1947 the most serious of all. In fact, crankcase explosions have occurred in all types of enclosed crankcase engines, including steam engines. Explosions occur in both trunk piston types and in engines with a scraper gland seal on the piston rod. Much research has been done in this field but the difficulties of full experimentation utilising actual engines under normal operating conditions is almost impossible to attempt. The following is a simplified presentation based on the mechanics of cause of explosion, appropriate DoT regulations and recommendations and descriptive details of preventative and protective devices utilised.

Mechanics of a crankcase explosion

1. A hot spot is an essential source of such explosions in crankcases as it provides the necessary ignition temperature, heat for oil vapourisation and possibly ignition spark. Normal crankcase oil spray particles are in general too large to be easily explosive (average 200 μm). Vapourised lubricating oil from the hot source occurs

at 400°C, in some cases lower, with a particle size explosive with the correct air ratio (average 6 µm). Vapour can condense on colder regions; a condensed mist with fine particle size readily causes explosion in the presence of an ignition source. A lower limit of flammability of about 50 mg/l is often found in practice. Experiment indicates two separate temperature regions in which ignition can take place, that is, 270–350°C and above 400°C.

2. Initial flame speed after mist ignition is about 0.3 m/s but unless the associated pressure is relieved this will increase to about 300 m/s with corresponding pressure rise. In a long crankcase, flame speeds of 3 km/s are possible, giving detonation and maximum damage. The pressure rise varies with conditions but without detonation does not normally exceed 7 bar and may often be in the range of 1–3 bar.

3. A primary explosion occurs and the resulting damage may allow air into a partial vacuum. A secondary explosion can now take place, which is often more violent than the first, followed by a similar sequence until equilibrium.

4. The pressure generated, as considered over a short but finite time, is not too great but instantaneously is very high. The associated flame is also dangerous. The gas path cannot ordinarily be deflected quickly due to the high momentum and energy.

5. Devices of protection must allow gradual gas path deflection, give instant relief followed by non-return action to prevent air inflow and be arranged to contain flame and direct products away from personnel.

6. Delayed ignition is sometimes possible. An engine when running with a hot spot may heat up through the low-temperature ignition region without producing flame because of the length of ignition delay period at low temperatures. Vapourised mist can therefore be present at 350–400°C. If the engine is stopped, the cooling may induce a dangerous state and explosion. Likewise, air ingress may dilute a previously too rich mixture into one of dangerous potential.

7. Direct detection of overheating by thermometry offers the greatest protection but the difficulties of complete surveillance of all parts is prohibitive.

8. A properly designed crankcase inspection door, preferably bolted in place, suitably dished and curved with, say, a 3-mm thickness of sheet steel construction, should withstand static pressures up to 12 bar, although distorted.

9. There are many arguments for and against vapour extraction by exhauster fans. There is no access of free air to the crankcase and the fan tends to produce a slight vacuum in the crankcase. On balance, most opinion is that the use of such fans can reduce risk of explosion. The danger of fresh air drawn into an existing over-rich heated state is obvious. On the practical aspect, leakage of oil is reduced.

Crankcase safety arrangements

The following are based on specific guidance to surveyors from a leading Flag State – Maritime Coastguard Agency (MCA):

1. Means should be adapted to prevent danger from the result of explosion in crankcases with forced lubrication systems.

2. Crankcases and inspection doors should be of robust construction. Attachment of the doors to the crankcase (or entablature) should be substantial.

3. One or more non-return pressure relief valves should be fitted to the crankcase of each cylinder and to any associated gear or chain casing.

4. Such valves should be arranged or their outlets so guarded that personnel are protected from flame discharge with the explosion.

5. The total clear area through the relief valves should not normally be less than 115 cm^2/m^3 of gross crankcase volume.

6. Engines not exceeding 250-mm cylinder bore but larger than 200-mm bore with strongly constructed crankcases and doors may have two valves, usually fitted at the crankcase ends. Similarly constructed engines not exceeding 200-mm cylinder bore or having a crankcase volume of less than 0.6 m^3 need not be fitted with relief valves.

7. Lubricating oil drain pipes from the engine sump to the drain tank should extend to well below the working oil level in the tank.

8. Drain or vent pipes in multiple engine installations are to be so arranged that the flame of an explosion cannot pass from one engine to another.

9. In large engines having more than six cylinders, it is recommended that a diaphragm should be fitted at near mid length to prevent the passage of flame.

10. Consideration should be given to means of detection of overheating and injection of inert gas.

Preventative and protection devices

The first line of defence in preventing crankcase explosions, as well as many other problems, will be good watchkeeping. However, the watchkeepers are not everywhere at all times, especially on ships that operate an unmanned machinery space (UMS) system, and therefore methods of continual monitoring of the 'critical' areas is essential for keeping the ship's staff as safe as possible. The following are regarded as the methods of providing constant monitoring of potential danger. These items are: explosion relief doors (preventing a secondary explosion), flame protection, explosive mist detection and bearing temperature detection.

Crankcase explosion door

A design is shown in figure 11.1. The sketch illustrates a combined valve and flame trap unit with the inspection door insertion in the middle. The internal section supports the steel gauze element and the spider guide and retains the spindle. The external combined aluminium valve and deflector has a synthetic rubber seal. Pressure setting on such doors is often 1/15 bar (above atmospheric pressure). Relief area and allowable pressure rise vary with the licensing insurance authority but a metric ratio of 1:90 should not normally be exceeded based on gross crankcase volume and this should not allow explosion pressures to exceed about 3 bar.

Flame trap

Such devices are advisable to protect personnel. The vented gases can quickly be reduced in temperature by gauze flame traps from, say, 1500°C to 250°C in 0.5 m. Coating on the gauzes, greases or engine lubricating oil greatly increases their effectiveness. The best location of the trap is inside the relief valve when it gives a more even distribution of gas flow across its area and liberal wetting with lubricating oil is easier to arrange. A separate oil supply for this action may be necessary. The explosion door in figure 11.1 has an internal mesh flame trap fitted.

Flame traps effectively reduce the explosion pressure and prevent two-stage combustion. Gas-vapour release by the operation of an oil-wetted flame trap is not

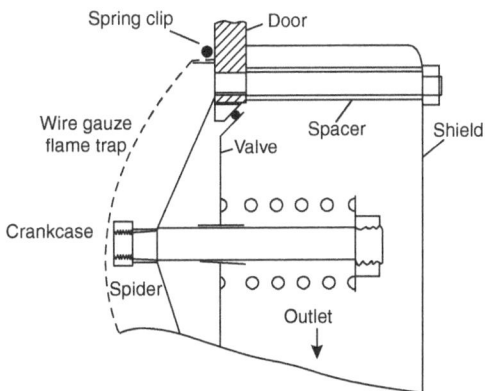

▲ **Figure 11.1** *Crankcase explosion relief door*

usually ignitable. Typical gauze mild steel wire size is 0.3 mm with a 40% excess clear area over the valve area.

Crankcase oil mist detector

The major cause of crankcase explosions is condensed oil mists and the cost-effective protection is with photoelectric analysis of the oil mist in the different parts of the crankcase. This should give complete protection; however, if the crankcase spray is explosive, the mist detection will only indicate a potential source of ignition.

The general principle of operation and the design of a typical detector should be fairly clear from figure 11.2. The photocells are normally in a state of electric balance, that is, measure and reference tube mist content in equilibrium. Out-of-balance current due to a rise of crankcase mist density can be arranged to indicate on a display following its measurement from electronic devices. The result can also be connected to continuous chart recording and auto visual or audible alarms. The suction fan draws a large volume of slow-moving oil–air vapour mixture in turn from various crankcase selection points. Oil mist near the lower critical density region has a very high optical density. The alarm is normally arranged to operate at 21/2% of the lower critical point, that is, assuming 50 mg/l as the lower explosive limit then warning at 1.25 mg/l.

Operation

The fan draws a sample of oil mist through the rotary valve from each crankcase sampling pipe in turn, then through the measuring tube, and delivers it to atmosphere. An average sample is drawn from the rotary valve chamber through the reference tube and delivered to atmosphere at the same time. In the event of overheating in any part

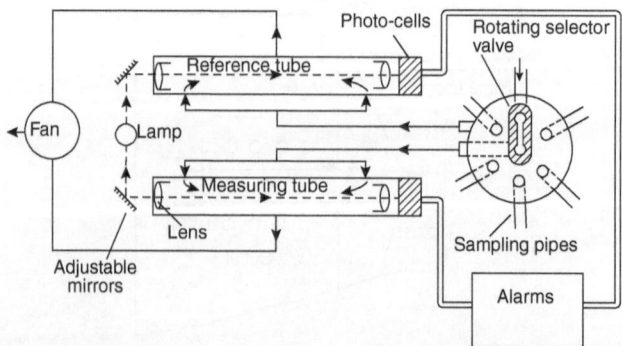

▲ **Figure 11.2** *General arrangement of the crankcase oil mist detector*

of the crankcase, there will be a difference in optical density in the two tubes, hence less light will fall on the photocell in the measuring tube. The photocell outputs will be different and when the current difference reaches a predetermined value, an alarm signal is operated and the slow-turning rotary valve stops, indicating the location of the overheating.

Normal oil particles as spray are precipitated in the sampling tubes and drain back into the crankcase.

CO_2 drenching system

Thirty per cent by volume of this inert gas is a complete protection against crankcase explosion. This is particularly beneficial during the dangerous cooling period. Automatic injection can be arranged at, say, 5% of critical lower mist density but in practice many engineers prefer manual operation. When the engine is opened up for inspection and repair at hot source, it will of course be necessary to ensure proper venting before working personnel enter the crankcase.

Bearing temperature detection

The principal bearings in an engine have the potential to become the cause of a crankcase explosion. If they are becoming overheated for some reason, the localised 'hot spot' causes vapourisation of the oil to form the oil mist and thereby sets up the conditions described above in 'The mechanics of a crankcase explosion'. The very same bearing can also act as the hot spot, initiating the start of the explosion.

One of the problems is that until recently it has been very difficult to measure the big end or crosshead bearing temperatures accurately. Systems in the past consisted of catching the oil that had sprayed from the big end bearing and measuring the temperature of the oil to give an indication of the temperature of the bearing.

Compared to the indirect measurements of a conventional oil mist detector, direct continuous monitoring of the crankpin bearing temperature permits earlier detection of a bearing overheating. This prevents major failures of critical and highly costly engine components and protects against the direct consequences of the non-operational availability of the engine. It also avoids any extra costs related to unplanned expensive maintenance operations. To be able to continuously detect and monitor the temperature of rotating bearings in an accurate and reliable manner, Wärtsilä has recently developed an innovative wireless temperature-sensing device.

The operating principle is to directly measure the temperature of the connecting rod big end bearing using a temperature sensor fitted as close as possible (within a few millimetres) to the bearing surface. The sensor then keeps in touch with the outside world by using a patented surface acoustic wave (SAW) radar technology, which has been proven to be the most reliable technology for real-time wireless temperature monitoring. The signal processing unit (SPU) generates a radio wave pulse, which is picked up by the stationary antenna and converted into an acoustic wave, which is sent to the rotating sensor. This acoustic wave propagates along the surface of a SAW chip fitted with multiple reflectors, thus permitting the sensor to reflect back a pulse train to the stationary antenna; the time delay between echoes depends on the temperature of the SAW chip.

The wireless temperature sensors are installed in the rotating connecting rod big end. The stationary antennas are screwed to a custom-designed bracket fixed inside the engine block in such a way that the sensors and antennas pass within a fixed distance of each other at each rotation of the engine crankshaft. The resultant signal is then transmitted via a thin cable passing through the engine block to the SPU fixed to the engine, and from there to the control room cabinet placed in the engine room.

The Environmental Agenda of Shipping

Marine fuels and emissions

In the century that has passed since the diesel engine first showed its promise as a power source for ships, it has come to dominate the marine scene. Steam power, once dominant, is reduced to niches such as LNG carriers and nuclear warships. Gas turbines have penetrated the marine propulsion market to a limited extent, primarily in warships and some cruise vessels and fast ferries. Natural gas is starting to make inroads in marine propulsion, while petrol and LPG are restricted to small craft. Thus, in terms of fuel used and installed power, the fuels and emissions question centres on the diesel engine.

There are four main constituents of diesel engine exhaust that are of environmental concern: NOx, SOx, CO_2 and particulates (soot). Until recently, CO_2 was seen as a harmless naturally occuring gas, neither controlled nor legislated against. Concern about NOx grew out of city smog problems. Sulphur has long been seen as a problem with fossil fuels. Columns of smoke, once seen as a positive sign of economic activity, are now frowned on at sea or ashore.

Cutting these emissions is not just an engineering matter, though some engineering solutions have been developed and others are under development. Looking at the broader picture, exhaust emissions from ships are a complex blend of politics, economics and engineering. Some of the engineering solutions are considered in more detail below but first it is instructive to look at the question of marine fuels.

Early in its history the diesel engine showed that with suitable design and adjustment it could burn a wide variety of liquid fuels. Sixty years ago there began a move towards using heavy fuels of the type used in boilers. This proved a successful and much cheaper solution than burning lighter distillate fuels, particularly in deep sea merchant vessels covering long distances.

Then, 30 years ago, the marine industry received warning from the oil industry that the quality of residual fuels was going to deteriorate substantially. Changing refinery processes in the quest for more valuable fractions for road and land use would leave a residue that the oil industry considered might be tough for the diesel engine to swallow. However, the engine manufacturers tackled the problem successfully, overcoming combustion troubles and also excessive wear caused by catalytic fines (abrasive particles) in the fuel. Ships' crews had to exert a much higher level of supervision to ensure that fuel purification systems were run correctly, and also in fuel management, since the latest load of fuel to be bunkered might not be compatible with the previous lot. In addition came the problem of disposal routes for sludge. Both large main engines and small-bore auxiliary engines were perfected to run on heavy fuel of dubious quality where viscosity and density were usually the only two parameters closely controlled. The incentive was low price; IF380 fuel, for example, is a fraction of the price of the light distillate marine gas oil.

Exhaust emissions

CO_2 is a product of combustion and the amount will depend on the chemical composition of the fuel, natural gas having a lower value than typical liquid fuels. Poor combustion will produce soot and unburned hydrocarbons in the exhaust.

NOx is a function of high-temperature combustion in the engine cylinders where the nitrogen in the air reacts with the oxygen (figure 11.3). Current and forthcoming

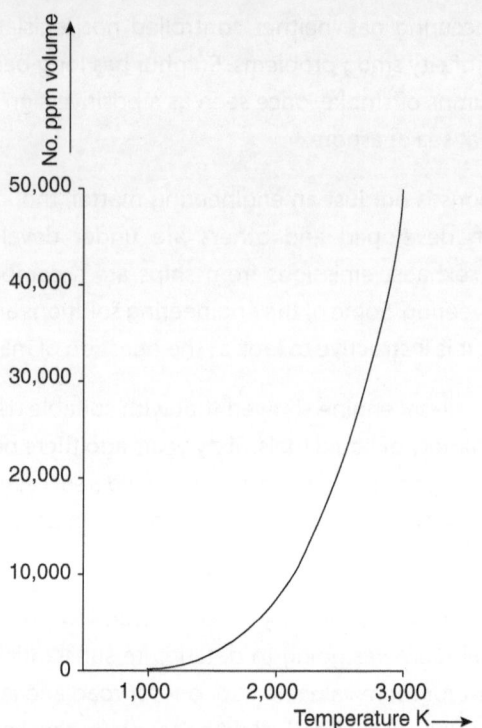

▲ **Figure 11.3** *Nitrogen chemical activity increasing with temperature*

legislation limits NOx emissions and this is one of the engine designers' biggest challenges. Apart from controlling NOx production within the engine cylinder, the exhaust stream can be treated by selective catalytic reduction to reduce final NOx emissions to very low levels.

SOx emissions in the exhaust stream represent the sulphur that was present in the fuel. They can be removed from the exhaust by scrubbing, or by eliminating sulphur from the fuel in the refining process before it is supplied to the ship. Intertanko and various other organisations have advocated the latter course. If this were done on a worldwide basis, vessels everywhere would burn an LSF. After all, the principle is applied on land. Vehicles, aircraft, heating systems and other consumers are supplied with LSFs. The proposal has been met with strong reactions – refineries are not geared to do this in the quantities needed, costs would be incurred and reorganisation required. However, to a limited extent this is happening with the introduction of ECAs in, for example, the Baltic, North Sea and some other coastal regions. There, sulphur in fuel is restricted to 1.5%, and this level will be successively reduced to 1% then even further in the course of the next few years. Residual fuel sold for ship bunkers outside the sulphur-controlled zones continues to have a high sulphur content.

Marine fuels are covered by an ISO specification and there are controls on various parameters. The standard also specifies that fuel shall not contain any added substance or chemical waste that jeopardises the safety of ships or adversely affects the performance of machinery or is harmful to personnel or contributes overall to air pollution. Although the limited routine bunker acceptance tests do not make a detailed chemical analysis, one specialist analysis company has reported that 2% or 3% of fuels analysed have evidence of chemical waste.

Reducing emissions

As seen, there are various sources of marine exhaust emissions, some of which are under the direct control of the ship operator. The challenge is both to meet emissions regulations covering the next few years and further to 'future proof' vessels, accounting for stricter emissions controls that will come into force in due course.

The International Council on Combustion Engines is one of the leading discussion groups for the development in internal combusion engines. This group suggests that emission control from diesel engines can be grouped into:

- *Pre-treatment*
  - o Denitration of fuel
  - o Alternative fuels
  - o Water addition to the fuel
- *Primary methods*
  - o Modification of the combustion process
  - o Modification of the air intake system
  - o Water injection
  - o Exhaust gas recirculation
  - o Humid air motor
- *Secondary methods*
  - o Re-burning
  - o SCR
  - o Plasma reduction systems
  - o Emission abatement systems
    - □ Wet systems
    - □ Dry systems

One effective solution is to use less fuel for the amount of work done. Here, Ro-Ro is very active in its ship design side in developing new propulsion solutions and also offshore vessel hull designs that reduce the amount of fuel needed to carry out a given operational profile. Its work also covers engine design itself. The thermal efficiency of marine diesel engines is already very high compared with prime movers on land and in the air, but design development is still producing small increments in efficiency. At the same time the focus is on reducing NOx emissions by careful design of the combustion system.

NOx emissions are covered by IMO regulations where progressively stiffer limits are being phased in. Tier II was implemented for vessels built from 2011, and the very much tougher Tier III applies to new buildings from 2015. The actual permitted emissions vary somewhat with engine size.

Clean design

Bergen engines meet Classification Society Clean Design notation now being widely specified by responsible shipowners. Among other things, Clean Design requires that NOx in the exhaust gases is reduced to a minimum of 20% less than the IMO's permissible limits that apply today. This NOx reduction is achieved without loss of efficiency, meaning that the specific fuel consumption – the amount of fuel burned per kilowatt of power output – is not increased. The means is application of the Miller cycle in combination with an increase in CR. To avoid low-load smoke and poor transient load behaviour in the low-load range, which is a consequence of the Miller cycle, the engines are equipped with VVT mechanisms by which the Miller cycle inlet air valve timing may be turned off for low-load running. The control of the VVT system is exercised by the engine's control logic.

Bergen lean-burn gas engines using natural gas as fuel (bunkered as LNG) already comply with the future Tier III requirements. Much of the NOx remaining in the diesel engine exhaust stream can, if required, be removed by selective catalytic reduction using urea. Many such systems are on the market.

For the common liquid fuels, CO_2 emissions are largely related to the efficiency of the engine and ship. They can be reduced by about 20% if natural gas is used as fuel, mainly because of the chemical composition. Ro-Ro marine gas engines in service have demonstrated this, together with a reduction in NOx of about 95% without further exhaust clean-up and an absence of SOx and particulates. Ro-Ro aero-derivative gas turbines for marine propulsion can also burn natural gas very efficiently with even less unburned hydrocarbons.

As noted, LSFs can be either specified or mandated, but if economic forces promote HSFs for marine use, a number of proprietary scrubber systems are now being approved by IMO. These are claimed to remove a high percentage of sulphur compounds and also particulates from the exhaust.

Energy efficiency

IMO has introduced an Energy Efficiency Design Index (EEDI) to indicate the number of grams of CO_2 per tonne mile that a particular design of ship will emit. It is intended that this will be refined over the coming years. The index is a formula that covers main engines plus auxiliary engines, shaft generators and electric motors set against the vessel's carrying capacity and speed. Among its uses is as a tool in the operational management of the ship, for assessing levies on bunkers or as an input to possible emissions' trading schemes. For some types of merchant ship the calculations would be fairly straightforward. The difficulty in developing an equitable index is where vessels habitually use only a proportion of their installed power; examples include offshore vessels and tugs.

Ships are also required by IMO to have a Ship Energy Efficiency Management Plan (SEEMP). It is intended that the SEEMP should be developed by individual shipping companies recognising that no two companies will be alike, therefore the owner will have to take into account parameters such as the ship's operation, type, machinery, charterer, routes and crew. The SEEMP is intended to improve a ship's energy efficiency by considering four stages: planning, implementation, monitoring, and self-evaluation and improvement. These components play a critical role in the continuous cycle to improve ship energy management. After each cycle it may be necessary to change some elements of the SEEMP, while others may remain in place as before.

The IMO members have agreed that the total carbon intensity of all shipping should be reduced by 40% by 2030. This is to be compared to the baseline measurement determined in 2008. The idea is to reduce the CO_2 emissions per transport work completed.

To help with this achievement two more recent regulations that have been introduced by the IMO, which are the Energy Efficiency Existing Ship Index (EEXI) and the Carbon Intensity Index (CII). These two regulations require owners to embark on a plan of continual energy improvement for their current ships.

The EEXI applies to all ships over 400gt and reductions in CO_2 will be calculated for the criteria set for each ship type. Obviously, the work of a cargo ship will be different from the operation of a passenger ship and therefore the calculations for the work completed will be slightly different.

The CII is a measure of the annual level of CO_2 emissions reduction that a vessel is required to make. The ratings go from A to E where A is the highest rating. A ship's actual level is documented in a "Statement of Compliance" and a plan for improvement should be set out in the Ship Energy Efficiency Management Plan (SEEMP).

The main strategies for improving a ship's CO_2 emissions are:

- Converting to an alternative fuel
- Ensuring that the hull is clean, and a suitable coating applied
- Fitting variable speed drives to constant speed electric motors
- More efficient Heating, Ventilation and Air Conditioning (HVAC) systems
- Energy efficient lighting
- Use of 'clean' technology such as wind or solar
- Energy efficient deck machinery
- Speed and route optimisation techniques

Ships should be moving toward achieving a grade level C and are required to submit plans for achieving this level if they are underachieving year on year.

A very important step in the process is to ensure that each flag state is fulfilling its responsibilities to check that each ship within its jurisdiction is carrying out the necessary updates and improvements. The IMO has developed a system for different member states to undertake an audit of other member states. The mechanism for the member states to audit each individual shipping line is set out in IMO resolution MEPC 347(78).

Planning

Here, the ship manager should recognise that there are a number of strategies to use. These could be speed and trim optimisation, just in time operation, weather routing and hull and machinery maintenance plans. This will take into account the business of the ship and company and will be different depending upon the ship type, routes, cargoes, etc.

There may be company internal structures to consider, such as early communication between operators, charters, ports and traffic management services. Staff development might well plan a role. IMO are suggesting that goals should be set so that progress can be measured, although these goals do not have to be widely publicised.

Implementation

When this stage has been designed by the senior managers, it should be clearly communicated to the company staff, especially the ship's staff.

Monitoring

The monitoring system used must be objective and quantitative and should be according to international standards. The IMO's Energy Efficiency Operational Index (EEOI) is such a monitoring tool and can be used for this monitoring task. Continuous, consistent and reliable data collection should be at the core of any system; however, IMO do not wish for another burden to be placed on the ship's staff and they suggest that the recording process is carried out ashore.

Self-evaluation and improvement

This part of the process is to identify the measures that work and the ones that do not. The next planning stage should be changed or modified in the light of findings from the first cycle.

Engineering strategies

IMO say that the following need to be considered by the company as a way of improving efficiency through better use of the propulsion machinery:

- Use of condition based maintenance (CBM)
- Use of fuel additives
- Adjustment of cylinder lubrication
- Improvements in inlet and exhaust valves due to the developments in material science
- Torque analysis
- Automated engine monitoring systems
- WHR (retrofit where cost effective)
- On-board energy management systems
- Different fuel types
- Renewable energy sources
- Regular engine management software upgrades.

Related to world trade levels

With the media and governmental spotlight firmly on the marine industry as a polluter, it has to be borne in mind that, until recently, the shipping industry has been a 'reactive one'. IMO has introduced 'Goal Based Standards', which is a system based on being proactive about managing risk.

If the amount of cargo decreases, the ships will either never be built, be laid up or will go slowly to save fuel. While air freight has captured some high-value cargoes such as cut flowers, fresh vegetables and high-value consumer electronics, some 90% of world trade goes by sea.

The cost and emissions per tonne mile for transporting, say, a container on a container ship is a small fraction of that incurred transporting the same container by road. With the current downturn in the world's economies, many ships are resorting to slow steaming. Speed at sea costs money since the power requirement for a given vessel increases disproportionately as the speed rises.

Throughout the history of the powered ship, transit speeds in given trades have been determined by the ruling balance between fuel cost, other operational costs, capital invested in ships, the cost of capital invested in goods in transit and the acceptable time from dispatch to arrival of a particular consignment of goods. When this balance is upset, slow steaming may be one solution. Direct fuel costs will be reduced but may be partly cancelled out by the other factors.

For the coming years, ways will be found of reducing emissions from ships. All the major engine manufacturers including Kongsberg, MAN Diesel & Turbo and Wärtsilä will be playing their part in this, and also in the broader overall picture of increasing efficiency and ensuring that the least possible amount of fuel is burned for the given amount of work done.

Other methods of reducing exhaust gas emissions

1. Adding water to the fuel. This works in two ways: it helps to create a homogeneous fuel/air mixture and also reduces the temperature of combustion. The injection of water requires an increase of 20–30% in fuel pump capacity. Precautions must also be taken to maintain the oil/water emulsion in a stable condition and to prevent corrosion of the fuel system components. Wärtsilä have favoured the water injection system where a controlled amount of water is injected alongside the fuel.

2. The use of SCR.

Selective catalytic reduction

SCR technology was developed for land-based installations and is being developed for main applications, and involves injecting small amounts of a single atom nitrogen-based additive, such as ammonia [NH_3], into the exhaust. Due to the difficulties and dangers of handling ammonia onboard ship, a safer, more easily handled ammonia compound called urea {$2[NH_2]CO$} is being used. The principle is to combine the nitrogen atoms of the NOx and NH_3 compound to form a stable nitrogen [N_2] molecule, which is the main constituent of air.

The principal catalytic reduction process of the ammonia compound is according to the following chemical reaction:

$$2NO + 2NH_3 + 1/2O_2 + CO_2 \xrightarrow{\text{catalysis}} 2N_2 + 3H_2O + CO_2$$

To accomplish this at the temperatures encountered in exhaust gas – 250–450°C, a catalyst is required. The catalyst is an oxide of vanadium carried on a heat-resistant honeycomb of ceramic. In order to minimise the pressure drop across the reactor, the gas passages of the honeycomb core must be of sufficient CSA.

The urea is mixed with water and metered into the exhaust gas upstream of the reactor at a rate dependent upon engine load (figure 11.4). The system is best suited to steady, high-load conditions and is less suited to low-load conditions. If the temperature is too low, the reaction rate will also be low and condensation of ammonia will poison the catalyst; and if the temperature is too high, NH_3 will burn rather than react with the catalyst.

Exhaust gas recirculation (EGR)

Of all the post-combustion treatment systems currently being placed on engines to meet the IMO Tier III legislation, the EGR system is fast becoming the most popular. Already a popular emission control technique for motor vehicles, the technology is now starting to mature.

The theory of diesel engine combustion will have been covered in Volume 3 of the Reeds series. Here, students will have learned that for complete combustion the engine is supplied with more air than is required. The additional air is called excess air. EGR technology replaces some of that excess air with recycled exhaust gas. This has the effect of lowering the peak temperatures in the combustion space and therefore the NOx formed is also lower.

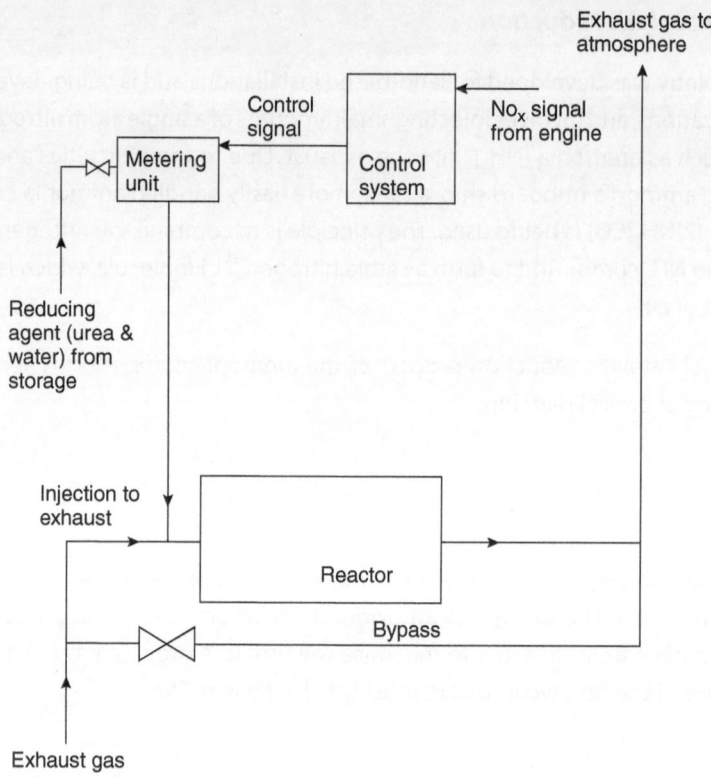

▲ **Figure 11.4** *Selective catalytic reduction (SCR) apparatus*

The technique used to accomplish this process is a valve working in the exhaust gas path. This could lead to the valve becoming clogged, which would stop it from working properly. Therefore, this will have to be considered an important maintenance item in the future because the vessel can be fined if the emissions from the funnel are not correct.

Developments in propulsion efficiency (including upgrading)

A study of Volume 4 of the Reeds series will determine that a propeller's function is to drive a vessel as far forward as it can with every revolution. Working in a solid, the distance travelled would match the pitch of the blade, which would be regarded as the maximum achievable, giving an efficiency index of 1.

Working in water, however, the propeller must generate a pressure difference, automatically leading to an acceleration of the water. The reactive force is then transmitted through the shaft line and thrust block to the hull of the vessel. The distance travelled by the vessel from the force produced by each revolution will not achieve the theoretical maximum for a number of reasons that are external to the propeller. These influences will be laminar flow of water into the propeller, cavitation produced and wake-field interaction as a result of the propeller disc rotating in a turbulent inflow of water following the water's disturbance having just passed over the hull of the vessel.

Optimising the function of a propeller so that it performs as close to the theoretical maximum as possible means that the designer should not consider the propeller in isolation. She/he needs to make reference to its working environment and match the propeller to the design of the aft end of the hull. This is because the flow of water around the hull of a vessel and then into the propeller is not uniform. The hull and/or a number of fittings or appendages will disturb the water before it reaches the propeller and for this reason the conventional propeller blade will revolve in water of different inflow speeds, which translates as vibration that will be transmitted to the hull of the vessel.

One answer to this problem has been the development of the skewed propeller, which due to its skewed propeller blade sections being located at different radii does not enter this so-called wake peak. The blades of propellers travel in a different direction across the wake field. At the top they cross in one direction and at the bottom in the other. This again can cause fluctuating pressure pulses, leading to inefficiency and increased vibration as a result of cavitation, and again the resultant vibration is transmitted back to the vessel's hull through the propeller shaft and bearings.

One of the most important design considerations is the lift/drag equation. Each propeller blade is an 'areofoil' shape and therefore a larger blade area will not only set up larger 'lifting' forces but will also cause more resistance 'drag' due to friction. As a consequence, larger blades may not necessarily increase efficiency and the shape of the blade has an effect on the cavitation produced.

It used to be thought that vibration occurrence at a specific combination of vessel speed and propeller revolutions per minute was due to the resonance between the blade and the natural frequency of the hull, coupled with the clearances of the propeller. As a consequence, it was sometimes possible to make improvements by changing the propeller for one with a different number of blades or changing the clearances between the blades and the hull of the ship. It has now been realised that any improvement in reducing vibration and increasing efficiency came about due to the new propeller being placed in a more favourable position in the vessel's wake field.

Hull to propeller tip clearance is also important for the prevention of vibration. Doubling the tip clearance reduces the pressure pulses and the integrated force by a factor of 2. Achieving a smooth inflow of water to each radial and circumferential position of the rotating propeller disc to enable an even force to be generated over the blade's surface, without creating cavitation, is not an easy task. However, research is now focused in this area, as MAN Diesel & Turbo explain in their report on the hydrodynamics of ships' propellers.

The Hydrodynamic Research Centre (HRC) originally built by the Swedish propeller manufacturer Kamewa, which is now part of the Kongsberg group, has celebrated over 40 years of operation. The cavitation tunnels are today a tool for analysing hydrodynamic performance, complementing the fast-developing field of CFD.

During their ownership Rolls-Royce gathered good hydrodynamic comparisons between fixed pitch propellers (FPP) and CPP. Generally, in hydrodynamic comparison an FPP is 0.5–1.0% more efficient in open water. The difference can be reduced by increasing the manufacturing accuracy of the CPP and its installation.

MAN Diesel & Turbo and Rolls-Royce hold the view that the major difference in efficiency between an FPP and a CPP is mostly related to the smaller hub of the former. A typical value for a non-ice-class propeller is in the range of 1–1.5% from this effect alone. However, the CPP will regain most of this difference due to its flexibility in adapting to the various conditions that a vessel experiences.

There is still a need for observations to be made on vessels while they are in operation. These observations required the installation of a window in the hull of the vessel during dry-docking. Lloyds Register (LR) have developed the system during the past 10 years and window-based observation techniques have evolved from still cameras, strobe lights and high-speed cine film to low-cost camcorders and high-speed digital video operating in natural daylight.

The latest technique, however, is a 'keyhole' observation technique which has been pioneered by LR. The bore holes are tapped holes made in the hull of the vessel, which can be carried out without the vessel being dry-docked.

The techniques have been enhanced by the use of low-light digital cameras operating at frame rates of 25 Hz and shutter speeds of up to 1/1,000 of a second. Although this is the main tool for propeller observations at LR, a further development is the use of high-speed video systems. Using digital technology, the cameras can be activated by a signal from the propeller shaft, which can also be synchronised with vibration and pressure pulse readings.

The resultant information can be brought together to better understand the relationship between the time-related cavity collapse event and the resultant information used to confirm the cause of erosion on propellers or surrounding fixtures.

With fuel costs now accounting for a high proportion of the running cost of a ship and the need to reduce CO_2 emissions from sea transport, developments in propulsion technology that save fuel and consequently reduce emissions can be very attractive.

Senior engineers will be interested in retrofit developments such as the integrated propulsion system Promas developed by Rolls-Royce, which could be one option in achieving efficiency goals as part of the SEEMP described elsewhere in this chapter.

Promas is an integration of the conventional propeller and rudder configuration of a typical single- or twin-screw vessel. A special hubcap is fitted to the propeller that smooths the flow of water from the propeller onto a bulb forming part of the rudder, while the spade rudder itself has a special form with a twisted leading edge. The resulting improvement in efficiency is made up of several components. One is the reduced loss in the hub region of the propeller. The second is that the shape of the rudder converts some of the swirl energy in the propeller slipstream, which is normally lost, into additional forward thrust. The third component is that the shape of the rudder gives a much higher side force for a given rudder angle in the ±5° range normally used for course corrections. Increased side force at low speeds also improves manoeuvrability.

The best results are achieved on blunt single-screw vessels with a block coefficient greater than 0.8 and a design speed in the 14–16 knot range. Here, efficiency gain can be as much as 6–9% compared with conventional solutions. Faster and slenderer single-screw vessels such as car carriers can have an efficiency gain of 2–5%. In a well-designed twin-screw vessel there is less improvement to be had, but even so 1–3% can represent a substantial amount of money at today's fuel prices, and the increased cost of the propulsion equipment can be quickly recovered.

Development work was carried out on many vessel types and a variety of rudder shapes and bulb configurations. Initial work was done using CFD methods, and promising solutions were then verified at model scale at the Hydrodynamic Research Centre.

In particular, the interactions between the hubcap and the curved leading edge of the bulb were studied at different loading conditions and rudder angles. In theory, the gap between the hubcap and the forward part of the rudder bulb should be as small as possible. In practice, there has to be a gap sufficient to allow for the structural deflection under load of the propeller aperture and rudder, and this Ro-Ro system also takes account of the tolerances that can be realistically achieved under real shipbuilding conditions.

In addition to achieving a high efficiency, the aim with Promas propulsion systems is always to reduce noise and vibration. With Promas, pressure pulses from the propeller can be reduced. Due to the presence of the hubcap and bulb, the risk of hub vortex cavitation is removed. Consequently, the radial distribution of hydrodynamic loads on the propeller blades can be modified, increasing the loading in towards the hub and reducing the loading at the tips, which helps to cut the intensity of blade pressure pulses.

Kongsberg also provides a simplified version of Promas (Promas Lite) for upgrading existing vessels. In this case the ship's existing rudder can be retained, but is fitted with a bulb, while the propeller is equipped with the special hubcap and new blades. Several ropax ferries that have had this treatment have shown dramatic improvements in efficiency, with payback times of 1–2 years. Often the vessels still have worn original propellers optimised for sea trial speed. Fitting new propellers, or new blades in the existing CPP hubs, optimised using the latest design techniques for the actual service speed, give a substantial cut in fuel consumption, helped by the other components, even though the rudder shape is not ideal.

Another Kongsberg propulsion product, the Azipull azimuthing thruster, draws on the same knowledge base as Promas with regard to hydrodynamic and mechanical interactions. Work on Azipull began in 1997, with a view to combining the theoretical advantages of pulling propellers and expertise in conventional azimuth thrusters to give a new concept. Azipull provides the shipowner with a very efficient propulsor that at the same time uses well-proven gear technology and is a low-risk solution. The inboard gearhouse and steering system is the same design as is used in other Kongsberg azimuth thrusters, and either direct diesel mechanical transmission can be specified or an electric motor coupled to the upper gear unit as part of a diesel electric transmission. CPP or FPP can be specified to suit the drive system and application.

The innovative part of the concept is the underwater unit, which has the lower bevel gear set enclosed in a streamlined housing behind a pulling propeller. The leg is hydrodynamically optimised to convert into an additional forward thrust some of the swirl energy in the propeller slipstream, which would otherwise be wasted. As the propeller works in a clean flow of water, provided the hull design is suitable, efficiency is increased and noise and vibration reduced. The shape of the underwater unit provides a much larger rudder effect than conventional azimuth thrusters, which in practice cuts fuel consumption since less thrust vectoring is needed to maintain a steady course.

The aim of the Azipull concept was to extend azimuth thruster technology into a higher-speed range. This has been achieved. The first units of this design have been very successful in a ferry operating at a 22-knot service speed. The advantages also hold good at much lower speeds, in the 13–15-knot range, where hundreds of these thrusters have been applied to offshore platform supply vessels, coastal and short sea cargo vessels and other types.

Application areas have been extended further, for example to short sea tankers, where twin Azipull azimuth thrusters are each coupled directly to a diesel engine, which also drives a generator set. The result is much better manoeuvrability than traditional solutions, and an enhanced level of safety through propulsion redundancy. Bergen Star is an excellent example, operating on the Norwegian coast.

Passenger vessels are another Azipull application, where efficiency and manoeuvrability combined with low levels of noise and vibration are key factors. A 12,700GT vessel has the distinction of being the first cruise ship to specify Ro-Ro Azipull thrusters for main propulsion, and the installation has proved very successful. Manoeuvrability is reported to be very good and vibration levels exceptionally low. Two Azipull AZP120 thrusters with pulling propellers provide propulsion and steering. Each has a rated input power of 2,310 kW supplied by a four-engine diesel electric system. Due to the operating area, the thrusters have LR ice class 1A notation.

Promas and Azipull can be combined in a single vessel. A ferry ordered in 2011 has a centreline CPP in a Promas configuration, flanked by two electrically driven Azipull thrusters. The thrusters manoeuvre the vessel in and out of the confined harbours and control it through a strong current setting across the piers, then at about 80% of service speed the main propeller provides the extra power for full speed.

The Professionalism of Marine Engineers

Initial education and training and the need for professional development

One of the primary aims of this book is to help qualified engineers and engineers under initial education and training in preparing for the motor engineering knowledge part of the certificates of competency as sea-going engineering officers on merchant vessels of all types and size.

The primary issue we face is that the skill set required of a marine engineering officer is changing all the time. Just a few short decades ago, the industry was faced with steeply rising capital and running costs, very little research and development, difficult recruitment patterns and an ageing workforce.

Not all of these problems have been solved but one of the most important has changed dramatically. The industry is now at the cutting edge of technology, with our knowledge of material science often holding back the pace of development. This is having the effect of projecting a more positive image of the industry and making it an exciting place to work and therefore attractive to young persons.

Modern marine engineers have to spend more of their time as 'systems engineers' than they did in the past. The maintenance requirement of the job role is still important but has to take a more secondary role to the watchkeeping for the following reasons:

- The machinery is becoming much more reliable and therefore does not need the service intervals of the older machinery.
- Reduction in overall staffing levels means that the current engineering officers working on board have less time to spend on maintenance activities.
- Due to increases in technology the engineers must devote a larger proportion of their time to ensure that they understand the theory behind the machinery's operation.
- Education and training systems around the world are not all as good as each other in producing the calibre of engineering officer required to operate the modern fleet and maintenance-induced failure is a growing trend.

Therefore, it is vital that the modern marine engineer continues his/her education and development throughout their career.

The impact of Standards for the Training and Certification of Watchkeepers (STCW)

We all appreciate that a major driver in setting the global minimum criteria for the education and training of modern ships' engineers is the IMO Standards for the Training and Certification of Watchkeepers (STCW). It has been a slow process from the original version but finally, with the 2010 Manila amendments, it looks like we will have a reasonable international base from which to build.

A second important influence on the education of seafarers is cost. Shipowners operate ships to make money and therefore cost-effective staffing arrangements are essential. Therefore, a shipowner or management company offering cadetships in the United Kingdom, for example, will be obliged to use the established education systems that are available. To step outside of the 'standard' system would mean that students would not achieve national qualifications and it would also be more expensive.

A recent EU funding project (Unification of Marine Education and Training – UniMET), aimed at sharing best practice throughout Europe, has praised the UK officer cadetship system. The partnership formed by the Merchant Navy Training Board, the Maritime Coastguard Agency, industry and representatives from academic institutions is an example of a system that other administrations could learn from.

The qualifications are now also linked to foundation degrees that must give the students the opportunity to complete a full honours degree, although this could be after the completion of their cadetship. Importantly, the new qualification structure reflects an up-to-date syllabus required by national and international regulatory authorities.

However, it is important that the education and training evolves to meet the changing requirements of the tasks of the seagoing engineer. IMO has now given a commitment to review the STCW criteria every five years.

Officer cadetships

In the United Kingdom the mix of technical content, skills and experiential learning within the officer cadetship programmes are also continually being reviewed by the members who make up the Maritime Skills Alliance. This is the employer-led organisation that has responsibility for setting the standards and keeping maritime skills, education and training relevant to the job role.

In the United Kingdom, the Merchant Navy Training Board (MNTB) considers that it is vitally important that all parts of the industry work together. This way, issues are identified, discussed and agreed, their involvement and 'buy-in' is secured and industry needs are met. This approach recognises and values all parts of the industry and their contribution to it.

The members of the Institute of Marine Engineering, Science and Technology (IMarEST) have a chance to play a part in the development of the UK's Education and Training system. This is because IMarEST has a place at IMO and through the correct mechanisms can influence decisions. It is also a member of the International Association of Maritime Institutions (IAMI), which is an academic discussion group and also has an important role to play in the UK decision-making process.

Initial education and training in the United Kingdom might start with the requirements of STCW but more importantly we are teaching the engineers to expand upon their understanding and to think for themselves in difficult situations. This takes the UK marine engineer beyond the basic requirements determined by IMO and is a feature that is so important for the successful management of modern tonnage. The technological advancements that are due in the next few years will be difficult to understand and the consequences of operational mistakes will be more severe.

For example, if the EGR valve on the main engine starts to malfunction (as happens all too often on a modern road car) then the exhaust gas emissions will not be correct and if the vessel is in an Emission Control Area then the engineers will have to be ready with a detailed knowledge of the system and all the problem-solving skills that they can muster. If not, the owner will be faced with administrations looking to recover costs for environmental damage.

Increases in electronic systems at sea are placing more emphasis on the work of the Electro Technical Officer (ETO) and the latest problem with the *Queen Mary 2* investigated by the MAIB shows the growing need for understanding complex control systems, and places a strong emphasis on the requirement for quality professional development of staff. Towards this end, Videotel has plans to expand its online academy and Viking Recruitment has just built a new training centre and is planning an expansion that will include a conference centre to be used for staff development discussions about different technical issues, such as 'future fuel' or which ballast water technology will be the most cost effective.

Advances in control engineering mean that in the near future mechanical components will be able to communicate with centralised systems. The increase in information will be such that engineers will have much more information available than at present to diagnose faults or poor performance. However, this increase will require engineers to apply their problem-solving skills and logical thinking across complicated systems that may be interacting and dependent upon each other.

With the increased reliability of machinery comes another problem. How many senior staff will we soon have that have not seen a main engine unit opened up? Manufacturers are predicting that main engines will run from dry dock to dry dock. This would lead to only staff present at the dry dock being involved with the main engine overhaul. If this is the case then possibly it is time for shipowners and managers to regard dry-docking as a major opportunity for the staff development of the fleet's most senior officers.

Shipping companies' staff development programmes and individual engineers' professional development are going to make the difference between the successful technical management of ships and struggling to stay out of trouble. Therefore, the ongoing professional development of marine engineers, beyond completing their class one certificate of competency, is essential.

Membership of an institute such as IMarEST is a very good method of progressing professional development. The technical publications that accompany membership supply quality and up-to-date information about developments in marine equipment and infrastructure. Many websites and discussion groups give product information free of charge and more and more are supplying technical information. Engineers can also subscribe to newsletters, free of charge, that give a brief overview of what is going on within the industry without the cost of a full subscription.

However, focused staff development should also be used as a powerful management tool available to senior managers. Conferences and seminars are expensive and it is not cost effective to send the entire staff on one seminar. However, any person who does have the opportunity should then be asked to complete a briefing note for the chief engineers of the fleet. This would be a cost-effective way to keep everyone up to date.

In the near future, marine engines will all carry equipment such as SCR systems or EGR valves. They will all have electronic combustion control, VVT, and the performance and condition will be monitored closely by various computer-controlled systems.

Thinking a little further on in time, the equipment will need to be so efficient that it will be totally enclosed with very little indication that it is even operating. This is a very different working environment from the one just a few years ago, when touching a hot surface or being allowed close to rotation machinery was a real possibility. This new world will be an environment needing a sophisticated skill set from the engineers in charge.

Watchkeeping duties

The maritime industry is full of tradition and often for a good practical reason. The maritime way of life, procedures and routines have been developed over many years and they should not be set aside without careful consideration.

A ship at sea obviously has to operate 24/7 and it must also be staffed to the correct level to maintain the safety of the vessel and crew. Traditionally, the shift patterns, or watches, as they are called at sea, have been broken into three 4-h work periods for every 12-h period. The same pattern is repeated for the second 12-h period. This means that three people would have covered the watches as follows:

- 08.00 h to 12.00 h – This is the morning watch and the start of the working day. The watch will be very busy due to the work and interruptions from day-working staff and senior engineers undertaking their duties. The end of a 24-h period main engine room log is usually taken during the time just before midday.

- 12.00 h to 16.00 h – The afternoon watch. Not quite as busy as the previous watch due to the lunch break and afternoon tea. Not so many interruptions and therefore the engineer is left to complete his/her own work.

- 16.00 h to 20.00 h – This is a slightly difficult watch due to the timing of dinner on board, which is usually 18.00 h. This means that the watchkeeper must have a replacement while she/he goes for dinner or alternatively the engineer will have to wait until 20.00 h when she/he completes the watch.

- 20.00 h til midnight 00.00 h – Not a bad watch but usually the watchkeeper misses out on the social activities going on elsewhere on board.

- 00.00 h to 04.00 h – Probably the most difficult watch, mostly because it is difficult to achieve a long rest period. Breakfast at 05.00 h watching the sun come up is a rare treat on this watch.

- 04.00 h to 08.00 h – Usually regarded as the best watch. The only drawback is having to start work at 04.00 h but once that is over this is the work pattern that fits into the ship's daily routine the best.

- This brings us back to the start of the day. Three people complete the two 4-h work shifts, making up the 8-h day.

Unmanned machinery spaces

These were mentioned at the end of Chapter 6 due to the special considerations required with the technical arrangements to operate a vessel in such a way. Considerable investment has been put into the design and construction of vessels designed to run UMS. The return on that investment is a:

- more efficient method of operation
- system that is more aligned to natural rest patterns
- machinery plant that can be operated safely with fewer engineers.

The system is that three people are again required to carry out the watchkeeping. However, due to the automation and the special machinery space design (as described in Chapter 6), the watchkeeper does not have to be in the machinery space 24/7. She/he is still responsible for the watch and will have to respond to any machinery alarm within 90 s of its activation, no matter where they are on board.

The first of the three watchkeepers is in charge for 24 h, after which she/he hands over to the next watchkeeper, who carries out their 24-h shift. If the machinery is running at a steady state with no malfunction and the watchkeeper has completed all his/her routine tasks then there should be no alarms and the watchkeeper should be able to leave the machinery space unattended for a period of up to 8 h.

During the 48-h period when an engineer is not responsible for the watch, that engineer falls back on to day work and has maintenance and up-keeping duties to complete. The UMS system of operation is generally regarded by the engineers as the better system because it fits into a daily routine better, they can get a good night's rest, especially on their two nights when not looking after the alarm, and a lot of the old watchkeeping chores are carried out by the automation, leaving the job actually more interesting than before.

Tasks of the engineering watchkeeper

When an engineer is assigned to the machinery watch for a period of time, she/he must regard that as his/her primary task. IMO point out in STCW that the engineering officer is a representative of the Chief Engineer while on duty and she/he should use their judgement accordingly. They should not, for example, hand over a watch if they don't think their reliever is capable of taking over.

Taking over a watch

It is always ideal to arrive at the engine room early to prepare to take over the watch at the allocated time. There are three primary reasons for this, which are as follows:

- So that the engineer taking over can familiarise him/herself with the current status of the machinery plant, including the level of any fluids in the bilge. This may not have changed much since the last time the engineer was in the engine room but it is a very important aspect of taking over a watch.
- Since the incoming engineer, at that stage, is additional to the current watch, it gives him/her the luxury of being able to walk around the plant and check for early signs of any defects in the operating plant. This should include a careful examination for any leak of fluids.
- Early arrival puts the current watchkeeper at ease as she/he may be tired following a busy spell on duty.

If it is during the daylight hours, the engineer taking over should start by going outside and having a look at the flue gases and making a note of any visible smoke from the funnel. Record the colour of the smoke and identify the engine that it is coming from.

Entering the engine room high up and working down, the engineer can inspect any WHR boiler plant that may be situated in the engine exhaust. Inspect for exhaust or water leaks. The engineer should then follow the trunking down to the turbochargers, which are viewed next for safe operation. The turbochargers will be placed close to the engine tops, which is where the exhaust valves will also be sited on a two-stroke engine. Older types of engine may well have the operation of rocker gear and physical springs to view but the latest engines will have all the moving parts totally enclosed. However, the engineer should be listening for abnormal noises or feeling for unusual vibration or heat.

If the cylinder heads are viewable at this level then each one should be inspected for the following:

- Fuel leaks from the high-pressure pipework
- Hydraulic leaks from the exhaust valve actuation
- Air leaks from the air return spring system on the latest engines
- Combustion leaks from the indicator cocks
- *Air start valve leaks* – checking for this fault is very important as leaking air start valves have in the past been a source of serious fire in engine rooms. The check is made by feeling the temperature of the air start line leading to the cylinder head. If the air start valve is leaking then a high temperature will be detected in the air start pipe adjacent to the valve.
- Check for loose supports and indeed any loose guardrails.

Down one level on the engine and the engineer would see the high-pressure fuel delivery pumps, cylinder lubricators and hydraulic pumps for the exhaust valve actuation. Around the back of the engine could be the charge air coolers and pipework for the engine services such as the cooling water and lubricating oil.

This level might also be a main engine room level giving access to the auxiliary machinery and the points to watch out for will be covered after the main engine section. It might also be the point to visit the machinery control room and let the outgoing watchkeeper know that you are there and are preparing to take over the watch.

At the lower level of the main engine, the engineer should be checking the crankcase explosion relief doors to ensure that they are not leaking oil. If they are then the rubber

'O' ring needs replacing. The lower level will also be the place to inspect any reduction gearing, shafting and/or thrust block, etc.

The correct function of the auxiliary machinery is vital for the efficient operation of the machinery plant and the engineer should conduct a comprehensive review of the auxiliary machinery, making a note of which pumps are running and which generators are supplying the electrical load. Generators should be checked for any fuel, water or oil leaks.

It is very important to check the pressure of compressed air inside the starting air receiver. This is vital for starting the main and auxiliary engines and if the pressure is low then it may be difficult to get the ship going again following a breakdown. Another very important inspection is the daily running tanks or service tanks. If these run low then there is a danger that the main engine of the auxiliaries will stop operating.

The auxiliary boiler will always be a source of concern because if they are neglected or are not understood well they can be a source of extreme danger. The Flag State examiners know this and will want to satisfy themselves that engineers gaining a certificate of competency are safe to operate auxiliary boilers.

While inspecting auxiliary boilers, engineers should look for efficient operation. As the boiler starts to work there should be a short purge cycle and then when the fuel is admitted the boiler should light up straight away. If it is struggling to do this then adjustments or cleaning may be needed. However, if the incoming watchkeeper sees a boiler working inefficiently then she/he would do well to pay attention to the boiler as soon as possible because it is likely to let him/her down during the next watch period.

Examiners will also be keen to find out that students understand the dangers from fire associated with boilers. Therefore, the engineer coming on watch should look out for any oil or fuel that may have accumulated around the boiler's burner. Engineers must check the boiler gauge glasses for the correct level of water. Low water levels have in the past been a major cause of accidents involving boilers.

While walking the machinery space, the engineer should make time to visit the steering flat. It might be remote from the main machinery space and therefore could be difficult for the main watchkeeper to inspect without first arranging someone else to cover the main space. Checks here would be correct and prompt operation, no abnormal noise and no oil leaks. Operation of stand-by pumps will be undertaken at set times so it would not be necessary to run these at every change of watch.

On the way back to the MCR the engineer should be listening for abnormal noise from running pumps, air compressors, purifiers and other machinery. Once back in the MCR the incoming engineer will be able to discuss the current status with the knowledge that she/he has just reviewed the machinery in operation. Once in the MCR the engineer can compare any temperatures and pressures taken locally with the ones that are showing remotely in the MCR and discrepancies can be discussed as part of the handover procedure.

It is important that both engineers discuss any issues that may have come up during the previous watch and also discuss anything that might need attention during the next watch. Anything that needs attention should be undertaken as soon as possible and not left to the next watchkeeper to deal with. This just leads to bad feelings and poor working relationships.

Depending upon the type of ship and the current operations, the watchkeeper might need to contact or work with other people in the ship's company, for example if the vessel has a large refrigeration plant or there is a need for inert gas generation or large electrical power for discharging cargo.

Engineers as officers

The overall management of ships requires an activity beyond the efficient 'technical' operation of the vessel. International shipping operates in a global trading environment and therefore it will have the opportunity to choose its staff from different places around the world. This means that owners recruiting staff for their ship will probably end up with crew from several different nations and cultures.

The careful management of staff from very different backgrounds is part of the responsibility of the engineering officers. Other important management functions they are responsible for include:

- Environmental
- Commercial
- Safety
- Strategic
- Training of ship's staff, including cadets

The management of a team to achieve the best safe-working practices and ensure the most cost-effective running of the ship will be a vital test of all the officers in such a pressurised environment.

Knowing the capability of fellow crew members will be an important step in that process. Therefore, engineering officers should be talking to their team, discussing technical problems and allowing all staff to have an input into the decision-making process.

Promoting a 'no blame' culture will encourage all staff to feel comfortable when discussing difficult situations that may have led to a breakdown and of course 'politeness oils the wheels of conflict'. Whenever people are brought together in a team there will be disagreements but if everyone is polite in their discussions then the disagreements can be worked through, with the team leader having the final say and taking ultimate responsibility for the collective decision.

Staff Development

The industry is changing at a 'great rate of knots', both technically and with new regulations, making it even more essential that staff keep themselves up to date with the latest changes.

Once the concepts, science behind the working of the machinery and the detailed construction of the components have been learned to a high standard, it is not overly difficult for engineers to keep up with the latest developments in the design of the latest equipment.

It is, however, essential that engineers make the time to enquire about the changes that are happening. The technical changes, for example, might well be linked to changes in legislation and therefore the introduction and correct operation of a new technology might be a legal requirement – as with Ballast Water Treatment. This means that if all the engineering staff do not keep themselves up to date then they may find themselves in breach of the law.

There are many forms of professional development for engineering staff and one of the best is via membership of the Institute of Marine Engineering, Science and Technology (IMarEST). They have an excellent technical publication as well as many 'technical' lectures that can be attended free of charge – other than the yearly subscription.

The lectures are also recorded and can then be watched by members after the event at their own convenience.

TEST QUESTIONS

(**S** shows questions used in the past as examination questions)

Chapter 1 – Class One

1. (a) With reference to fatigue of engineering components, explain the influence of stress level and cyclical frequency on expected operating life.

   (b) Explain the influence of material defects on the safe operating life of an engineering component.

   (c) State the factors that influence the possibility of fatigue cracking of a bedplate transverse girder and explain how the risk of such cracking can be minimised.

2. With reference to engine performance monitoring, discuss the relative merits of electronic indicating equipment when compared with traditional indicating equipment.

3. As Chief Engineering Officer, how would you ascertain if the main engine is operating in an overloaded condition?

   If the engine is overloading, what steps would you take to ensure that the engine was brought within the correct operating range?

4. A set of indicator readings suggests that the power from each of the individual cylinders of the main engine is not balanced:

   Describe the action you would take to rectify the problem.

   Outline how you would check the accuracy of the cards.

5. Explain, by referring to the theoretical considerations, how the efficiency of an IC engine is dependent upon the Compression Ratio (CR).

   State why an actual engine power card is only an approximation to the ideal cycle.

Chapter 2 – Class One

1. (a) Evaluate the main causes of normal and abnormal cylinder liner wear.

   (b) State the ideal properties of cylinder oil for use with a slow-speed engine burning residual fuel and why the oil might need to be changed when low sulphur fuel oil is to be used.

(c) Explain the possible consequences of operating an engine with a cylinder liner worn beyond normally acceptable limits.

2. (a) Critically evaluate three methods of crankshaft construction, indicating the type of engine to which each method is most suited.

(b) State the nature of and reasons for the type of finish used at mating surfaces of a shrink fit.

(c) Explain each of the following:

    a) Why slippage of a crankshaft with a shrink fit can occur.

    b) How such slippage may be detected.

    c) How slippage may be rectified.

3. During recent months it has been necessary to frequently re-tighten some main engine holding down bolts as the steel chocks have become loose:

(a) Explain possible reasons for this.

(b) State the reasons why re-chocking using a different material might reduce the frequency of incidents.

(c) Explain the possible consequences if the situation is allowed to continue unchecked.

4. (a) State, with reasons, why engine air inlet and exhaust passageways should be as large as possible.

(b) Explain how such passageways can become restricted even when initially correctly dimensioned.

(c) Explain the consequences of operating an engine with:

    (i) Restricted air inlet passageways.

    (ii) Restricted exhaust passageways.

5. (a) Describe, using sketches if necessary, a main engine chocking system using resin-based compounds, explaining how such a system is installed.   **S**

(b) State the advantages and disadvantages of resin-based materials for use as chocks when compared with iron or steel.

Chapter 3 – Class One

1. Bunkers have been taken in a port and a sample is sent to a laboratory for analysis. The vessel proceeds to sea before results of the analysis are obtained. The analysis indicates that the fuel is off specification in a number of ways, but the fuel must

be used as there is insufficient old oil supply available to enable the ship to reach the nearest port. Explain with reasons what action should be taken to minimise damage and enable safe operation of the engine if the following fuel properties were above or below specified levels: S

(a) Viscosity.

(b) Compatibility.

(c) Sulphur.

(d) Ignition quality.

(e) Conradson carbon.

(f) Vanadium and sodium.

2. (a) Describe using sketches a Variable Injection Timing (VIT) fuel pump and explain how timing is varied while the engine is in operation.   S

(b) Explain why it is necessary to adjust the timing of fuel pumps individually and collectively.

3. With respect to residual fuel, explain the effects of EACH of the following on engine components, performance and future maintenance, stating any step that should be taken in order to minimise these effects:

(a) High Conradson carbon level.

(b) Aluminium level of 120 ppm.

(c) Low ignition quality.

(d) 450 ppm vanadium plus 150 ppm sodium.

4. (a) With reference to 'slow steaming nozzles' as applied to main engine fuel injectors, state with reasons when and why they would be used.   S

(b) State with reasons the engine adjustments required when changing to a fuel having a different ignition quality. Explain the consequences of not making such adjustments.

(c) State the procedures that should be adopted to ensure that main engine fuel injectors are maintained in good operative order, indicating what routine checks should be made.

5. With respect to fuel oil:

(a) Explain the meaning of the term 'ignition quality' and indicate the possible problems of burning fuels of different ignition quality.   S

(b) State how an engine may be adjusted to deal with different fuels of different ignition qualities.

(c) State how fuel structure dictates ignition quality.

Chapter 4 – Class One

1. (a) Sudden bearing failure occurs with a turbocharger that has been operating normally until that point. Explain the possible causes if the turbocharger has:
    (i) Ball or roller bearings.
    (ii) Sleeve bearings.
    (b) State with reasons the measures to be adopted to ensure that future failure is minimised.

2. With respect to turbochargers, indicate the nature of deposits likely to be found on EACH of the following and in each case state the possible consequences of operating with high levels of such deposits and explain how any associated problem might be minimised:
    (a) Air inlet filters.
    (b) Impeller and volute.
    (c) Air cooler.
    (d) Turbine and nozzles.
    (e) Cooling water spaces.

3. (a) State what is meant by the term *surge* when applied to turbochargers.   S
    (b) State why surging occurs and how it is detected.
    (c) Explain how the possibility of surging may be minimised.
    (d) State what action should be taken in the event of a turbocharger surging and explain why that action should not be delayed.

4. It is discovered that delivery of air from a turbocharger has fallen even though engine fuel control has not been changed. State with reasons:
    (a) The causes of such reduced delivery.   S
    (b) The effects of this reduced air supply on the engine.
    (c) The immediate action to be taken.
    (d) How future incidents might be minimised.

5. At certain speed, vibration occurs in a turbocharger.
    (a) State with reasons the possible causes.   S
    (b) Explain how the cause can be detected and corrected.
    (c) Explain how the risk of future incidents can be minimised.

Chapter 5 – Class One

1. **(a)** State the possible reasons for an engine failing to turn over on air, despite the fact that there is a full charge of air in the starting air receiver, and explain how the problem would be traced.

   (b) Explain how the engine could be started and reversed manually in the event of failure of the control system. **S**

   (c) Outline planned maintenance instructions that could be issued to minimise the risk of failure indicated in (a) and (b).

2. Describe the safety interlocks in the air start and reversing system of a main engine.

   What maintenance do these devices require?

   At what interval would they be tested?

3. As Chief Engineering Officer, what standing orders would you issue to your engineering staff regarding preparing the main engines for manoeuvring?

4. Routine watchkeeping reveals that a cylinder air start valve is leaking.

   What are the dangers of continued operation of the engine?

   What steps would you take if the vessel was about to commence manoeuvring?

5. Describe the main engine shutdown devices. How and how often would you test them?

   The shutdown system on the main engine fails, immobilising the engine. Checks reveal that all engine operating parameters are normal. What procedures would you, as Chief Engineering Officer, adopt to operate the unprotected engine to enable the vessel to reach port?

Chapter 6 – Class One

1. **(a)** Describe briefly the operation of an electrical or hydraulic main engine governor.

   (b) For the type described, indicate how failure can occur and the action to be taken if immediate correction cannot be achieved and the engine must be operated. **S**

2. Complete failure of the UMS, bridge control and data logging systems has occurred, resulting in the need for the main engine to be put on manual control and monitoring:

   (a) State with reasons six main items of data that require to be monitored and recorded manually.

   (b) Explain how a watchkeeping system should be arranged to provide for effective monitoring and control of the main engine.   **S**

   (c) Explain how the staff will be organised to allow the engine to be manoeuvred safely and state the items of plant that will require attention during such manoeuvring.

3. Discuss the relative merits and demerits of hydraulic and electronic main engine governors.

4. Describe, with the aid of a block diagram, a bridge control system for main engine operation.

   As Chief Engineering Officer, what standing orders would you issue to your engineering staff when the vessel was operating under bridge control?

5. Describe a jacket cooling water system temperature controller. When operating under low-load conditions for an extended period, how can cylinder liner corrosion be minimised?

Chapter 7 – Class One

1. During a period of manoeuvring it is noticed that difficulty is being experienced in maintaining air receiver pressure:

   (a) State, with reasons, possible explanations.

   (b) Explain how the cause may be traced and rectified.   **S**

   (c) State what immediate action should be taken to ensure that the engine movements required by the bridge are maintained.

2. (a) Explain why it is essential to ensure adequate cooling of air compressor cylinders, intercoolers and aftercoolers.

   (b) State, with reasons, the possible consequences of prolonged operation of the compressor if these areas are not adequately cooled.   **S**

3. (a) With reference to air receivers, explain:

   (i) Why regular internal and external inspection is advisable.

    (ii) Which internal areas of large receivers should receive particularly close attention. **S**

    (iii) How the internal condition of small receivers is checked.

  (b) Where significant corrosion is found during an internal inspection, what factors would you take into account when revising the safe working pressure?

4. It has been found that during recent periods of manoeuvring a number of air start valve bursting discs or cones have failed:

  (a) Explain the possible reasons for this.

  (b) Indicate how the actual cause might be: **S**

    (i) Detected.

    (ii) Rectified.

5. **(a)** State why starting air compressor performance deteriorates in service and how such deterioration is detected. **S**

  (b) Explain the dangers associated with some compressor faults.

Chapter 8 – Class One

1. **(a)** Explain the advantages and problems of using aluminium in the construction of composite pistons for medium-speed engines.

  (b) Briefly describe the removal, overhaul and replacement of a pair of pistons connected to a single crank of a vee-type engine, explaining any problems regarding the bottom end bearings.

2. **(a)** Explain the advantages of fitting highly rated medium-speed engines with double exhaust and air inlet valves.

  (b) State the disadvantages of double valve arrangements. **S**

  (c) Explain the possible causes of persistent burning of exhaust valves if it is:

    (i) General to most cylinders.

    (ii) Specific to a single cylinder.

3. Explain the problems associated with medium-speed diesel exhaust valves when operating with heavy fuel oil.

  How can these problems be minimised?

  (a) By design.

  (b) By maintenance.

4. Describe a suitable maintenance schedule for one unit of a medium-speed diesel engine operating on heavy oil.

5. Describe the torsional vibration of medium-speed diesel engine crankshafts.

   Describe, with the aid of sketches, a coupling that will aid the damping of torsional vibration.

Chapter 10 – Class One

1. As Chief Engineering Officer, what standing orders would you issue your engineering staff to ensure that the auxiliary boiler was operated in a safe and efficient manner?

2. Describe a waste heat plant that is able to produce sufficient steam to a turbogenerator to supply the entire ship's electrical load at sea.

   Due to trading requirements, the vessel is sailing at reduced speed.

   Describe the steps you would take to ensure the slowest ship's speed commensurate with supplying sufficient steam to the turbogenerator without allowing the boiler to fire or starting diesel generators.

3. Describe, with the aid of sketches, an auxiliary boiler suitable for use with a waste heat unit.

   Explain how the pressure of the steam plant is maintained when operating under low steam load conditions.

4. Sketch and describe a composite thimble-tube boiler. Describe how the thimble tubes are fitted and discuss burning of tube ends and other possible defects.

5. You are the Chief Engineering Officer of a motor vessel equipped with a steam plant incorporating a waste heat unit in the engine uptake. On passage, it is reported to you that the uptake temperature is rising.

   (a) What would this information indicate and what steps would you take?

   (b) How could you prevent a reoccurrence?

Chapter 1 – Class Two

1. (a) State the ideal cycle most appropriate to the actual operations undergone in the modem diesel engine.

(b) Give reasons why the actual cycle is made approximate to the ideal heat exchange process. **S**

(c) State how the combustion process in the actual cycle is made approximate to the ideal heat exchange process.

2. (a) State why bottom end bolts of four-stroke engines are susceptible to failure.

(b) Sketch a bottom end bolt of suitable design.

(c) Explain how good design reduces possibility of failure. **S**

(d) State how the possibility of failure is reduced by good maintenance.

3. (a) Explain why in large, slow-speed engines, power balance between cylinders is desirable.

(b) State why it is never achieved in practice.

(c) Describe how power balance between cylinders of a medium-speed engine is improved. **S**

(d) Describe how power balance in a slow-speed engine is improved.

4. (a) Give an example of each of the four types of two-stroke engine indicator diagrams, explain how each is taken and the use to which it is put.

(b) Illustrate two defects that can show up on a compression card. **S**

(c) How is cylinder power balance checked on a higher-speed engine?

5. (a) Explain how the power developed in an engine cylinder is determined:

(i) From indicator cards. **S**

(ii) By electronic means.

(b) State which of these is the most representative and why.

Chapter 2 – Class Two

1. (a) State TWO reasons why large crankshafts are of semi-built construction.

(b) State SIX important details of crankshaft construction that will reduce the possibility of fatigue failure. **S**

(c) List FOUR operational faults that may induce failure in a crankshaft.

2. (a) State the nature of the stresses to which crank webs of large diesel engines are subjected. **S**

(b) Explain how they are designed and manufactured to resist these stresses.

3. (a) State the reason for fitting crosshead guides to engines and explain why 'ahead' and 'astern' faces are required with unidirectional engines. **S**

   (b) Describe how crosshead guide clearance is checked and adjusted.

   (c) List reasons for limiting such crosshead clearance.

4. (a) State why bedplates of large engines are fitted with chocks rather than directly on foundation plates.

   (b) Sketch an arrangement of lateral chocking showing the position relative to the engine. **S**

   (c) State why such an arrangement is employed.

   (d) State the factors that determine the spacing of the main chocks.

5. With reference to auxiliary diesel engine machinery:

   (a) (i) State why this may be mounted on resilient mountings.

   (ii) State why such mountings have great flexibility. **S**

   (b) State why limit stops are provided.

   (c) State how the external piping is connected.

Chapter 3 – Class Two

1. (a) Describe, with the aid of sketches, a fuel pump capable of VIT.

   (b) State why injection timing might need to be changed. **S**

   (c) State how injection timing is adjusted while the engine is running.

2. (a) Sketch and describe a fuel valve for a diesel engine.

   (b) State FOUR factors that indicate that fuel valve(s) require attention. **S**

3. (a) State the factors that influence:

   (i) Droplet size during fuel injection.

   (ii) Penetration. **S**

   (b) State TWO methods of improving air turbulence.

4. (a) Sketch and describe a jerk-type fuel pump that is not helix controlled.

   (b) Explain how the pump may be timed. **S**

   (c) State TWO advantages of this type of pump.

5. (a) Sketch a main engine fuel pump of the scroll type.

(b) Explain how the fuel quantity and timing are adjusted. **S**

(c) To what defects is this type of pump subject and how is the pump adjusted to counter their effects?

Chapter 4 – Class Two

1. (a) Describe with the aid of sketches:

   (i) A pulse turbocharger system.

   (ii) A constant pressure turbocharger system.

   (b) State the advantages and disadvantages of each system in **S** Q.1(a) for use with marine propulsion engines.

   (c) In the event of turbocharger failure with one of the systems in Q.1(a), state how the engine could be arranged to operate safely.

2. (a) Sketch a simple valve timing diagram for a naturally aspirated four-stroke engine.

   (b) Sketch a simple valve timing diagram for a supercharged four-stroke engine. **S**

   (c) Comment on the differences between the above two diagrams.

3. (a) Sketch and describe a turbocharger with a radial flow gas turbine, showing the position of the bearings. **S**

   (b) State the advantages of radial flow gas turbines.

4. (a) State why turbochargers are used to supply air to an engine rather than expanding the gas further in the cylinder and then employing crank-driven scavenge pumps.

   (b) Explain what measures should be adopted to ensure safe operation of the engine should all turbochargers be put out of action. **S**

   (c) State why a two-stroke cycle engine relies upon a pressurised combustion air supply but a four-stroke cycle engine does not.

5. (a) Explain why air coolers and water separators are fitted to large turbocharged engines. **S**

   (b) Sketch a water separator, explain how it operates and indicate its positioning in the engine.

   (c) What are the defects to which coolers and separators are susceptible?

Chapter 5 – Class Two

1. (a) Sketch a starting air distributor used for a large reversible engine.
    (b) Explain how the engine may be started with the crankshaft in any rotational position. **S**
    (c) Explain how the engine is started on air in either direction.
2. (a) Sketch a pneumatically operated starting air valve.
    (b) Explain how the valve is operated. **S**
    (c) State what normal maintenance is essential and the possible consequence if it is neglected.
3. (a) Sketch and describe the reversing system for a large slow-speed diesel engine. **S**
    (b) List the safety devices fitted to the air start system.
4. (a) Explain why it is necessary to have air start overlap.
    (b) Show how air start timing is affected by exhaust timing. **S**
    (c) State why the number of cylinders have to be taken into consideration.
5. (a) Sketch an engine air start system from the air receiver to the cylinder valves and describe how it operates. **S**
    (b) List the safety devices and interlocks incorporated in such a system and state the purpose of each.

Chapter 6 – Class Two

1. With reference to a jacket water temperature control system:
    (a) Sketch and describe such a system.
    (b) (i) Explain how disturbances in the system may arise. **S**
        (ii) Describe how these disturbances may be catered for.
2. (a) Construct a block diagram, in flow chart form, to show the sequence of operations necessary for the starting of a diesel engine on bridge control. **S**
    (b) Identify the safety features incorporated in the system of Q.2(a).

3. (a) Sketch a cylinder relief valve suitable for a large engine.

   (b) State with reasons why such a device is required.

   (c) If the relief valve lifts, state the possible causes and indicate the rectifying action needed to prevent engine damage. **S**

   (d) State why the relief valve should be periodically overhauled even though it may never have lifted.

4. With reference to mechanical/hydraulic governors, explain:

   (a) Why the flyweights are driven at a higher rotational speed than the engine.

   (b) How dead band effects are reduced. **S**

   (c) How hunting is reduced.

   (d) How the output torque is increased.

5. Sketch and describe a hydraulic governor with proportional and reset action.

Chapter 7 – Class Two

1. (a) Sketch a jacket water cooling system.

   (b) State why chemical treatment of the jacket cooling water is necessary. **S**

   (c) Describe how the correct concentration of the chemicals in the jacket water cooling system may be determined.

2. (a) Explain how oil may become mixed with starting air and state the attendant dangers.

   (b) Describe how this contamination may be reduced or prevented. **S**

   (c) Outline the dangers of lubricating oil settling in air starting lines.

   (d) How may an air start explosion be initiated?

3. (a) Explain why air compression for starting air duties is carried out in stages and why those stages are apparently unequal. **S**

   (b) What is the purpose of an intercooler and explain why it is important that it is kept in a clean condition?

   (c) What is the significance of clearance volume to compressor efficiency?

   (d) What is bumping clearance and how is it measured?

4. (a) State why compressor suction and delivery valves should seat promptly.

   (b) Explain the effect on the compressor if the air is induced into the cylinder at a temperature higher than normal. **S**

(c) What would be the effect of the suction valves having too much lift?

(d) Explain why pressure relief devices are fitted to the water side of cooler casings.

5. (a) State why inhibitors are employed with engine cooling water even though distilled water is used for that purpose.

(b) State the merits and demerits of the following inhibitors used in engine cooling water systems: **S**

(i) Chromate.

(ii) Nitrite-borate.

(iii) Soluble oil.

(c) Briefly explain how each inhibitor functions.

Chapter 8 – Class Two

1. Describe, with the aid of sketches, an exhaust valve of a medium-speed diesel engine suitable for use with heavy fuel oil. Explain the procedure adopted when overhauling this valve.

2. Describe, with the aid of sketches, a piston suitable for use in a medium-speed engine. Why is aluminium being generally superseded for pistons on highly rated medium-speed engines?

3. Describe with the aid of sketches a system for main propulsion in which two medium-speed diesel engines are coupled to a single propeller.

4. Describe the advantages and disadvantages of medium-speed diesel engines compared to large slow-running engines.

5. Explain why lubricating oil consumption is greater in medium-speed engines than in slow-running diesels and the steps taken to minimise the consumption.

Chapter 10 – Class Two

1. (a) Describe, with the aid of sketches, an arrangement for producing electricity using steam generated from waste heat.

(b) State how electricity can be generated with the system in Q.1(a) when the engine is not operating. **S**

(c) State the circumstances that could lead to an emergency shutdown of the steam plant in Q.9(a) and the use of diesel engines for electrical generation.

2. Describe the inspection of an auxiliary boiler.

   What precautions should be taken prior to entering the boiler?

3. Describe, with the aid of sketches, a boiler which may be alternatively fired or heated with main engine exhaust gas in which the heating surfaces are common. Describe the changeover arrangements and state any safety devices fitted to this gear.

4. What are the precautions that should be taken before and during the 'flashing up' operation of an auxiliary boiler?

   State the checks carried out on the boiler when a fire is established.

5. What are the advantages and disadvantages of forced circulation and natural circulation multi-boiler installations?

   How can the steam pressure of the waste heat plant be controlled when operating on exhaust gas?

6. Describe the dangers of dirty uptake in the waste heat unit. Explain how these dangers are minimised.

SPECIMEN QUESTIONS

(**S** shows questions used in the past as examination questions)

Class One

1. (a) Define the term *hot spot*.

   (b) State SIX specific areas in a diesel engine where hot spots have occurred.   **S**

   (c) State other factors that may contribute to the occurrence of a crankcase explosion.

2. With reference to crankcase explosions, state:

   (a) The conditions that may initiate an explosion.

   (b) What may cause a secondary explosion.   **S**

   (c) How a crankcase explosion relief valve works.

3. (a) State the basic processes leading up to a crankcase explosion and explain how a secondary explosion can occur.   **S**

   (b) List with reasons the precautions that can be taken to minimise the risk of a crankcase explosion occurring.

4. (a) Explain how a primary crankcase explosion is caused and how it may trigger a secondary explosion.

   (b) Indicate the possible benefits or dangers of the following features on the likely development of a crankcase explosion: **S**

      (i) Oil mist detector.

      (ii) Inert gas injection.

      (iii) Infrared heat detectors.

      (iv) Bearing shells having a layer of bronze between the white metal and steel backing steel.

5. (a) Describe, using sketches if necessary, the procedure for complete inspection of a propulsion engine main bearing and journal.

   (b) State the possible bearing and pin defects that might be encountered.   **S**

   (c) State what precautions should be taken before returning an engine to service following such bearing inspection and adjustment.

6. (a) Explain the reason for fitting crossheads and guides to large, slow-speed engines.

   (b) Explain:

      (i) Why guide clearance is limited.   S

      (ii) How guide clearance is adjusted.

      (iii) How guide alignment is checked.

7. (a) During an inspection it is noticed that tie-rods of certain main engine units have become slack. State with reasons the possible causes of this.

   (b) Explain how correct tension is restored and the risk of future slackness minimised.   S

   (c) A tie-rod has fractured and cannot be replaced immediately. State with reasons the course of action to be adopted in order to allow the engine to be operated without further damage.

8. (a) Explain the term *fuel ignition quality* and indicate how a fuel's chemical structure influences its value.

   (b) State, with reasons, the possible consequences of operating an engine on a fuel with a lower ignition quality than that for which it is timed.   S

   (c) (i) Explain how an engine might be adjusted to burn fuel of different ignition quality.

      (ii) State what checks can be carried out in order to determine that the engine is operating correctly.

9. (a) Describe the phenomenon of surging as applied to turbochargers.

   (b) Explain why turbochargers are not designed to completely eliminate the possibility of surging.   S

   (c) State with reasons the possible consequences of allowing a turbocharger to continue to operate while it is surging.

10. With reference to turbocharger systems, state how deposit build-up might be detected on the following parts and explain the consequences on turbocharger and engine operation of excessive deposits:

   (a) Suction filter.

   (b) Impeller.

   (c) Turbine nozzle and blades.

   (d) Air cooler.

11. Difficulty is experienced in starting an engine even though there is full air pressure in the air receivers and fuel temperature is correct. Explain how the cause of the problem can be:　　　　　　　　　　　　　　　　　　　　　　　　　　　**S**

    (a) Detected.

    (b) Rectified.

12. With reference to piston ring and liner wear:

    (a) State, with reasons, the causes of abnormal forms of wear known as cloverleafing and scuffing (micro-seizure).

    (b) Explain how cylinder lubrication in terms of quantity and　　　　**S** quality can influence wear.

    (c) Describe the procedure for determining whether piston rings are suitable for use.

13. With reference to main engine holding down studs/bolts:

    (a) Explain the causes of persistent slackening.

    (b) State, with reasons, the likely consequences of such slackening.　　**S**

    (c) Describe how future incidents of slackening might be minimised.

14. (a) Inspection of an engine indicates an unexpected increase in cylinder liner wear rate. State with reasons the possible causes if:

    (i) The problem is confined to a single cylinder.

    (ii) The problem is common to all cylinders.　　　　　　　　　　　　**S**

    (b) Explain how cylinder wear rate may be kept within desired limits and indicate the instructions to be issued to ensure that engine room staff are aware as to how this can be achieved.

15. Cracks have been discovered between the crankpin and web on a main engine crankshaft:

    (a) Describe action to be taken in order to determine the extent　　　**S** of the cracking.

    (b) Explain the most likely reasons for the cracking.

    (c) State, with reasons, the action to be taken in order that the ship may proceed to a port where thorough inspection facilities are available.

16. It is found that tie-rods are persistently becoming slack:

    (a) State, with reasons, the possible causes.

    (b) State, with reasons, the likely effects on the engine if it is　　　　**S** allowed to operate with slack tie-rods.

    (c) Explain how this problem can be minimised.

17. As Chief Engineering Officer, explain the procedure to be adopted for the complete inspection of a main engine cylinder unit, emphasising the areas of significant interest. **S**

18. (a) The water jacket on a turbocharger casing has fractured, allowing water into the turbine side. State possible reasons for this.

    (b) Explain how the engine may be kept operational and the restrictions now imposed upon the operating speed. **S**

    (c) State how the fracture can be rectified and how future incidents can be minimised.

19. (a) State the conditions that could result in a fire in the tube space and/or uptakes of a waste heat boiler.

    (b) State how such conditions can occur and how the risk of fire can be minimised. **S**

    (c) State how such fires can be dealt with.

20. As Chief Engineering Officer, explain the procedure to be adopted for the survey of an air compressor on behalf of a classification society. **S**

21. (a) Identify, with reasons, the causes and effects of misalignment in large, slow-speed engine crankshafts.

    (b) Describe how the alignment is checked. **S**

    (c) State how the measurements are recorded and checked for accuracy.

22. (a) Explain why side and end chocking arrangements are provided for large direct-drive engines.

    (b) State, with reasons, why non-metallic chocking is considered superior to metallic chocking. **S**

    (c) State why top bracing is sometimes provided for large engines and explain how it is maintained in a functional condition.

23. (a) As Chief Engineering Officer, describe how a complete inspection of a main engine turbocharger may be carried out, indicating, with reasons, the areas requiring close attention. **S**

    (b) Describe defects that may be found during inspection and their possible cause.

24. The main engine has recently suffered problems related to poor combustion and inspection indicates that a number of injector nozzles are badly worn:

    (a) Explain the possible causes of the problem and how they may be detected. **S**

    (b) State how future problems of a similar nature can be minimised.

25. With reference to fuel pumps operating on residual fuel:

    (a) (i) State, with reasons, the defects to which they are prone.

        (ii) Explain the effects of such defects on engine performance.　　　**S**

    (b) State, with reasons, corrective action necessary to restore a defective fuel pump to normal operation.

    (c) Suggest ways in which the incidence of these defects might be minimised.

Class Two

1. Describe the routine maintenance necessary on the following components in order to obtain optimum performance from a main engine turbocharger:　　　**S**

    (a) Lubricating oil for ball bearings.

    (b) Air intake silencer/filter.

    (c) Turbine blades.

    (d) Diffuser ring.

2. (a) List the advantages of multi-stage air compression with intercooling compared with single-stage compression.

    (b) Explain the faults that may be encountered during overhaul of the HP stage and indicate how they may be rectified.　　　**S**

3. (a) Outline the problems associated with air compressor cylinder lubrication, indicating why it should be kept to a minimum.

    (b) State why a restricted suction air filter might make the situation worse and lead to the possibility of detonation in the discharge line.　　　**S**

    (c) Explain why the compressor discharge line to the air receiver should be as smooth as possible with the minimum number of joints and connections.

4. (a) Explain the need for additives in engine jacket water cooling systems.

    (b) State what factors determine the choice of chemicals used.　　　**S**

    (c) State why chromates are seldom used.

5. (a) Give a simple line sketch of a jacket water cooling system.

    (b) Describe a control system capable of maintaining the jacket water temperature within close limits during wide changes in engine load.　　　**S**

6. (a) Sketch an arrangement for securing turbocharger blades to the blade disc.

    (b) How is blade vibration countered?

    (c) What is the cause of excessive turbocharger rotor vibration?   **S**

    (d) Briefly describe an in-service cleaning routine for the gas side of a turbocharger.

7. (a) Describe with sketches a scroll-type fuel pump.

    (b) Explain how the quantity of fuel is metered and how the governor cut-out functions.   **S**

    (c) State how this type of pump is set after overhaul.

    (d) State the reasons that necessitate pump overhaul.

8. (a) Sketch a fuel injector.

    (b) Explain how it operates and what determines the point at which injection occurs.   **S**

    (c) Describe the defects to which injectors are prone.

    (d) How can injection be improved when a low-speed engine is to operate at prolonged low load?

9. With reference to turbocharging:

    (a) (i) Explain the terms 'pulse system' and 'constant pressure system'.

         (ii) List the advantages of each.

    (b) State how in a pulse system the exhaust from one cylinder may be prevented from interfering with the scavenging of another.   **S**

    (c) State why electrically driven blowers are usually fitted in addition to turbochargers.

10. (a) Show how combustion forces are transmitted to the cross members of the bedplate.   **S**

    (b) Describe TWO means by which the stresses within the cross members can be accommodated.

11. (a) Describe how crankshaft alignment is checked.

    (b) Identify, with reasons, the causes of crankshaft misalignment.   **S**

    (c) State how the measurements are recorded.

12. (a) Sketch a cross section of a main engine structure comprising bedplate, frames and entablature, showing the tie-bolts in position.

    (b) Explain why tie-bolts need to be used in some large, slow-speed engines. **S**

    (c) Explain in detail how the tie-bolts are tensioned.

13. Give reasons why, when compared to the other bearings of large, slow-speed engines, top end bearings: **S**

    (a) Are more prone to failure.

    (b) Have a greater diameter in proportion to pin length.

14. (a) State how engine cylinder power is checked and approximate power balance is achieved.

    (b) Explain why the methods of checking may differ between low- and high-speed engines. **S**

    (c) State why perfect cylinder power balance cannot be achieved.

    (d) State the possible engine problems resulting from poor cylinder power balance.

15. (a) Describe, with sketches, the monobox frame construction that is being used to replace the traditional A-frame arrangement for some crosshead engines. **S**

    (b) State why this form of construction is considered to be more suitable than one using A-frames.

16. (a) State TWO reasons why large crankshafts are of semi-built or fully built construction.

    (b) State SIX important details of crankshaft construction that will reduce the possibility of fatigue failure. **S**

    (c) State FOUR operational faults that may induce fatigue failure.

17. (a) Define the cause of corrosive wear on cylinder liners and piston rings.

    (b) Explain the part played by cylinder lubrication in neutralising this action. **S**

    (c) State how the timing, quantity and distribution of cylinder oil is shown to be correct.

18. With reference to large fabricated bedplates, give reasons to explain:

    (a) Why defects are likely to occur in service and where they occur. **S**

    (b) How these defects have been avoided in subsequent designs.

19. (a) Define the cause of cylinder liner and piston ring wear. **S**

    (b) Describe how cylinder liner wear is measured and recorded.

    (c) Explain the possible consequences of operating a main engine with excessive cylinder liner wear.

20. (a) Sketch a main engine holding down arrangement employing long studs and distance pieces.

(b) Explain why the arrangement sketched in Q.6(b) may be employed in preference to short studs. **S**

(c) Describe, with the aid of sketches, how transverse movement of the bedplate is avoided.

21. (a) Briefly discuss the relative advantages and disadvantages of oil and water for cooling.

(b) Sketch a piston for a large two-stroke crosshead engine indicating the coolant flow. **S**

(c) State the causes of piston cracking and burning, and how it can be avoided.

22. (a) Sketch the arrangement of a large two-stroke engine, cylinder liner in position in the cylinder block.

(b) Describe how jacket water sealing is accomplished between the liner and cylinder block. **S**

(c) For the liner chosen, illustrate the directions of cooling water flow, exhaust gas flow and combustion air flow.

(d) Explain how thermal expansion of the liner is accommodated.

23. (a) Give the reasons for progressive 'fall-off' of piston ring performance in service.

(b) State, with reasons, which ring clearances are critical. **S**

(c) State what effects face contouring, bevelling, ring cross section and material properties of rings and liners have on ring life.

24. (a) Sketch the arrangement for connecting a piston to the crosshead.

(b) State the type of piston coolant employed and show how the **S** coolant is directed to and from the piston.

(c) State the precautions to be exercised when lifting or overhauling the piston described.

25. (a) Explain the reasons for employing two air inlet and two exhaust valves for high-powered trunk piston four-stroke engines. **S**

(b) State the problems relating to tappet setting with such valves.

(c) Sketch a caged valve as fitted to a trunk piston engine.

INDEX

REEDS MARINE ENGINEERING AND TECHNOLOGY SERIES

Vol. 1 Mathematics for Marine Engineers
Kevin Corner, Leslie Jackson and William Embleton
ISBN 9781408175552

Vol. 2 Applied Mechanics for Marine Engineers
Paul A Russell, Leslie Jackson and William Embleton
ISBN 9781472910561

Vol. 3 Applied Thermodynamics for Marine Engineers
Leslie Jackson, William Embleton and Paul A Russell
ISBN 9781408160749

Vol. 4 Naval Architecture for Marine Engineers
Richard Pemberton, E A Stokoe
ISBN 9781472947826

Vol. 5 Ship Construction for Marine Engineers
Paul A Russell, E A Stokoe
ISBN 9781472924285

Vol. 6 Basic Electrotechnology for Marine Engineers
Christopher Lavers, Edmund G R Kraal and Stanley Buyers
ISBN 9781408176061

Vol. 7 Advanced Electrotechnology for Marine Engineers
Christopher Lavers and Edmund G R Kraal
ISBN 9781408176030